THE GENERATION OF ELECTRICITY BY WIND POWER

THE GENERATION OF ELECTRICITY BY WIND POWER

The late

E. W. GOLDING

sometime Technical Secretary of the Wind-Power Committee of the Electrical Research Association

With an additional chapter by

R. I. HARRIS

Director of the Environmental Sciences Research Unit, Cranfield Institute of Technology

LONDON

E. & F. N. SPON LTD

A Halsted Press Book

John Wiley & Sons, Inc., New York

First published 1955
Reprinted, with additional material, 1976
Reprinted 1977
E. & F. N. Spon Ltd
11 New Fetter Lane, London EC4P 4EE

Printed and bound in Great Britain by
Redwood Burn Limited
Trowbridge & Esher

ISBN 0 419 11070 4

Distributed in the U.S.A. by Halsted Press,
a Division of John Wiley & Sons, Inc., New York

Library of Congress Cataloging in Publication Data

Golding, Edward William, 1902–1965
The generation of Electricity by wind power.

Bibliography: p.
Includes indexes.
1. Electric power-plants. 2. Wind power. I. Title.
TK1541.G6 1976 621.312'136 76–7907
ISBN 0–470–14986–8

PUBLISHERS' NOTE ON THE REVISED IMPRESSION

First published in 1955 *The Generation of Electricity by Wind Power* was soon recognised as the definitive account of the research on wind power undertaken up to that time.

In the immediate post-war years there was considerable interest, in Europe particularly, in wind-power research. The reasons for this, which Golding lists in his Introduction, sound all too familiar today: fears about the depletion of the stocks of fossil fuels, the rising cost of existing methods of electricity generation and the political importance of being self-sufficient in energy resources.

However, in the years following the publication of the book, interest in wind power began to fall away. The research programmes had raised more problems than the technology of the day could handle, atomic power seemed to offer the prospect of unlimited energy and new discoveries of oil and natural gas encouraged complacency about fuel reserves. Thus when the author died in 1965 it seemed inevitable that the book would remain out of print and forgotten, except as a scientific curiosity.

Instead, the 'energy crisis' of 1974/75 and the increasing interest in 'self-sufficiency' and 'alternative technology' produced a sudden demand for what was now seen as an essential reference. A reprint was planned to meet this demand but it seemed inappropriate to reissue the book without including an account of subsequent research and technical developments.

The publishers therefore decided to include an additional chapter outlining these developments contributed by a former research colleague of Golding, R. I. Harris, Director of the Environmental Sciences Research Unit at the College of Aeronautics, Cranfield Institute of Technology. Cranfield has long been a centre of wind-power research and much of the practical work described in the book was carried out there.

LONDON, *February* 1976

PREFACE TO THE FIRST EDITION

In this age we are so greatly dependent upon mechanical power for industry, for agriculture and in almost every aspect of our daily life and work, that the level of a country's energy consumption is now commonly accepted as a measure of its prosperity or its degree of development. Electricity plays an ever-increasing part in the provision of this much-needed power and widespread interest is taken in all the possible sources of energy from which it can be generated. Coal, oil and other fuels are being used up at an alarming rate, water power is being exploited wherever it can be brought into use economically and, of course, tremendous efforts are being made to apply nuclear energy for power production. But the possibilities of several inexhaustible sources of energy such as the wind, solar radiation, the tides and geothermal heat, are being considered seriously. It is clear that they have their peculiar advantages and can serve very useful purposes in many parts of the world; for even nuclear energy, when it is fully developed, is not likely to provide the solution to all of the world's power problems.

This book gives an account of the research and development work which has been done in Great Britain and other countries, particularly during the last few years, in an endeavour to make use of wind power on a significant scale. It outlines the technical and economic questions which arise and describes the steps being taken to provide answers to them.

The subject matter could be roughly divided into three parts dealing respectively with wind behaviour and its determination, with wind-driven machines and with the economic use of wind power under different conditions. Selected bibliographies are given at the ends of the chapters and these include most, if not all, of the references likely to be useful for a fuller study of wind power. Appendices give a comprehensive list of meteorological references and a three-language glossary of wind power terms.

In attempting to make acknowledgements the author is in some difficulty because of the very large number of people and organizations, throughout the world, from whom he has received valuable help during the last few years. At international meetings sponsored by O.E.E.C., UNESCO, and the World Power Conference and at private meetings, wind power problems have been freely discussed with exchanges of information from the experiences of workers in different countries. There has arisen, in fact, a world-wide group of people, interested in the subject, who have thus been drawn into loosely-organized contact and one object of the book is to assemble the information in an ordered form for their benefit.

The author has been privileged to act as technical secretary of the Wind Power committee of the Electrical Research Association and to lead the team of investigators whose work has been guided by that committee. In his writing he has, naturally, drawn freely on the experience gained and on the E.R.A. Technical Reports and he is glad to have this opportunity of acknowledging the most valuable and enthusiastic help which he has thus received from his colleagues. To his friends abroad, in particular Professor D. Dresden (Holland), J. Juul, H. Lykkegaard and O. V. Mörch (Denmark), P. Ailleret, A. Argand, L. Serra, G. Lacroix, P. Basiaux, J. B. Morel and the late J. Andreau (France), Dr. U. Hütter and H. Christaller (Germany), H. Munro (Ireland), Professor M. S. Thacker (India), J. Frenkiel (Israel) and Dr. P. C. Young and J. Swarbrick (UNESCO), the author expresses his gratitude for their continued help in many different forms, but there are others, too numerous to name, to whom he is also very grateful. He is glad to acknowledge also cordial assistance from the manufacturers of wind power plant in England. To the Ministry of Fuel and Power thanks are due for permission to include the information, on their design study, given in Chapter 16.

Acknowledgements for the loan of blocks and for permission to include illustrations

Plate I. From *Farmers of Forty Centuries* by F. H. King. (Jonathan Cape, Limited.)

Plates III & IV. From *Windmills in England* by Rex Wailes. (Architectural Press.)

Plates XV & XVII. R. W. Munro Limited.

Plate XVIII. Kelvin and Hughes Limited.

Plates XXVI and XXVII. Enfield Cables Limited.

Plate XXVIII. From *Power from the Wind* by P. C. Putnam (Copyright, 1948, D. Van Nostrand Company Inc.).

Fig. 2. From *The pneumatics of Hero of Alexandria*, Bennett Woodcroft. (Charles Whittingham.)

Fig. 4. From *The Story of Sprowston Mill* by H. C. Harrison. (Phoenix House.)

Fig. 28. The Director of the Air Ministry Meteorological Office.

Figs. 30, 61, 62, 76. Mr. J. Juul, S.E.A.S., Haslev, Denmark.

Figs. 36, 46. R. W. Munro Limited.

Figs. 39, 40, 45. The Controller of H.M. Stationery Office.
Fig. 47. ASEA Electric Limited.
Figs. 63, 64, 65. Studiengesellschaft Windkraft, E.V. Stuttgart.
Figs. 79, 85, 86. Joseph Lucas, Limited.
Figs. 83, 84. Messrs. Merz and McLellan.
Fig. 87. Allgaier-Werke, G.M.B.H.
Other acknowledgements are made in the text.

E. W. GOLDING

LONDON, *July* 1955

CONTENTS

PLATES

between pages 78 *and* 79

FIGURES

THE GENERATION OF ELECTRICITY BY WIND POWER

Chapter 1

INTRODUCTORY

THE wind has been used as a source of power—in sailing ships—for many centuries and, indeed, throughout history countries have owed their prosperity to their skill in sailing.

On land also, though not perhaps for so long a period, windmills have taken their place, with water mills, as power plant for such purposes as water pumping and corn grinding. Before the development of the steam engine in the second half of the eighteenth century, the energy of flowing water and of the wind were the only natural sources of mechanical power which were exploited in any significant degree. Water mills had to be located on the banks of streams and, although they were used in Great Britain long before windmills appeared, they suffered from this disadvantage of limited location. There was greater freedom in the choice of site for a windmill because, with virtually no competition from alternative methods of power production, and with constructional costs so much lower than they are today, it was much less important that the site should have a high average wind speed. Naturally, the windiest available site in the locality—on a hill if possible—was chosen by the old millwrights, but other factors, such as convenience of transport for the corn to be ground, or the position of the water to be pumped, had to be considered. Thus, in England the old windmills were located in the towns and in the eastern, corngrowing and low-lying, counties, not because these were very windy districts but because power was needed there.

During the last half century, the old corn mills have fallen into disuse, as a result of the reorganization of the flour-milling industry on the basis of large-scale, centralized operation rather than on a multiplicity of small independent plants. Except in Holland, the old windmills for pumping have been largely replaced by machines driven by less fickle sources of power than the wind though even now, in some parts of the world, extensive areas depend on a more modern form of windmill for their water supply.

With the development of electricity supply, there has been increasing use of small wind-driven generators to afford the amenity of electric lighting for isolated premises which cannot economically be connected to public supply networks. It is mainly the need for storage of energy—which may be expensive—to provide for periods of calm which has limited the use of such isolated plants to small sizes. There is also the question of the capital cost of wind-driven machines.

Inevitably one tends to think of power generated by the wind as free. In fact, while the source of the energy is certainly free, the energy itself is 'thinly distributed' in the air. Because of its low density, large volumes of air must be tapped to provide an appreciable output of power. Without close attention being paid to its design and construction, the plant required to capture this power, and to convert it to a usable form, may be expensive and the capital charges for interest and depreciation may be high enough to raise the cost, per unit of energy produced, to an uneconomic value. It is generally true that the cost per horse power of power-producing plant decreases with increasing size and no doubt this applies also to wind-driven plant, at least up to sizes greater than those which have so far been built. During the past few decades, in several countries, there have thus been many projects for exploiting the energy in the wind by means of large plant used on a scale sufficient to make a significant contribution to electricity generation. That so very few of these have materialized is explained by the difficulty which has been experienced in constructing such plants cheaply enough for their generating costs to compare favourably with those of large steam-driven generating stations using cheap coal.

Revival of interest in wind power

Since the war, increasing and widespread interest has been taken in the potentialities of the wind as a source of energy and, in several countries, private and Government-sponsored organizations have been established to investigate the matter. Some of the reasons for this interest are as follows:

(i) The rapidly increasing demand for electrical energy accompanied by the inadequacy—in some countries—of fuel supplies or of potential hydro-electric resources which can be economically developed;

(ii) high and rising costs of generation in steam-driven stations or in newly-constructed hydro-stations. Both the initial cost of construction of hydro-electric stations, and that for the transmission of the power generated by them, are now often increased by the fact that the more easily developed sources, near to load centres, have already been exploited;

(iii) difficult economic and political conditions of the post-war years tending to make countries depend upon their own resources for the generation of power rather than upon imported fuels;

(iv) the realization that coal and oil resources are being used up at an increasingly high rate and that they can be put to better use than burning them as fuels;

(v) the greatly increased knowledge of aerodynamics as applied to aeroplane construction, resulting from war-time research and development. This renders the problems to be faced in constructing large windmills less formidable than they were formerly;

(vi) the appreciation of the advantages of diversity in the availability of power from different sources connected to a widespread network;

(vii) the Smith-Putnam experimental aerogenerator, of 1250 kW capacity, built during the war on Grandpa's Knob in Central Vermont, U.S.A. Although this machine was not completely successful as an economic unit, its performance certainly indicated the practical possibility of employing large plants to generate electricity from wind power.

Future possibilities

One can envisage two ways in which wind power may be used economically on a significant scale. The first is by means of machines of medium size—up to perhaps 100 kW or 200 kW capacity—used in conjunction with other forms of generating plant, for the supply of electricity to isolated communities which cannot be supplied economically from a main network. Examples of such communities are to be found on, literally, hundreds of islands in many parts of the world. Usually there is no water power or other usable source of energy available locally and fuel for power generation has to be imported with heavy charges for transport. Quite often such islands have a high average wind speed which offers a good scope for wind-driven generators to supplement power supplies from firmer, though more expensive, sources. Many sets of this kind have been used in Denmark and there should be no great difficulty in developing them for much more widespread application at prices within the economic limit. Medium-size plant is economic under such circumstances, not because it is necessarily cheap to build, but rather because the alternative method of power generation is expensive.

The second possibility, which may be even more important in its contribution to total energy generated, lies in the employment of large wind-driven generators, each having a capacity of at least 1000 kW—the upper limit of size cannot be precisely determined and may, indeed, advance with increasing experience in construction. These units would be located on especially favourable windy sites, fairly close to supply networks, into which they would feed their output without employing any direct form of storage. Like the medium-size generators, when connected to networks supplied from thermal power stations they would act as fuel savers so that the cost of the energy generated by them must compare favourably with the fuel component of the generating cost when coal or oil is used. Some firm power might be obtained through diversity of wind régimes at widely scattered sites but the possibility of a general calm occasionally affecting all the installations cannot be ruled out. The average wind speed at a given site remains sufficiently constant, year by year, for an estimated annual output of energy to be relied upon within fairly close limits though there can be no certainty of power being available at any particular time.

Wind power could be combined with water power to supply a network. Then the cost of the wind-generated energy must be comparable with that of its production by augmenting the hydro-electric capacity installed; each kilowatt-hour generated by the wind would save water which could be used later for power production provided that the storage capacity of the reservoir were ample. The wind power would have the effect of an increase in rainfall, or in catchment area, to justify a larger water power installation.

The annual energy available in the winds over the earth's surface amounts to many millions of millions of kilowatt-hours (A. Parker [Ref. 8] gives 13 million million kWh) although it is only possible to envisage the utilization of a fraction of the enormous total. Nevertheless, it might prove possible in a windy country to develop wind power to an installed capacity of some ten to twenty per cent of the total generating capacity. Without some form of storage, or a load which could absorb large quantities of energy at random times, much more than this might create a difficulty in that, if full output were to occur simultaneously from all the wind-driven installations during summer night-time, the available wind power would exceed the normal demand on the system so that some would have to be wasted.

BIBLIOGRAPHY

(1) SKILTON, C. P. *British windmills and watermills.* Collins (London, 1947).
(2) AUBERT DE LA RÜE, E. *Man and the winds.* Hutchinson (1955).
(3) PUTNAM, P. C. *Power from the wind.* Van Nostrand (1948).
(4) WITTE, HANS. *Windkraftwerke.* Rudolf A. Lang (Possneck, 1950).
(5) AILLERET, P. L'énergie éolienne: sa valeur et la prospection des sites. *Revue générale d'électricité* (March 1946).
(6) EGERTON, A. C. G. Power and combustion. The Institution of Mechanical Engineers, *The twenty-seventh Thomas Hawksley Lecture*, Vol. 144, No. 3 (1941).
(7) EGERTON, A. C. G. *Civilization and the use of energy.* UNESCO/NS/74.

(8) PARKER, A. World energy resources and their utilization. The Institution of Mechanical Engineers, *The thirty-sixth Thomas Hawksley Lecture*, Vol. 160, No. 4 (1949).

(9) SIMON, F. E. *Energy in the future*. UNESCO/NS/79.

(10) THACKER, M. S. *The role of energy in under-developed areas*. UNESCO/NS/78.

(11) GOLDING, E. W. Large-scale generation of electricity in wind power – preliminary report. Electrical Research Association, *Technical Report*, Ref. C/T101 (1949).

(12) THOMAS, PERCY H. Harnessing the wind for electric power. *Proceedings, United Nations Scientific Conference on the Conservation and Utilization of Resources*, Vol. III, p. 130 (Lake Success, 1949).

(13) BRUN, E. A. and ONIGA, T. Utilizacão da Energia dos Ventos. *Dados gerais.* A situacão no Brasil. Instituto Nacional de Technologia (Rio de Janeiro, 1952).

(14) FARDIN, R. Windpower: its advantages and possibilities. *Proceedings, United Nations Scientific Conference on the Conservation and Utilization of Resources*, Vol. III, p. 322 (Lake Success, 1949).

(15) VEZZANI, R. Il problema Italiano dell'utilizzazione del vento. *Annali dei Lavori Pubblici*, Anno 1942, XX, Fasc. 3.

(16) LACROIX, G. Les Problèmes électriques soulevés par l'utilisation de l'énergie du vent. *Bulletin de la Société Française des Electriciens*, Vol. V, No. 103, p. 211–215 (Avril 1950).

(17) *The value of windmills in India*. Commonwealth Relations Office for India (Madras, 1903).

(18) ASTA, ANTONINO. Esperienze sull'utilizzazione dell'energia del vento per produzione d'energia elettrica. *La Ricerca Scientifica*, Anno 23, No. 4 (April 1953).

(19) JUUL, J. Investigation of the possibilities of utilization of wind power. *Elektroteknikeren*, Vol. 45, pp. 607–635, No. 20 (22nd October 1949).

(20) KASPAR, F. Větrné Motory a Elektrárny. *Elektrotechnický Svaz Československý* (Praha, 1948).

(21) LANOY, H. *Les Aéromoteurs Modernes*. Girardot et Cie (Paris, 1947).

(22) PARSONS, H. E. Wind power—history and present status. (Paper presented to the 66th Annual General and Professional Meeting of The Engineering Institute of Canada, at Vancouver, 9th May 1952.) *The Engineering Journal* (January 1953).

(23) SIL, J. M. Windmill power. *The Indian Journal of Meteorology and Geophysics*, Vol. 3, No. 2 (April 1952).

(24) PUTNAM, P. C. *Energy in the future*. Macmillan & Co. Ltd. (London, 1954).

(25) CAMBILARGIU, E. *La energía del viento en el Uruguay*. Instituto de Máquinas, Publicacon No. 12, Montevideo (Uruguay 1953).,

(26) HALDANE, T. G. N. Power from the wind. *The Times Review of Industry* (October 1949).

Chapter 2

THE HISTORY OF WINDMILLS

The precise date when man first used a 'machine' to assist him in his daily work would be virtually impossible to determine but it seems clear that the earliest machines were based on the principle of rotation as a means of providing continuous motion for routine tasks such as grinding corn or pumping water. Thus, there were the mills, driven by animal- or man-power, in which the rotating shaft was vertical and was driven by a long horizontal beam, fixed to it, and pulled or pushed round by the animal walking round-and-round in a circular path. Another form was the treadmill which was driven by treading or 'walking' on vanes or paddles attached radially to its horizontal shaft.

It is easy to understand how the water mill came to be invented: the pressure of a stream of water was simply substituted for the foot pressure of the treadmill. So far as can be discovered from references in early literature, water mills seem to have been used before windmills.

The earliest forms of windmill

These were first based on the vertical-shafted mill, rather than on the treadmill; a number of radially-mounted sails, pushed round by the wind, replaced the animal-drawn horizontal beam. The sails, or vanes, carried forward by the wind in one half revolution, move against it in the second, returning, half. To produce a net driving torque, therefore, either the vanes must be shaped so that the wind pressure on their front is less than on their back, or the back pressure must be removed in some other way. One method was to have the sails hinged so that they moved back against the wind in an edgewise position while, held by a stop, they presented their full surface to the wind as they moved with it. An alternative was to blank off one side by a cover to prevent the wind pressing on the returning sails, although a disadvantage of this method is that the blanking cover must move round to follow any changes in wind direction. The rotor with hinged sails, on the other hand, works independently of wind direction.

Illustrations of these early forms of windmill are given in Fig. 1 and Plate I. The first shows the construction of an old Persian windmill (Ref. 1) and the second shows a Chinese wind-driven brine pump

(Ref. 2) which, it is easy to believe, may be the descendant of machines of the same type built in China many centuries ago. A. Flettner (Ref. 1) states that the Babylonian emperor Hammurabi, in the seventeenth century B.C., planned to use windmills for his ambitious irrigation schemes.

Hero of Alexandria, who lived in the third century B.C., described (Ref. 3) a simple windmill, having a horizontal axis and four sails, which was used to blow an organ (see Fig. 2).

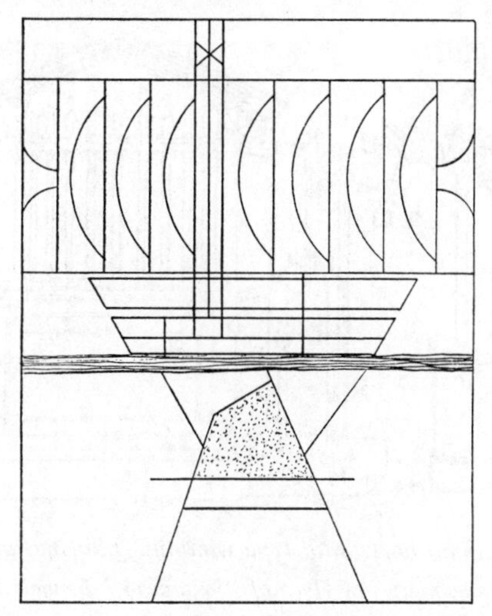

FIG. 1. *Diagram of old Persian windmill of the vertical axis type*

For how long the Persians had used windmills is uncertain but Washington Irving (Ref. 4) tells the story of a Persian slave at Medina who complained to the Caliph Omar because he was being taxed two pieces of silver a day. The slave was, however, 'a carpenter, and expert in the construction of windmills' and was thus considered well able to pay the tax. Unfortunately no details are given of the Persian windmills themselves but it is apparent that, in the middle of the seventh century A.D. windmill building was a well-recognized craft in that part of the world.

Crude and mechanically inefficient though these old machines undoubtedly were, as judged by modern standards, there can be no doubt that they served their purpose well through many centuries, as did their European successors, of a different and more efficient type, much later. They were made from locally-available material and probably by cheap labour. Maintenance could not have presented a serious problem and, with a free source of energy to be tapped, their low mechanical efficiency was unimportant. Their size was probably that which proved most convenient and economic in relation to the materials and labour available; more power needed means more windmills rather than larger ones.

The type still persists in, for example, the cup anemometer commonly used for wind measurements, and some present-day inventors are convinced that this simple construction should be

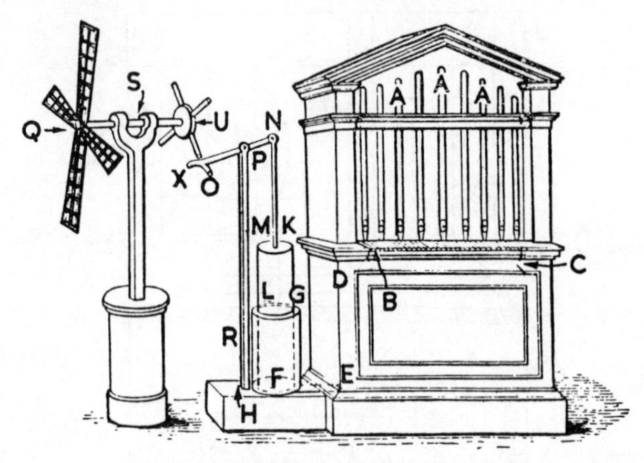

FIG. 2. *Small horizontal-type windmill, blowing an organ (From "The pneumatics of Hero of Alexandria," Bennett Woodcroft)*

followed in modern machines (see Ch. 12). Incidentally the Stastik windmill (see p. 198) is, in essence, a modern development of the treadmill.

European windmills

There is a long gap between the scanty historical records of the Persian windmills and those of the so-called European mills, having sails mounted on a horizontal axis, which have continued in use—though now in rapidly diminishing numbers—until the present day. It has been suggested that they were introduced into western countries by the Crusaders upon their return from the Middle East but it is doubtful whether windmills of the horizontal axis type were in use then in the Middle East. Possibly the English Crusaders came across such machines during their journeys in one of the European countries where they may have been used before they were in England.

The primitive windmills of the horizontal-axis type (although, in fact, the axis was slightly inclined to the horizontal to help in carrying the weight of the sails) had very simple canvas sails spread over a wooden framework (see Plate II). They had no arrangement for turning the sails into the wind so that they must have faced the direction of the prevailing wind.

Later, orientation was accomplished by mounting the mills on floats or by turning the body of the mill on a post on which it was carried. The latter construction—the post mill—is illustrated in

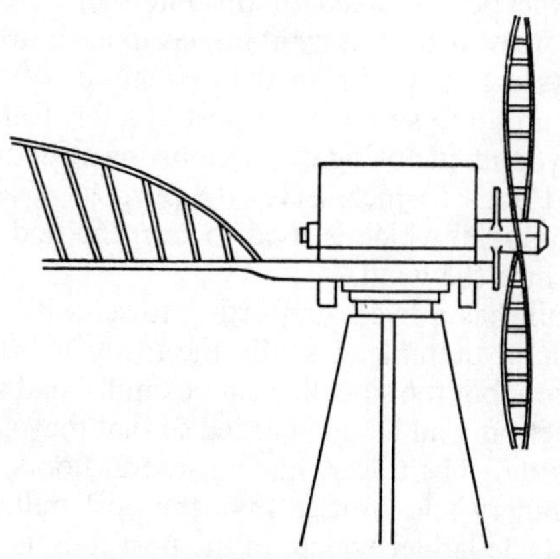

FIG. 3. *Parent's propeller-type windmill described by Belidor in "L'Architecture hydraulique"* (1737). *Probably the first of its type*

thirteenth century manuscripts and is said to have originated in Germany. Among the many sketches of all sorts of mechanical devices by Leonardo da Vinci (1452–1519) is one of a windmill with six sails.

The Dutch certainly did much to improve the early windmills and may have introduced the tower mill, having a fixed body (tower) and rotatable cap, which was first recorded in illustrations of the fourteenth century. Belidor (Ref. 5) describes, with an illustration, (Fig. 3) the construction of what appears to be the first propeller-type windmill and the forerunner of those now being developed for the generation of electricity.

English windmills

The earliest recorded English windmill dates from 1191 and the progress made in the construction and use of what we now choose to call the 'old-fashioned' windmill has been well described by C. P. Skilton (Ref. 6) and Rex Wailes (Ref. 7).

There are two main types, the older one being the 'post mill' and the other the 'tower mill.' These are illustrated in Plate III and Plate IV.

The post mill consists of a weather-boarded wooden body, housing the machinery and carrying the sails on an inclined 'wind-shaft' which projects through the top of one side. This body is built on a horizontal 'crowntree,' of up to some 2 feet square in section, socketed on top of the heavy oaken 'post' on which it can turn. The lower end of the post is fixed to the intersection of two massive wooden 'cross trees' placed at right angles to each other and carried on four piers (see Fig. 4). From the extremities of the cross trees, 'quarter bars' run up to support the post at a few feet above its base. Below the body, and enclosing this supporting structure, is the brick or stone 'round house' which serves the purpose of a store house.

There is a 'tailpole' which is used to turn the body of the mill to bring the sails into the wind.

The tower mill has a fixed supporting tower with a cap, carrying the sails, which is rotatable. Usually the tower is brick-built and is circular in section, but the so-called 'smock mills' had wooden towers, polygonal in section, and being boarded so that they gave the appearance of 'smocking.' The tower, having several floors, provides more space for machinery and storage than the post mill and avoids the need for an outside ladder which, in the post mill, is attached to the body of the mill and rotates with it. The lower end of the ladder rests on the ground but has to be raised when the mill is turned by means of the tailpole which actually passes through the ladder.

To save the miller the labour of pushing round the mill, or the cap, to make the sails automatically face into the wind as its direction changes, the 'fantail' was patented by Edmund Lee in 1745 (Ref. 7) (although Andrew Meikle has also been credited with its invention). This is a small wheel, having a number of radial vanes, which is fixed with its plane of rotation perpendicular to that of the main sails. If the latter are not facing directly into the wind the fantail is driven by it and, through gearing of about 3000 to 1 ratio, drives round the sails until they are facing correctly when the fantail itself is becalmed. On tower mills the fantail is at the tail of the top cap, while on post mills it is usually mounted on the ladder, or on the tail pole, and drives a wheeled carriage on a circular track round the outside of the mill.

Windmill sails

By far the commonest number of sails for windmills of the types just described is four although five-, six- and eight-sailed mills have

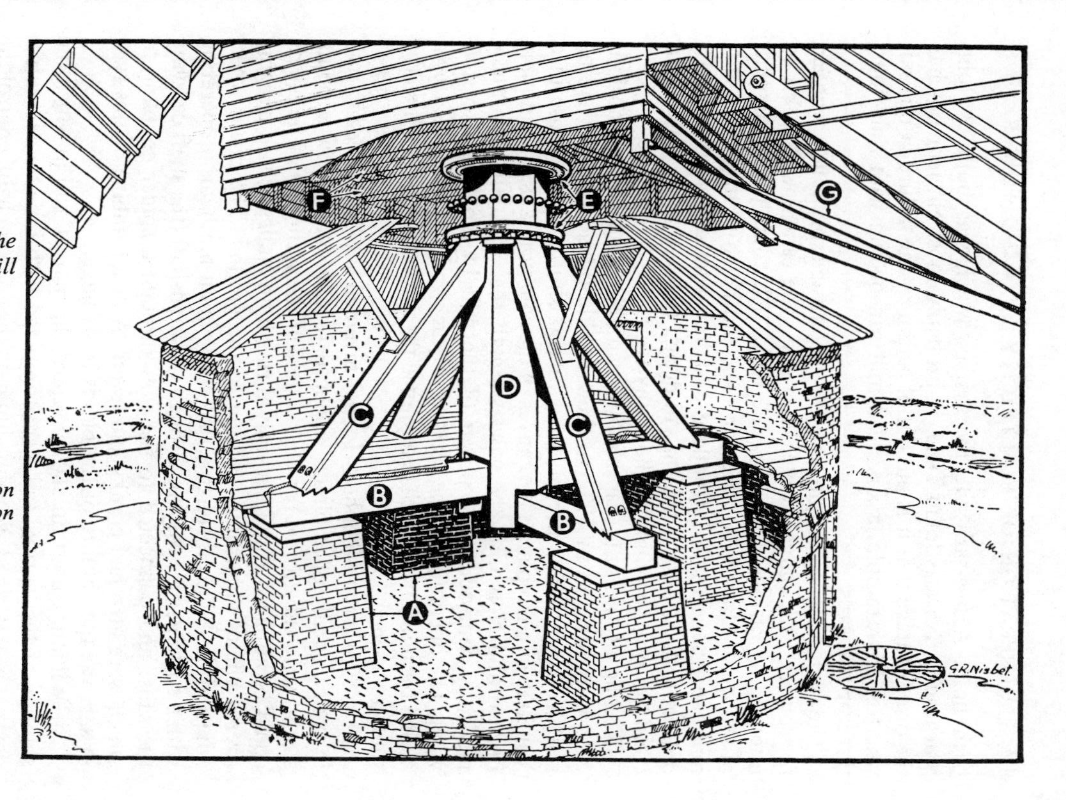

FIG. 4. *Construction of the base of an English post mill*

A Piers
B Cross trees
C Quarter bars
D Post
E Ball race (exploded)
F Sheer beams
G Tail pole

[*From "The Story of Sprowston Mill" by H. C. Harrison (Phoenix House, London)*]

been built. A five-sail mill, the first of which was built by John Smeaton at Newcastle-on-Tyne, is probably most efficient but has the disadvantage that it is put out of action by the failure of one sail whereas, for example, a six-sailed mill can be run with only four, three or two sails.

The oldest sails—'common sails'—consisted of a flat wooden framework with canvas stretched over it. The canvas was furled or unfurled to various degrees by the miller according to the strength of the wind. Until John Smeaton, who did much to improve windmill construction and introduced cast iron windshafts and wheels, did the first scientific work on sail design (see p. 13) the sails had a constant 'weather angle' without any twist.

Reefing of the sails, especially in variable winds, was laborious for the miller and in 1772 Andrew Meikle introduced the 'spring sail' in which a number of small hinged shutters, like those in a Venetian blind, replaced the canvas of the common sails. The shutters opened and closed, as the wind strength rose and fell, thus automatically controlling the power of the mill by 'spilling' excessive wind. The operation of the shutters on each sail was controlled by a spring the tension of which determined the wind strength at which the shutters opened.

In 1789, Captain Stephen Hooper invented the 'roller reefing sail' in which small roller blinds, attached to one another by webbing straps, were substituted for the shutters of the spring sail. The blinds could be adjusted simultaneously through an endless chain hanging over a chain wheel and reaching to the ground. The chain wheel adjusted the blinds through a rack-and-pinion mechanism operating with a striking rod, passing through the windshaft (the shaft carrying the sails) and connected to the sails through a spider coupling to which were attached cranks and levers. A disadvantage of this type of sail was the liability of the blinds to damage.

Sir William Cubitt, in 1807 invented the 'patent' sail in which the shutters of the 'spring' sail were combined with the remote control device just described. Weights, hung on the endless chain, control the opening and closing of the shutters and the size of the weights used determines the wind speed at which the shutters open and begin to spill wind. This method of control is used in the many Lykkegaard windmills, built during the last few decades, which are still working in many parts of Denmark.

The sails of the English windmills were up to about 40 ft long, in the tower type of mill, and up to 9 ft in width and the power generated by the larger mills has been stated as 30–40 h.p. in a good wind.

Wailes estimates that at one time there were some ten thousand windmills at work in Great Britain. These were located mainly in the eastern and south-eastern counties of England where their purpose was for corn grinding and for drainage—in the Fens where some two thousand were used. A few windmills were to be found also in the Midlands, in Lancashire and Cheshire and in Anglesey.

Smeaton's experiments

One of the earliest scientific investigators of windmill performance was John Smeaton who described his experiments 'On the Construction and Effects of Windmill Sails' in a classic paper read before the Royal Society in 1759 (Ref. 8).

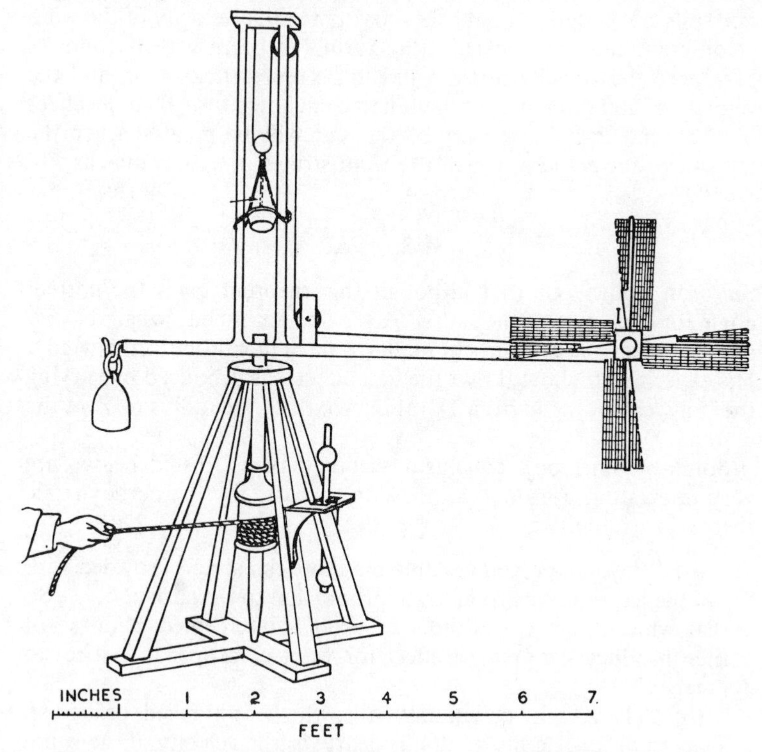

FIG. 5. *Smeaton's laboratory equipment for testing windmill sails*

Using the apparatus shown in Fig. 5, he mounted various forms and settings of windmill sails, in model size, at the end of an arm which was made to rotate by pulling a cord; the rotational speed was kept uniform by adjustment of the pull so that each half revolution of the arm corresponded with a vibration of the double-ball pendulum.

The sails, in rotating on their own axis, raised a weight which could be varied. The product of the weight and the number of turns of the sails in a given time, when corrected for friction in the apparatus, was used as a measure of the 'effect' (actually the power) produced by the windmill.

By this means, Smeaton determined the best form and 'weather' for the sails. (In old windmill practice the term 'weather' meant the angle between the plane of the sail and its plane of rotation.) He refers to the work of Maclaurin on the effect of the wind upon windmill sails taking into account the varying relative velocity of the wind at different radii on the sail. Maclaurin's theorem was as follows: "Suppose the velocity of the wind to be represented by a, and the velocity of any given part of the sail to be denoted by c, then the effect of the wind upon that part of the sail will be greatest when the tangent of the angle in which the wind strikes it, is to radius as

$$\sqrt{2 + \frac{9c^2}{4a^2}} + \frac{3c}{2a} \text{ to 1."}$$

Smeaton points out that although this theorem gives the correct angle for each part of the sail, it leaves undetermined "what velocity any one given part of the sail ought to have in respect to the wind". His experiments showed that the best effect was produced by varying the sail weather angle from 18° at one-sixth of the radius to 7° at the extremity.

Some of Smeaton's conclusions and 'maxims', listed below, are very interesting, forming, as they do, a basis for the aerodynamic theory of windmills:

(*a*) "Beyond a certain degree the more the area is crowded with sail the less effect is produced in proportion to the surface. . . . So that when the whole cylinder of wind is intercepted, it does not then produce the greatest effect for want of proper interstices to escape."

(*b*) "The velocity of windmill sails, whether unloaded, or loaded, so as to produce a maximum, is nearly as the velocity of the wind, their shape and position being the same."

(*c*) "The effects of the same sails at a maximum are nearly, but somewhat less than, as the cubes of the velocity of the wind."

(*d*) "In sails of a similar figure and position, the number of turns in a given time will be reciprocally as the radius or length of the sail."

Smeaton, in addition to investigating the ratio of the velocity of the sail extremity to that of the wind for Dutch and English type mills and showing that it varied between 2·2 and 4·3, determined the power output for a given wind velocity and size of mill. Thus, for a mill having a sail length of 30 ft, the sails being of the best design, he computed the output as "the power of 18·3 men or of $3\frac{2}{3}$ horses, reckoning 5 men to a horse" for a wind speed of $8\frac{2}{3}$ m.p.h. His definition of 1 man-power makes his horse-power 32,000 ft lb/min so that the actual output for this wind speed would be 3·5 h.p. as compared with 12·5 h.p. available from the wind in the area swept by the sails.

One may conclude therefore, since Smeaton checked his computations by observations on a mill with 30 ft sails in normal operation, that the old English mills of this size would produce about 44 h.p. in a 20 m.p.h. wind, the overall efficiency being some 28 per cent.

Another piece of information given in the same paper relates to the pressure of the wind at different wind velocities. Smeaton gives a table of pressures which had been sent to him by a contemporary investigator, Mr. Rouse ("an ingenious gentleman of Harborough in Leicestershire") from which the Smeaton-Rouse formula

$$P = 0{\cdot}00492\ V^2$$

can be stated and which applied particularly to winds up to 50 m.p.h.

P = perpendicular force in pounds per square foot
V = wind speed in miles per hour

In 1755 Smeaton visited Holland and wrote a Diary (Ref. 9) in which he describes Dutch practice in windmill construction.

Danish windmill experiments

Denmark, with negligible energy resources in the form of water power or coal, has used wind power extensively right up to the present day. The Danish windmills are, in the main, less picturesque than the Dutch mills because many of them are of a more modern type which resulted from, or at least was largely influenced by, the work of Professor P. La Cour. He was placed in charge of a windmill experimental station established by the Danish State at Askov (between Kolding and Esbjerg) in 1891 and worked there until his death in 1907. His researches were aimed at the improvement of windmill performance and the development of a type which could be economically constructed and, especially, the generation of electricity for application to various mechanical drives on Danish farms. La Cour was thus the first to undertake systematic investigations on 'aero-generation' and he certainly succeeded in popularizing wind power for electro-agricultural purposes.

His experimental mill (see Plate V) having slatted and remote controlled sails, had a rotor diameter of 22·8 m (75 ft) and a sail width of 2·5 m ($8\frac{1}{4}$ ft). It was oriented by two small fantails and it drove two 9 kW dynamos.

La Cour's work led to his establishing certain guiding principles for the construction of the 'ideal windmill'. The most important of these are:

(*a*) It should have four sails with low resistance surfaces, especially at the tips:

(*b*) the sails should be parallel-sided and of width one-fifth to one-quarter of the length;

(*c*) the sail surface should begin at a distance of about one-quarter of the length from the axis of rotation;

(*d*) the sail profile near the tip should not be straight but bent at one-sixth to one-quarter of the width from the front edge;

(*e*) the weather angle should be 10° at the tip and should increase regularly to be 15° at two-thirds of the length from the axis and 20° at one-third of the length;

(*f*) the tip speed should be 2·4 times the speed of the wind from which the maximum amount of energy is to be extracted;

(*g*) the power produced by such a windmill, in g-m/sec. is $60 \times (\text{total sail area}) \times (\text{wind speed})^3$ the sail area and wind speed being expressed in square metres and metres per second respectively.

The results obtained at Askov were put into practice by the Danish windmill manufacturers such, for example, as Lykkegaard of Ferritslev (Fünen) one of whose machines is shown in Plate VI.

In 1920 a commission was established by the Danish Board of Trade for the purpose of comparing the performance of several types of Danish-made windmills. There were tests at the Askov station and also tests on established windmills in service in different parts of the country (Ref. 11).

During the second World War, to assist in remedying electric power shortages due to the scarcity of oil fuel, a new propeller type of windmill was developed by the firm of F. L. Smidth of Copenhagen (see p. 221).

Danish wind power experiments have continued to the present day (see p. 214 and Refs. 11, 12, 13). The Association of Danish Electricity Works (Danske Elværkers Forening) has set up a wind power committee.

Other modern developments

Present-day researches on the development of wind power are to be described in later chapters but, to complete this historical outline one should mention some of these briefly.

In Holland, the Society for the preservation of Dutch windmills (De Hollandsche Molen), founded in 1923, has worked towards the preservation, improvement of performance and the more effective use of the Dutch mills. It has restored some 750 old mills. Under its auspices Dekker—a Dutch millwright—introduced a special roller bearing for the windshaft and applied aerofoils to the leading edges of the sails. Another organization (the Prinsenmolen Commissie) for the improvement of the old Dutch windmills was started in 1936. With support from the Schieland Catchment Board who provided a big poldermill—the 'Prinsenmolen'—the committee has carried out an important investigation on the performance and improvement of the mill (Ref. 14).

The Netherlands government and tourist organizations have given financial support to windmill preservation. The total number of windmills in the country in 1953 was 1143 of which 643 were corn-mills and 468 poldermills; in 1750 there were between 6000 and 8000. Experiments are now being made to improve their utilization by using the mills for electricity production at times when they are not needed for pumping (Refs. 15, 16).

Germany is another European country in which windmills have been in common use for centuries, especially in the north, and it is not surprising to find there evidence of sustained interest in wind power. H. Witte (Ref. 17) gives the numbers of windmills in use in Germany as 18,242 in 1895, 17,000 in 1907, 11,366 in 1914 and between 4000 and 5000 in 1933. In the province of Brandenburg some 400 windmills are still in use for corn grinding. During the past three decades there have been many ambitious projects for wind-driven electric generators of up to tens of thousands of kilowatts in capacity. These have not materialized but a number of small plants, of ten or twenty kilowatts, and of different types have been developed These have been described fully by Witte and by J. W. van Heys (Ref. 18). We owe to Germany also the classic work of A. Betz of Göttingen (Refs. 19, 20, 21, 22) and that of Bilau (Refs 23, 24, 25, 26, 27).

Anton Flettner was responsible for the development of the rotor ship (Ref. 1).

More recently U. Hütter and others working in Wurttemburg (Refs. 28, 29, 30) have studied the construction of propeller type windmills of medium size (see also p. 213).

Date		*Reference*
B.C.		
2000	(?) Chinese and Japanese windmills in use.	1
1700	(?) Hammurabi reported the use of windmills for irrigation in Babylon.	1
circa 200	Hero of Alexandria describes a small windmill.	3
134	Arabian explorer Istachri mentions windmills in Persian province of Segistan.	1
100	(?) Windmill in use in Egypt.	1
A.D.		
7th century	Persian windmills in use (vertical axis type).	4
1105	French document permitting construction of windmills	1
1191	First reported windmill in England.	6 and 7
circa 1270	Windmill Psalter containing earliest illustration of a windmill (horizontal axis, sail type)	6 and 7
13th century	Manuscript Aristotle's Physica; illustration of windmill with tailpole.	6
1327	Deed referring to a windmill at Lytham St. Anne's (Lancashire).	6
circa 1340	Illustration of a windmill in the Luttrell Psalter.	6
1349	Flemish brass illustrating a windmill in St. Margaret's Church, Kings Lynn.	6
1390	Picture of a windmill on a rug in the Germanic Museum of Nuremberg.	1
1393	Records in the chronicles of the city of Speyer tell of an engineer from the Netherlands being called in to build a windmill.	1
14th century	English illustration of a windmill with four sails in a Decretal of Gregory IX.	6
1439	First corn-grinding windmill built in Holland.	1
circa 1500	Sketch by Leonardo da Vinci (1452–1519) of windmill construction.	
1506	Woodcut showing a windmill in 'Expositio Sequentiarum'.	6
1665	Construction of postmill at Outwood, Surrey. This is still working.	
1737	Illustration in Belidor's 'Architecture Hydraulique' Tome II, Livre 3e, Ch. II, of French windmill with primitive form of propeller having two blades.	5
1745	Edmund Lee patented a method of turning mills into wind automatically.	7
1750	Andrew Meikle invented the fantail.	6
1759	John Smeaton awarded a gold medal by the Royal Society for his paper on windmills and water mills.	8
1772	Andrew Meikle introduced the 'spring sail'.	7
1789	Stephen Hooper invented the 'roller reefing sail'.	7
1807	Sir William Cubitt invented the self-reefing or 'patent' sail.	6 and 7
1891	Establishment of windmill experimental station at Askov, Denmark, under Professor P. La Cour.	
First half of 20th Century	Development of windmills for the generation of electricity and for water supply to individual premises. This period is that of the changeover from sail to propeller-type windmills.	
Post World War II	Researches, in a number of countries, on the possibilities of large-scale utilization of wind power.	

In Russia, where power is needed for large numbers of widely scattered agricultural communities, the Central Wind Power Institute of Moscow was established by the government at the end of World War I.

The work of this institute resulted in the building of a 100 kW pilot plant for electricity generation near Balaclava, Crimea, in 1931 and in the introduction of several types of small wind-driven generators.

Recent work has been concentrated on the development of wind-driven machines of small-to-medium size.

At the Sectional meeting of the World Power Conference in Brazil in 1954 a delegate from U.S.S.R. gave the present number of wind power plants operating in Russia as 29,500 with an aggregate capacity of 167,000 h.p.

In France, G. Darrieus, L. Constantin, J. Andreau, G. Lacroix and others have worked on wind power development for electricity generation during the last three decades (Refs. 41–46) and, in 1943, a wind power committee to be responsible for wind power research in France and French overseas territories was established (see also p. 71).

During World War II the possibilities of generating electricity on a large scale by wind power were studied in the United States of America. Several comprehensive reports (Refs. 31–35) were issued and a 1250 kW wind-driven generator was built and tested at Grandpa's Knob, Rutland, Vermont (Ref. 36 and pp. 213, 220).

Work on the subject in Italy by R. Vezzani (Refs. 36, 37) during and since World War II has led to the formulation of a project for a 500 kW wind turbine of novel design incorporating wind ducts.

This brief summary does not mention all the work now in progress towards wind power development but it may be sufficient to bring this review of the history up-to-date.

BIBLIOGRAPHY

(1) Flettner, A. *The story of the rotor*. F. O. Willhoft (New York, 1926).
(2) King, F. H. *Farmers of forty centuries*. Jonathan Cape Ltd. (1927).
(3) *The pneumatics of Hero of Alexandria*, translated for and edited by Bennett Woodcroft. Charles Whittingham (London, 1851).
(4) Irving, Washington. *Lives of the successors of Mahomet*. Henry G. Bohn (York Street, Covent Garden, London, 1850).
(5) Belidor. De la théorie des machines mues par le vent, et de la manière d'en calculer l'effet. *Architecture Hydraulique*, Tome II, Livre 3, Chapitre II (1737).
(6) Skilton, C. P. *British windmills and watermills*. Collins (London, 1947).
(7) Wailes, R. *Windmills in England*. The Architectural Press (London, 1948).

(8) SMEATON, J. On the construction and effects of windmill sails. *Royal Society* (London, 1759).

(9) *John Smeaton's diary of his journey to the Low Countries* 1755. Extra Publication No. 4, Newcomen Society (1938).

(10) LA COUR, P. *Die Windkraft und ihre Ausnutzung zum Antrieb von Elektrizitätswerken*. M. Heinsius Nachfolger (Leipzig, 1905).

(11) ARNFRED, J. Th. and PRYTZ, K. Arbejdsprøve med vindmøller og med enkelte vandløftningsredskaber i aarene 1921–24, 36. Beretning fra Statens Redskabsudvalg, August Bangs, Boghandel, København.

(12) JUUL, J. Undersøgelse af muligheder for vindkraftens udnyttelse. *Elektroteknikeren*, Nr. 20, 45 Aarg, p. 607 (22 Oktober 1949).

(13) JUUL, J. Report of results obtained with SEAS experimental wind power generator. *Elektroteknikeren*, Vol. 47, p. 5 (7 January 1951).

(14) JUUL, J. Supplement til beretning om resultater opnået med SEAS' forsøgsmolle. *Elektroteknikeren*, Nr. 4, 48 Aarg, p. 65 (22 February 1952).

(15) *Het prinsenmolenboek*. H. Veenman en Zonen (Wageningen, 1942).

(16) Wind energie. Medeling Prinsenmolen-Commissie, *De Ingenieur*, No. 45, 1951 (Algemeen Gedeelte 45).

(17) WITTE, H. *Windkraftwerke*. Rudolf A. Lang, Verlag (Possneck, 1950).

(18) HEYS, J. W. VAN. *Wind und Windkraftanlagen*. Georg Siemens (Berlin, 1947).

(19) BETZ, A. Das Maximum der theoretisch möglichen Ausnutzung des Windes durch Windmotoren. *Zeitschrift fur das gesamte Turbinen wesen*, V. 17 (20 September 1920).

(20) BETZ, A. Windmills in the light of modern research. (U.S. National Advisory Committee for Aeronautics, Technical Memorandum No. 474, August 1928, p. 27, 2 plates.) *Die Naturwissenschaften*, Vol. XV, N. 46 (18 November 1927).

(21) BETZ, A. Der Magnuseffekt, die Grundlage der Flettnerwerke. *V.D.I.*, Bd. 69, No. 1, 3. I (1925).

(22) BETZ, A. *Windenergie und ihre Ausnutzung durch Windmühlen*. Vandenhoek and Rupprecht (Göttingen, 1946).

(23) BILAU, K. Die Ausnutzung der Windkrafte. *Zentralblatt der Bauverwaltung*, V. 45, p. 393–5 (19 August 1925).

(24) BILAU, K. Schnellaufende Windmotoren. *Elektrotechnische Zeitschrift*, V. 46, p. 1405–7 (10 September 1925).

(25) BILAU, K. *Die Windkraft in Theorie und Praxis*. Paul Parey (Berlin, 1927).

(26) BILAU, K. *Die Windausnutzung für die Krafterzeugung*. Paul Parey (Berlin, 1942).

(27) BILAU, K. Windmühlenbau einst und jetzt. Verlag der Wochenschrift, *Die Muhle* (Leipzig, 1933).

(28) HÜTTER, U. Die Entwicklung von Windradern hoher Leistung. *Bericht*, Nr.4, der Ventimotor G.m.b.H. (Weimar, 1941).

(29) HÜTTER, U. Beitrag zur Schaffung, von Gestaltungsgrundlagen für Windkraftwerke. *Dissert* (Weimar, 1942).

(30) HÜTTER, U. Der Einfluss der Windhauligkeit auf die Drehzahlabstimmung von Windkraftanlagen. *Zeitschrift fur Elektrotechnik*, 1948, Heft 6, und 1949, Heft 1.

(31) Final report on the wind turbine. *Research Report PB* 25370, Office of Production, Research and Development, War Production Board (Washington, D.C., 1946).

(32) THOMAS, PERCY, H. *Electric power from the wind*. Federal Power Commission (1945).

(33) THOMAS, PERCY H. *The wind power aerogenerator, twin wheel type*. Federal Power Commission (1946).
(34) THOMAS, PERCY H. *Aerodynamics of the wind turbine*. Federal Power Commission (1948).
(35) THOMAS, PERCY H. Harnessing the wind for electric power. *Proceedings of the United Nations Scientific Conference on the Conservation and Utilization of Resources*, Vol. III, p. 310 (Lake Success, 1949).
(36) PUTNAM, P. C. *Power from the wind*. Van Nostrand (1948).
(37) VEZZANI, R. Il problema Italiano dell'utilizzazione del vento. *Annali dei Lavori Pubblici*, XX, Fasc., 3 (Anno 1942).
(38) VEZZANI, R. Study of a project of a wind power generating station of medium power driving a pumping station for hydraulic accumulation. *Elettrotecnica*, Vol. 37, pp. 398–419 (September 1950).
(39) HARRISON, H. C. *The story of Sprowston Mill*. Phoenix House (London, 1949).
(40) PARSONS, H. E. Wind power—history and present status. Paper presented to the 66th Annual General and Professional Meeting of The Engineering Institute of Canada, at Vancouver, May 9, 1952. *The Engineering Journal* January 1953).
(41) *Procès-verbal des séances du congrès du vent*, Carcassonne (1946).
(42) CONSTANTIN, L. Le vent. *La Nature* (21 June 1924).
(43) CONSTANTIN, L. Le Problème de la propulsion des navires par le vent; les rotors Flettner et les turbines éoliennes. *La Nature*. (1 January 1928)
(44) LACROIX, G. Les Moteurs à vent; les éoliennes électriques Darrieus. *La Nature* (15 December 1929).
(45) ANDREAU, J. Utilisation de l'énergie du vent. Société des Agriculteurs de France. Les journées d'études sur l'utilisation de la force motrice dans l'enterprise agricole, p. 48 (21 et 22 mai 1947).
(46) CHAMPLY, R. *Les Moteurs à vent* (Paris, Dunod, 1933).
(47) WAILES, R. *The English windmill*. Routledge & Kegan Paul Ltd. (London, 1954).

See also meteorological references listed in Appendix I.

CHAPTER 3

ESTIMATION OF THE ENERGY OBTAINABLE FROM THE WIND

WIND is merely air in motion. The air has mass—though its density is low—and when this mass has velocity the resulting wind has kinetic energy which is proportional to $\frac{1}{2}$[mass $\times$ (velocity)^2].

Let ρ = the mass per unit volume of air = air density,
V = velocity of the wind,
A = an area through which the wind passes normally.

The mass of air passing in unit time is ρAV and the kinetic energy passing through the area in unit time is

$$P = \tfrac{1}{2} \,.\, \rho AV \,.\, V^2 = \tfrac{1}{2}\rho AV^3.$$

This is the total power available, in the wind, for extraction by a wind-driven machine; only a fraction of this power can actually be extracted.

As for the units to be used in the expression for the power, for calculations in British units the most convenient formula is that connecting power, in kilowatts, with ρ in pounds per cubic foot, A in square feet and V in miles per hour; but so many different units can be used that it might be well to give alternatives.

Thus, taking throughout $\rho = 0{\cdot}08$ lb/ft^3 = 1290 g/m^3 (*i.e.* grams per cubic metre) and using the general formula $P = K \,.\, AV^3$, we have the different values of the constant K given in Table I for various units of P, A and V.

TABLE I

Unit of power P	Unit of area A	Unit of velocity V	Value of K
Kilowatts	Square feet	Miles per hour	0·0000053
Kilowatts	Square feet	Knots	0·0000081
Horse-power	Square feet	Miles per hour	0·0000071
Watts	Square feet	Feet per second	0·00168
Kilowatts	Square metres	Metres per second	0·00064
Kilowatts	Square metres	Kilometres per hour	0·0000137

Air density. The density ρ of the air varies with altitude and with the atmospheric conditions. For dry air;

$$\text{Density (in g/m}^3) = 348{\cdot}8 \times \frac{\text{atmospheric pressure in millibars}}{\text{temperature in degrees } (C) \text{ absolute}}$$

The standard density—at an atmospheric pressure of 1000 millibars (29·53 in.) and a temperature of 290° absolute—is given, by this expression, as 1201 g/m³.

The density decreases slightly with increasing content of water vapour but corrections for this are negligibly small in wind power calculations. During the year the air density may vary, due to pressure and temperature changes, by 10–15 per cent.

Putnam (Ref. 1) gives a curve for the variation of air density with altitude in the mountains of New England. This shows a decrease from about 1300 g/m³ at 1000 ft above sea level to little more than 900 g/m³ at 10,000 ft. The value for 2000 ft is 1200 g/m³.

In calculations on available wind power in Great Britain, where the hills selected as potential wind power sites range in altitude from about 500–2000 ft, the value 1290 g/m³ used is thus probably a reasonable figure.

Power extracted by a wind-driven machine

A. Betz, of Göttingen, showed in 1927 that the maximum fraction of the power in the wind which could be extracted by an ideal aeromotor was $\frac{16}{27}$, or 0·593 (see also Ch. 12). If this fraction is applied, the formula for the maximum possible power theoretically obtainable is $0{\cdot}593KAV^3$.

To give some idea of the possibilities, Table II has been drawn up showing power values for different wind speeds and different areas swept by the rotor of the wind-driven machine.

TABLE II

($P = 0{\cdot}593KAV^3$)

Wind speed (m.p.h.)	Power in kilowatts from circular areas				
	dia = 12·5 ft	*dia = 25 ft*	*dia = 50 ft*	*dia = 100 ft*	*dia = 200 ft*
10	0·38	1·5	6·0	24	96
20	3·08	12·3	49·2	196	784
30	10·4	41·6	166·4	666	2,664
40	24·6	98·4	393·6	1,574	6,296
50	48·2	192·8	771·2	3,085	12,340
60	83·2	332·8	1,331·2	5,325	21,300

Because of aerodynamic imperfections in any practical machine and of mechanical and electrical losses, the power extracted is less than that calculated as above so that, in practice, the multiplying factor may not be greater than about 0·4 (instead of 0·593) and may be less. Its actual value will depend upon the type and detailed design of the machine and on the operating conditions (see Ch. 10 and 12).

One of the most important points in the design of a wind-driven generator is the 'rated wind-speed', i.e. the lowest wind speed at which full output is produced; at higher wind speeds the output is limited, by the controlling mechanism, to this full rated value.

Small wind-driven units, for the supply of electricity to isolated premises, often have propellers of about 12 ft diameter. Thus, their maximum power outputs will be of the same order, though rather less than, those in column 2 of Table II, i.e. they will vary from (say) 1·5 to 6 kW according to the rated wind-speed for which they are designed. This is usually less than 30 m.p.h.

Medium-sized units, with propeller diameters of around 50 ft, can have an output of the order of 100 kW while the largest size, with about 200 ft diameter, could be rated at between 2000 and 3000 kW.

As will be shown later, it is uneconomic to design for rated wind speeds much greater than 35 m.p.h. so that the higher outputs given in Table II are not obtained in practical designs.

Wind speed and power data

The fact that the power in the wind is proportional to the cube of its speed makes this speed a prime consideration in aeromotor design. Its annual average and distribution, or frequency of occurrence, is clearly of great importance in assessing the energy potentialities of a site or district.

The most essential information required when considering these potentialities is that relating to the annual duration of wind speeds of different magnitudes. Wind speed measurements should thus determine hourly mean speeds throughout the year. These can then be analysed and displayed in the form of (*a*) a velocity-duration curve or, (*b*) a velocity-frequency curve.

Velocity- and power-duration curves. The velocity-duration curve is one showing, as ordinates, the range of wind speed and, as abscissae, the number of hours in the year for which the speed equals or exceeds each particular value.

As examples, Fig. 6 shows three velocity-duration curves. Curve A is drawn from measurements made on the summit of Rhossili Down, a hill in South Wales having an altitude of 633 ft. The long-term, annual average wind speed there is 24 m.p.h. Curve B is plotted from Meteorological Office records for the observation station at St. Ann's Head, at 142 ft above sea level, some forty miles distant from Rhossili Down, and having an annual average wind speed of 16·2 m.p.h. Curve C is for the Meteorological Office station at Leicester, at 267 ft, with an average wind speed of 6·2 m.p.h. These curves show the great difference which exist between wind régimes at an inland site (Leicester) and a coastal site (St. Ann's Head), and also the advantage which can be gained by choosing a good site (Rhossili

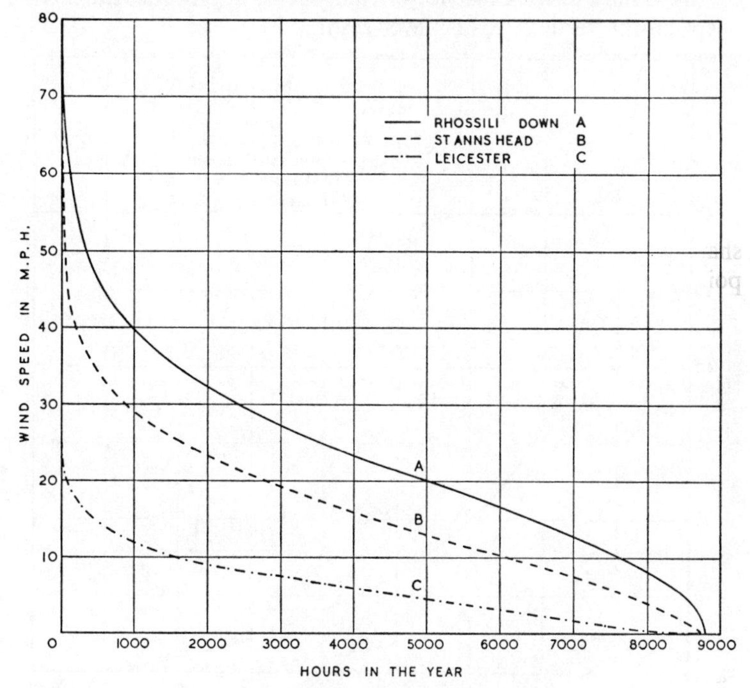

FIG. 6. *Velocity duration curves*

Down) at only a short distance away from a reasonably windy, though not specially selected, site (St. Ann's Head).

In Fig. 7 the three curves are converted to power-duration curves by cubing their ordinates which are then proportional to the power in the wind for a given swept area. The differences between the sites, as potential sources of energy, now become much more marked, especially when it is realized that the areas under the curves are proportional to the annual amounts of energy in the wind.

Velocity-frequency curves. These curves show the annual hours of duration of various wind speeds plotted vertically against the wind speeds horizontally. Fig. 8 shows the velocity-frequency curves for the sites whose velocity-duration curves are given in Fig. 6.

Putnam (Ref. 1) gives frequency curves for some New England sites and these have shapes similar to those in Fig. 8; the general shape is probably typical of the frequency curve for any site. He points out that:

(*a*) Owing to the existence of calm spells, at any site, the intercept on the vertical axis is never zero;

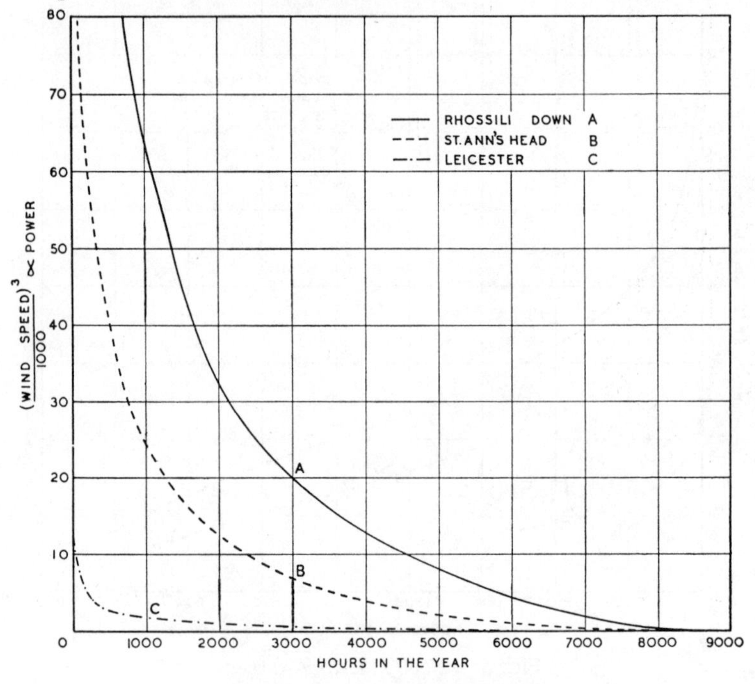

FIG. 7. *Power duration curves*

(*b*) the most frequent velocity is always lower than the mean velocity but varies with it;

(*c*) the duration of the most frequent velocity decreases as the mean velocity increases. (This can be observed in Fig. 8.)

Another point is that the area under the curves is constant, since it must total 8760 (the number of hours in one year).

Fig. 9 (*a*) shows (full line) Putnam's curve of the most frequent velocity plotted against mean velocity for sixteen sites in New England, while the dotted curve is for a number of sites in Great Britain. It can be seen that the most frequent wind speed is much closer to the annual mean value at the British sites than at the American ones. Fig. 9 (*b*) compares the American and British figures for the duration of the most frequent velocity for various values of the mean velocity.

Attempts have been made (*e.g.* see Ref. 2, which obtains the expression for the frequency

$$f = \frac{1}{\sqrt{\pi}} \int_z^\infty \varepsilon^{-z^2} . dz$$

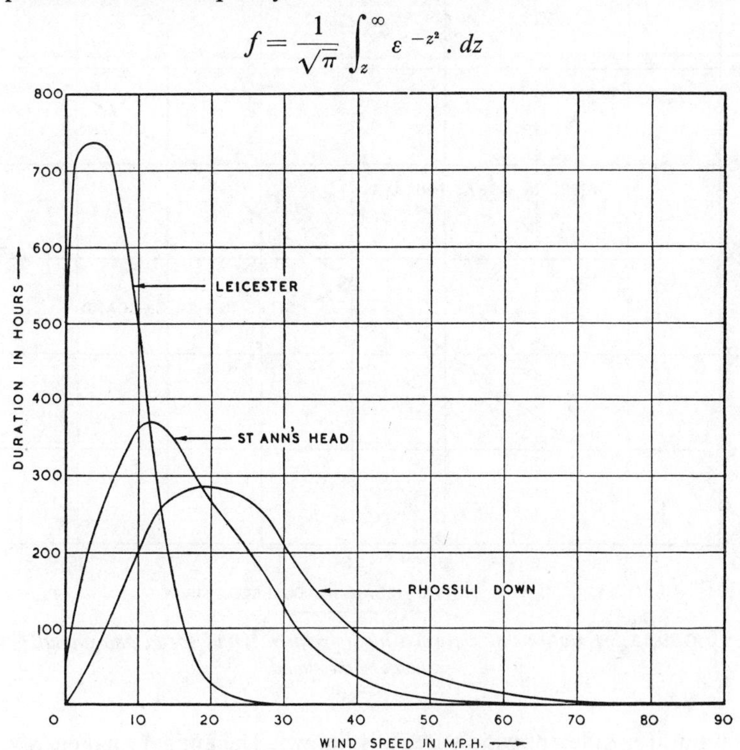

FIG. 8. *Velocity frequency curves*

where z is a function of the wind speed v and can be written $z = av^2 + bv + c$, a, b and c being constants) to obtain a mathematical expression for the velocity-frequency curve at a given site so that it may be constructed from a limited amount of wind data for the site. It appears doubtful, however, whether this is worth while since the determination of the site constants to be used in the expression involves almost as much effort as to obtain the data for plotting the curve itself.

In Great Britain frequency curves have not been much used, the velocity-duration curve being more favoured as a means of presenting the wind data. But they may be rather more convenient than the latter when it is a question of calculating the annual energy which might be obtained from a wind-driven generator whose actual power

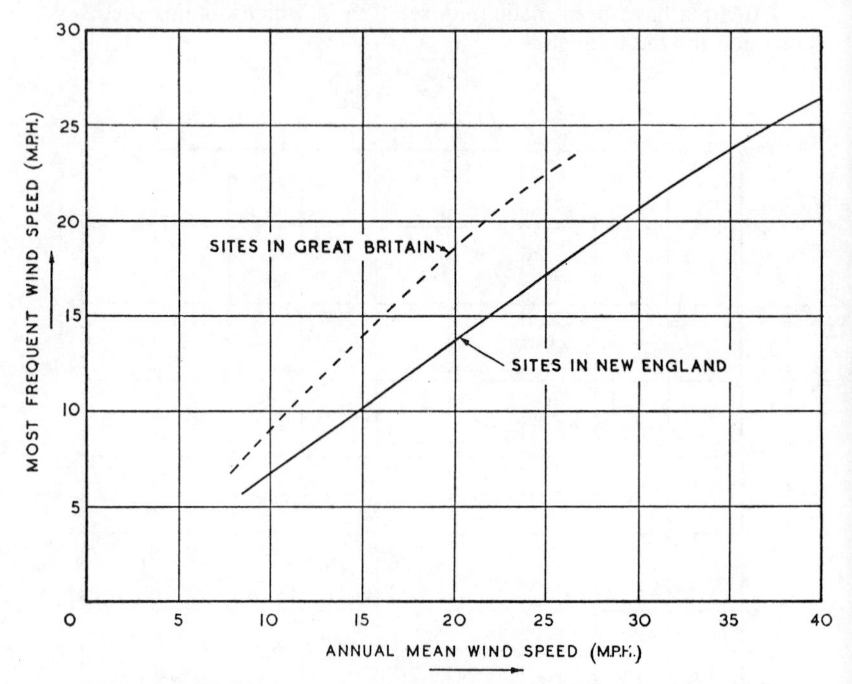

FIG. 9 (a). *Relationship between most frequent wind speed and annual mean wind speed*

output, for different wind speeds, is known. The annual total energy is then calculated by adding together the increments of energy for each duration of wind speed at 1 m.p.h. intervals.

Energy pattern factors

This is a factor which has been introduced, in one form or another, in American literature on the subject (see Refs. 3 and 1). It may be defined as,

$$K_e = \text{Energy pattern factor}$$

$$= \frac{\text{Total energy available in the wind}}{\text{Energy calculated by cubing the mean wind speed}}$$

or, mathematically expressed,

$$K_e = \frac{\int_0^T v^3 \, . \, dt}{T\left(\frac{\int_0^T v \, . \, dt}{T}\right)^3}$$

where v = the wind speed at any time
and T = the total time

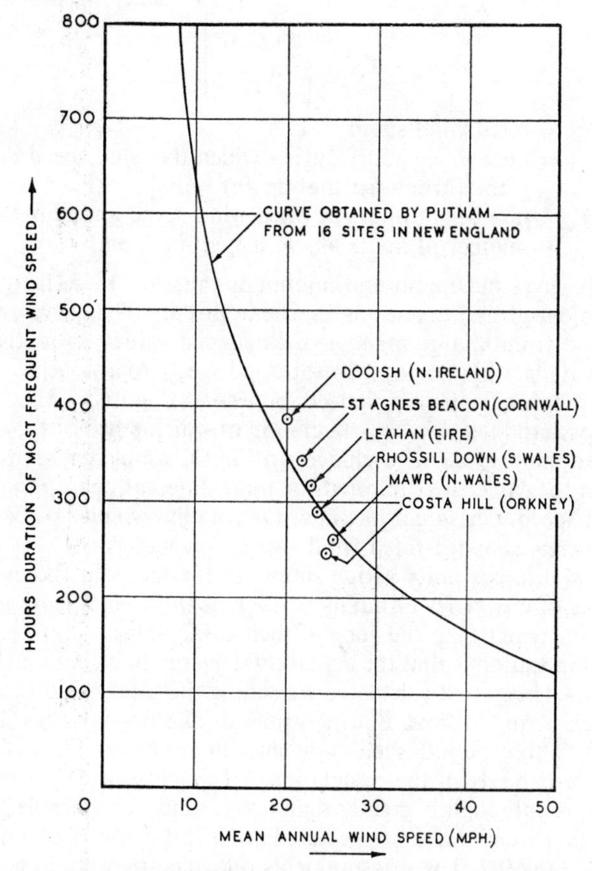

FIG. 9 (b) *Duration of most frequent wind speed*

The numerator of this expression takes into account all the energy available in the wind throughout the period considered. If, however, some arbitrary limits are chosen for the operating range of a wind-driven generator of given design, then a value for the 'usable' energy may be calculated.

The 'usable energy pattern factor' (Ref. 4) obtained by replacing 'total energy', in the definition of K_e, by 'usable energy' will always be less than the value calculated from the total energy in the wind. This factor can be expressed as

$$K_{eu} = \frac{V_R^3 (T_2 - T_1) + \int_{T_2}^{T_3} v^3 \, . \, dt}{T \left[\frac{\int_0^T v \, . \, dt}{T} \right]^3}$$

where V_R = rated wind speed,
T_1 = number of hours during which the wind speed exceeds the furling (or shut-down) value;
T_2 = number of hours at rated wind speed and above;
T_3 = number of hours above the cut-in speed.

Fig. 10 gives maximum and minimum values of these factors for Meteorological Office stations in Great Britain. The curves, which are drawn from a large number of calculated values, show that the factor K_e falls, with increasing mean wind speed, to a value not much greater than unity. It may, indeed, be less than unity for the 'usable energy pattern factor' in the calculation of which much of the energy from very high speeds is excluded. In Fig. 11 usable energy pattern factor at the same site is plotted for three different operating ranges of wind speed. These curves show that, at the windier places, this factor increases as the rated wind speed is increased.

That K_e exceeds unity is due simply to the fact that the mean of the cubes of any series of numbers, such as those representing wind speeds, is greater than the cube of their mean value.

The implication is that the actual total energy to be obtained from the wind exceeds that calculated by cubing the mean wind speed for any given period of time. But the values of K_e shown in Figs. 10 and 11 are for a long period, such as one month or one year, using hourly mean wind speeds in the calculations. If much shorter periods are taken K_e is only slightly greater than unity. Thus, for example, values of energy pattern factors calculated for short periods of time are given in Table III. The measurements on which they are based were made with,

(*a*) Gust anemometers and quick response recorders (for one second periods).

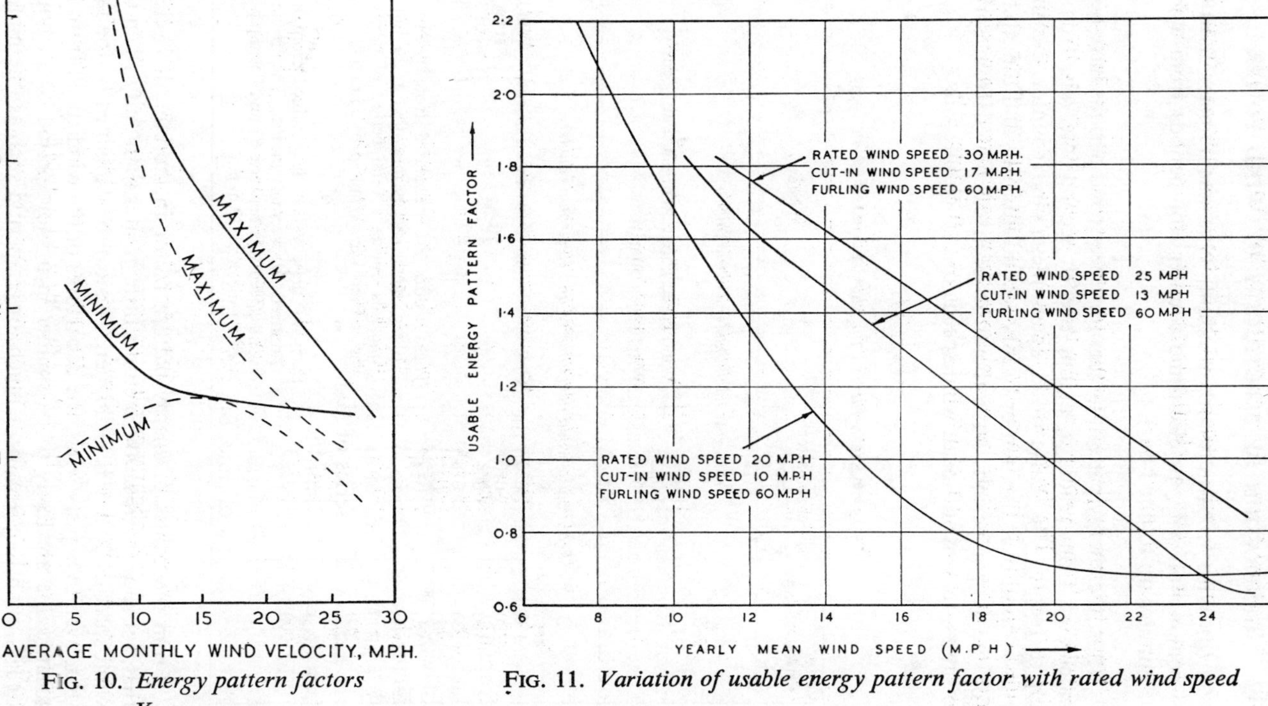

FIG. 10. *Energy pattern factors*

K_e ———
K_{eu} - - - - - -
$V_r = 30$ m.p.h.

FIG. 11. *Variation of usable energy pattern factor with rated wind speed* (*b*) With a balsa wood, windmill type anemometer used in conjunction with an electronic counter (for periods from a few minutes up to one hour).

Thus, the energy pattern factors for one second periods have been calculated from readings, at one-twelfth second intervals, from a chart record. The balsa-wood anemometer and counter measure run-of-wind over five, ten or twenty second intervals and the e.p.f.'s for the longer periods have therefore been calculated from series of measurements of mean wind speeds over such intervals.

TABLE III

Energy pattern factors for short periods of time

Period of time	Energy pattern factor	Notes
		From gust anemometer measurements
sec.		
1	1·037	Wind speed about 30 m.p.h. Fairly gusty. (applies to both rows)
1	1·022	
1	1·002	Wind speed 50 m.p.h.
		From balsa-wood anemometer and electronic counter
sec.		
115	1·020	Calculated from 5 sec. mean speeds. — Wind speed 46 to 52 m.p.h. (applies to this and next two rows)
230	1·010	,, ,, 10 ,, ,, ,,
460	1·004	,, ,, 20 ,, ,, ,,
200	1·018	These values are calculated from twenty, 10 sec. mean speeds measured at 10 min. intervals during a total period of rather more than 1 hr. Wind speed 55 m.p.h. (applies to this and next seven rows)
200	1·007	
200	1·014	
200	1·005	
200	1·008	
200	1·012	
200	1·002	
1000	1·020	Each of these values is obtained from a hundred 10 second means. The mean wind speed was between 63 and 75 m.p.h. with a mean of 70 m.p.h. (applies to both rows)
1000	1·023	

From the values of energy pattern factor in the table it is clear that, for short periods, the factor is little greater than unity when the wind speed is 30 m.p.h. or higher. This can be explained by the fact that, even in a gusty wind, the departures of the wind speed from its mean value are likely to be smaller in a short period of an hour, or less, than they are for a much longer period such as one month or one year.

Making some simplifying assumptions it is possible to obtain an indication of the value of the energy pattern factor under varying conditions as follows:

Referring to Fig. 12, suppose that, when the wind speeds for the time T are replotted in ascending order of magnitude, the graph is

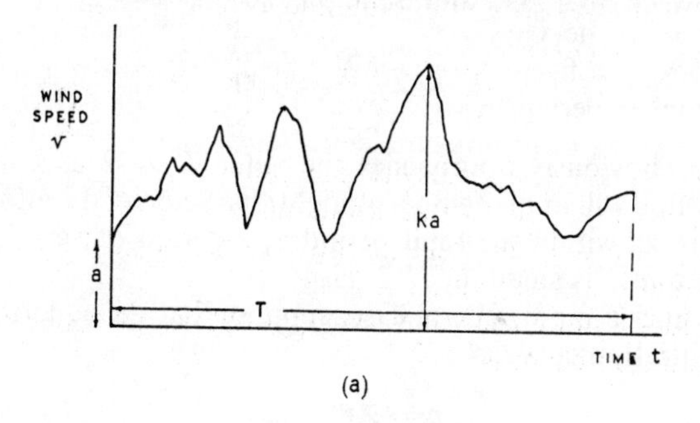

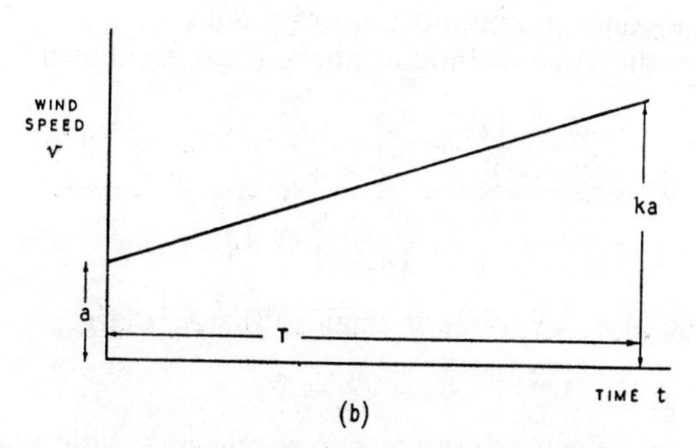

FIG. 12. *Derivation of energy pattern factor*

the straight line velocity duration curve shown in the lower diagram and that the ratio of the maximum to minimum speed is k: 1. Then,

$$\text{Energy pattern factor} = \frac{\int_0^T v^3 \,.\, dt}{T\left[\frac{1}{T}\int_0^T v \,.\, dt\right]^3} = K_e$$

and, since $$v = \left(\frac{ka - a}{T}\right) t + a$$

in this straight-line case,

$$K_e = \frac{2(k^3 + k^2 + k + 1)}{(k + 1)^3}$$

Thus, for example, if the wind speed during the period considered varies between 30 m.p.h. and 60 m.p.h. so that $k = 2$.

$$K_e = \frac{2(15)}{27} = 1{\cdot}11$$

The factor obviously approaches the value $K_e = 2$ as k increases and this value will apply (with a straight line velocity duration curve) when there is, within the total period T, a period of very low wind speeds so that a is small and k is large.

To consider a more general case, suppose that the equation of the velocity duration curve is

$$v = At^n + a$$

where A is a constant and n is a positive index.

Following the same method as above it can be shown that, now,

$$K_e = \frac{\dfrac{(k-1)^3}{3n+1} + \dfrac{3(k-1)^2}{2n+1} + \dfrac{3(k-1)}{n+1} + 1}{\left(\dfrac{k-1}{n+1} + 1\right)^3}$$

For example, if $n = 2$ and a is small so that k is large,

$$K_e = \tfrac{27}{7} = 3\tfrac{6}{7}$$

The annual velocity duration curves for sites with a low mean wind speed (less than 10 m.p.h.) may, in fact, be covered fairly closely by this example which explains the high energy pattern factors shown in Fig. 10.

The value of the energy pattern factor for a long period varies considerably, as shown in Fig. 10, with the shape of the monthly, or annual, velocity-duration curve so that it may lie anywhere between the envelope curves for any given mean wind speed. For short periods, of one hour or less, the value of the factor will depend upon the magnitude and frequency of the wind speed fluctuations during the period, but it will not vary between such wide limits neither will it usually be much greater than unity.

One may thus conclude that the main use of the conception of an energy pattern factor is to show how fallacious it is to calculate available energy from long-period mean wind speeds. Such calculations are only useful when based upon the velocity-duration curve for the site.

Again, the fact that the energy pattern factor is usually so small when it applies to a short period of time, justifies its being neglected entirely in estimates of available energy.

What has been called the 'cube factor' has sometimes been used instead of the energy pattern factor. This factor, K_c, is the ratio of the cube root of the mean cube of the wind speeds to the mean wind speed, i.e.

$$K_c = \frac{\sqrt[3]{\dfrac{\int_0^T v^3 \, . \, dt}{T}}}{\dfrac{\int_0^T v \, . \, dt}{T}}$$

Obviously $K_c = \sqrt[3]{K_e}$ but some confusion has occasionally been caused by insufficiently precise definition of the two factors.

Taking into account the overall power coefficient, i.e. the ratio $\frac{\text{electrical power output}}{\text{power in the wind}}$, the power output curve, between zero and full output, is more nearly parabolic than cubic in shape so that the usable energy pattern factor to be applied over this operating range should perhaps take the form

$$\frac{V_R{}^3 (T_2 - T_1) + \int_{T_2}^{T_3} v^2 \, . \, dt}{T \left[\dfrac{\int_0^T v \, . \, dt}{T} \right]^3}$$

BIBLIOGRAPHY

(1) PUTNAM, P. C. *Power from the wind.* Van Nostrand (1948).

(2) BOIS, CHARLES. Sur les fréquences des vents dont la vitesse a dépassé une valeur donnée d'après les observations de Casablanca. Note transmise par M. Charles Maurain. *Comptes Rendus des Séances de l'Académie des Sciences*, Tome 229, No. 10, pp. 522–524 (5 Septembre 1949).

(3) THOMAS, PERCY H. *Electric power from the wind.* Federal Power Commission (1945).

(4) GOLDING, E. W. and STODHART, A. H. The potentialities of wind power for electricity generation (with special reference to small-scale operation). Electrical Research Association, *Technical Report*, Ref. W/T16 (1949).

(5) Bhatia, K. L. Energy available for windmills in India. *Journal of Scientific and Industrial Research*, Vol. 11A, No. 8, pp. 329–333. New Delhi, India (August 1952).

(6) *Climatological atlas of the British Isles*. Air Ministry, Meteorological Office, H.M. Stationery Office (1952).

(7) Golding, E. W. Large-scale generation of electricity by wind power—preliminary report. Electrical Research Association, *Technical Report*, Ref. C/T101 (1949).

(8) Golding, E. W. and Stodhart, A. H. The selection and characteristics of wind power sites. Electrical Research Association, *Technical Report*, Ref. C/T108 (1952).

(9) Pick, W. H. *A short course in elementary meteorology*. Air Ministry, Meteorological Office, H.M. Stationery Office (1938).

(10) Studiengesellschaft Windkraft e.V., Stuttgart. *Windkraft Mitteilungen*, No. 3 (1st April 1954).

See also meteorological references listed in Appendix. I

CHAPTER 4

WIND CHARACTERISTICS AND DISTRIBUTION

It would be quite inappropriate here to attempt to usurp the functions of the text books on meteorology which deal fully with the many aspects of wind behaviour. Some of these are mentioned in the bibliography to which the reader is referred. We shall be concerned in this chapter mainly with the distribution, in space and in time, of the speed and direction of the wind as related to its possible use for power generation.

Some notes on the causes and general nature of wind may, however, be helpful as a basis for the more specific considerations which follow.

Nature and occurrence of wind

Differences of air density, due to temperature variations, cause the air to move from one latitude to another and the resulting wind direction is influenced considerably by the velocity of the earth's surface which falls from about 1000 m.p.h. at the equator to zero at the poles. This influence is superimposed on that of atmospheric pressure differences which produce the air flow from a high pressure point to one at a lower pressure. For average conditions throughout the year, two belts of high pressure surround the earth between approximately 30° N. and 40° N. and 30° S. and 40° S. with a low-pressure belt lying in the equatorial regions between them. Hence the 'trade' winds which blow in a north-easterly direction between 30° N. and the equator and south-easterly between 30° S. and the equator. These high- and low-pressure belts also give rise to the persistent westerly winds between 40° and 60° latitude in both North and South Hemispheres.

Thus, it is possible to predict the prevailing wind direction in many parts of the world with some confidence. Nevertheless, the single-track trade winds are often disturbed by storms obscuring the prevailing direction, though this direction is always apparent from continuous records. Cyclonic depressions, which may move in any direction but which have certain established direction tendencies, are superposed on the general system of atmospheric pressure.

A knowledge of the prevailing direction of the wind is important in considering the exposure of a potential wind power site but it is the speed of the wind which bears more directly on the economic possibilities of its use. Frictional effects, caused by trees, broken ground or other obstructions, and local temperature variations, have a great influence on the wind at low altitudes (see Ch. 7) but its speed is, of course, otherwise proportional to the steepness of the atmospheric pressure gradient. Observations of this pressure, continuously made at a large number of meteorological stations throughout the world, enable the 'isobars' i.e. lines joining points having the same atmospheric pressure (at sea level and corrected for temperature and latitude) to be drawn for any particular time. From these the 'gradient wind' may be calculated although, owing to surface friction, the calculated value is not likely to agree with the actual speed at altitudes less than about 1500 ft.

Wind characteristics and power production

When considering the use of wind for power purposes important questions are:

(i) In what parts of the world is there sufficient wind to be economically useful?

(ii) What annual amounts of wind energy can be expected?

(iii) How is the wind distributed, in time, during the day, month or year, or over even longer periods?

(iv) What are the probable durations of very high wind speeds, or of calm periods, during any given period?

These questions are discussed below.

World distribution of wind and its economic use

Two important factors enter the economic question: (*a*) the annual mean wind speed and (*b*) the cost of power generation by alternative methods. The economic utilization of wind power, discussed in Ch. 16, is governed by these factors, together with the annual capital charges for (and, therefore, with the initial cost of) the wind power plant designed for optimum performance under the particular climatic conditions at the place considered. An annual mean wind speed which would be economically useful in an area where the cost of power generation is high, might be quite otherwise in another where this cost is lower.

Fig. 13, plotted from the constructional cost data given in Ch. 15, shows a curve for the probable variation of the costs of wind power generation with annual mean wind speed. Two scales are given—one for large scale generation and one for medium scale—which implies that costs for the latter are about 2·5 times those for the former. The curve shows that large-scale generation is only economic at places with an annual mean wind speed of 10 m.p.h. if the fuel component of generating costs by alternative means is greater than 1·5 pence/kWh but that, if the mean wind speed is 25 m.p.h., wind power can compete with a fuel component of generating cost as low as 0.25 pence/kWh.

Only by a very comprehensive study of the economy of electricity generation throughout the world would it be possible to state with

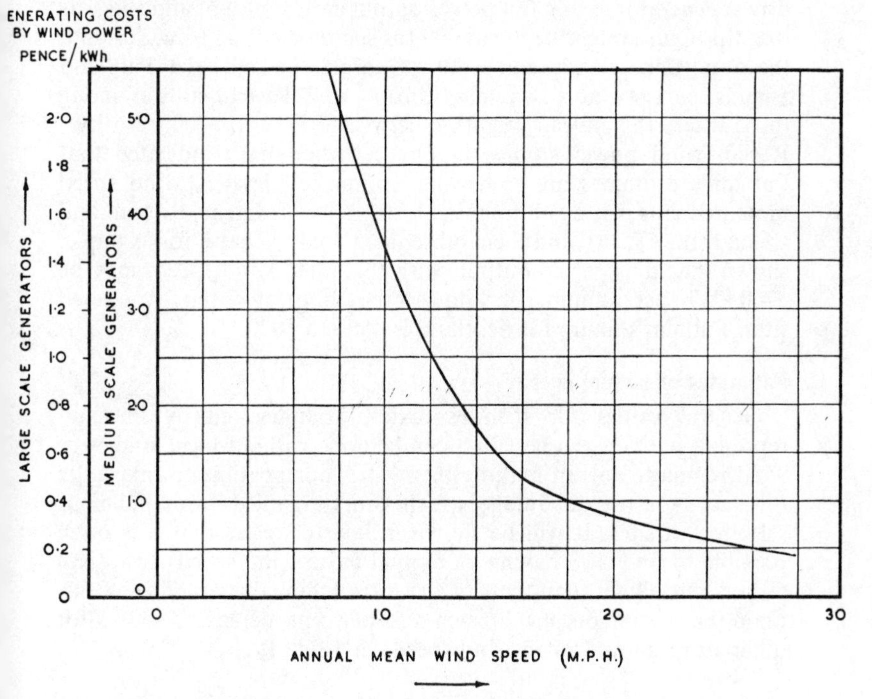

FIG. 13. *Curve of generating costs for large and medium scale wind power*

precision where, and on what scale, wind power generation would be most profitable. Information exists, however, to form a basis for such a study.

Long-term measurements of wind speed and direction, usually for the standard height of 10 m (33 ft), have been made covering most of the world and these are obtainable from the meteorological services of individual countries (see Meteorological References). British Meteorological Office charts Nos. 483, 484, 518 and 519 give information, respectively, for the Atlantic, Western Pacific, Eastern

Pacific and Indian Oceans. The observations on which they are based cover the period 1855–1938.

P. C. Putnam (Ref. 1) indicates, on a world map, the coastal areas which are particularly windy and gives a list also of windy continental regions—mainly the mountainous areas. He assesses windiness in terms of the probable specific output (kilowatt-hours per annum per kilowatt installed) which would be obtained from a large wind-driven generator giving full power output at 32·5 m.p.h. and suggests that this figure might be over 6000 for such regions as New Zealand, the Sandwich Islands and South Georgia. Iceland, the Falkland Islands, parts of north western Europe, and Tasmania, are among those where this output might lie between 5000 and 6000 kWh/kW. Recent wind power studies in Great Britain have indicated that Putnam's estimates are somewhat optimistic; detailed wind speed measurements in western coastal districts of Great Britain and Ireland (see Ch. 10) and their subsequent analysis have, for example, shown that the specific output, with this rated wind speed, may be 3700 kWh per annum per kilowatt but only at specially selected sites. Putnam's figure for Scotland is 4500 to 5000.

Estimates of annual energy

General values for such estimates, as suggested by Putnam, represent perhaps the best that can be done without local studies of wind régimes. But, in making them, it is not possible to take fully into account the advantage which can be gained from judicious selection of sites. It will be shown in later chapters that it is often possible to find sites having an annual mean wind speed more than 50 per cent higher than that for the surrounding district. The effects, upon the specific output, of such a 50 per cent increase is shown for different mean and rated wind speeds in Table IV.

TABLE IV

Annual mean wind speed V m.p.h.	Ratio $\frac{\text{Specific output for mean speed } 1{\cdot}5V}{\text{Specific output for mean speed } V}$		
	Rated wind speed 30 m.p.h.	*Rated wind speed 25 m.p.h.*	*Rated wind speed 20 m.p.h.*
10	3·0	2·5	2·05
15	2·1	1·8	1·5
18	1·85	1·5	1·3

Thus, for example, if the wind-driven generator has a rated wind speed of 30 m.p.h. its specific output will be three times greater at a site with an annual mean wind speed of 15 m.p.h. than at one with 10 m.p.h. Although the ratio falls as the mean wind speed rises, and as the rated wind speed falls, the importance of careful selection of site is obvious.

Variations of mean wind speed with time

These variations must be considered for (*a*) daily, (*b*) monthly, (*c*) yearly and (*d*) long-term periods.

(*a*) *Diurnal variations.* In the daytime, especially in summer, the decrease in temperature with increasing altitude causes thermal convection with consequent interlocking of the air at different levels. Some of the momentum of the upper air, moving at a higher velocity, is transmitted to the surface layers, causing an increased wind speed. Hence the tendency towards a higher surface wind speed during the day than at night, although the effect is often masked, in countries such as Great Britain, by changes of atmospheric pressure which can occur at any time. In some countries, with lower wind speeds, the effect is more regularly experienced. Thus, for example, at Madras (SE. India) the diurnal variation of wind speed throughout the year is shown by the four curves in Fig. 14. These are taken from a report by V. D. Iyer (Ref. 2) who gives similar curves for many places in India. The increased wind speed between 12.00 and 18.00 hours is very noticeable. E. Kidson and M. E. Ewart (Ref. 3) show curves for the diurnal variation of wind speed at three stations in New Zealand. These closely resemble the Indian ones in shape but the peaks occur almost at noon throughout the year.

Hourly records of wind speed for eighteen stations in South Africa (Ref. 4) show that, throughout the year, there is a marked increase during the hours 10.00 to 18.00.

Up to some 15 to 20 miles inland from the coast in Great Britain and countries with a temperate climate, land and sea breezes may occur due to the more rapid increase in temperature of the land during the day—causing an inflow of air from the sea to take the place of the ascending warmed air—and a less rapid cooling of the sea at night which causes a wind from land to sea. Such winds regularly occur, for instance, in Israel and other countries on the eastern Mediterranean seaboard. They are especially noticeable when the land surface is dry: the land temperature is kept down by evaporation if the surface is wet or well covered by vegetation. With hilly country near the coast, land and sea breezes are increased. The sea breeze may extend to as far as 80–100 miles inland in the tropics (Ref. 5).

The curves of diurnal variation in wind speed for two stations in

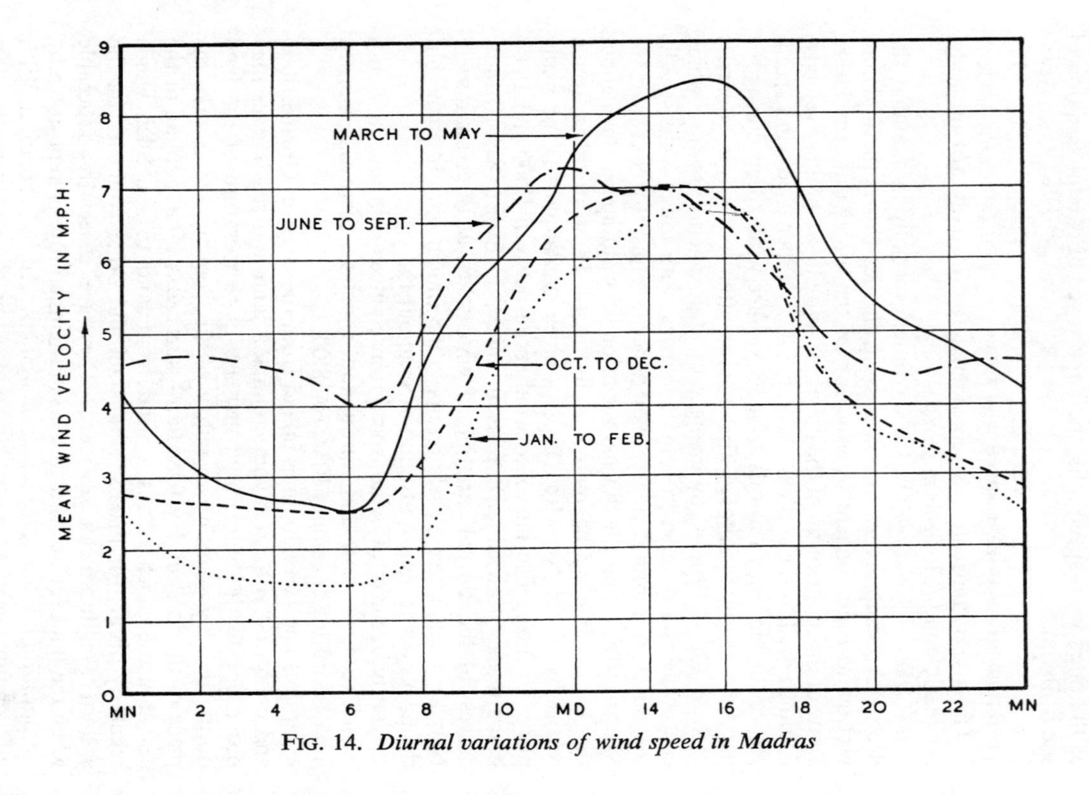

FIG. 14. *Diurnal variations of wind speed in Madras*

British Somaliland are shown in Fig. 15. For the coastal station, Berbera, there is a clearly marked daily peak occurring at about 14.00 hours during nine months of the year but, during June, July and August, when the SW. monsoon blows, there are high winds in the early morning hours with a gradual decrease throughout the day. At Hargeisa, 60 miles inland, the daily curves for most months are less

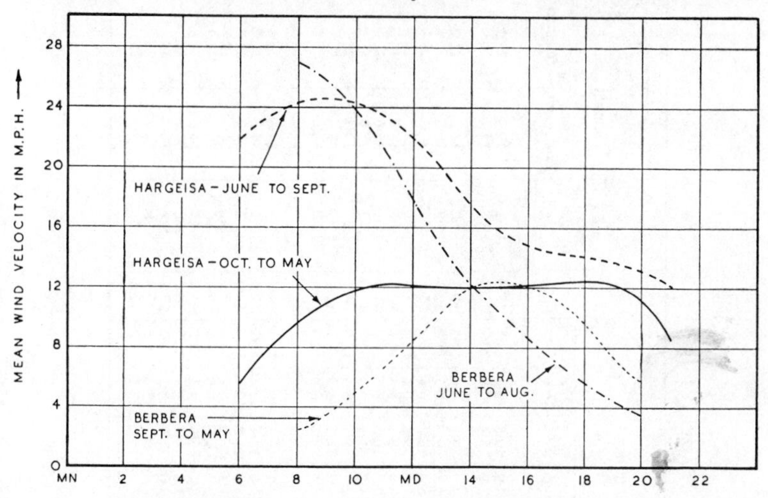

FIG. 15. *Diurnal variations for Hargeisa and Berbera*

peaked than at Berbera but there is a distinct peak at around 09.00 hours during the monsoon period.

As might be expected from the island's surface area and its position in the South Atlantic Ocean, the figures of wind speed for St. Helena (see Table V) show no significant variation during the day.

TABLE V

Longwood (St. Helena)

Hour of day	0	1	2	3	4	5	6	7	8	9	10	11	12	13	14	15	16	17	18	19	20	21	22	23
January	20	21	21	22	22	22	22	22	23	23	22	21	20	19	19	19	19	19	19	19	19	20	21	21
July	17	17	16	17	16	18	18	18	19	19	20	18	18	17	17	17	17	17	17	17	17	17	17	17
Year	18	18	18	19	19	20	20	20	20	20	21	19	18	18	18	18	18	18	18	18	18	18	18	18

(*b*) *Monthly variations.* The day-to-day variations of wind speed and direction in Great Britain and countries with a similar climate

Table VI

Place or region	Monthly mean wind speed in m.p.h.												Yearly mean speed (m.p.h.)	Percentage departures from the annual mean		No. of years of observations	Height of anem. above ground (feet)
	Jan.	*Feb.*	*Mar.*	*Apr.*	*May*	*June*	*July*	*Aug.*	*Sep.*	*Oct.*	*Nov.*	*Dec.*		*Maximum*	*Minimum*		
Northern hemisphere																	
England and Wales*	14·4	13·8	12·5	12·3	10·8	10·7	10·6	9·9	10·8	12·8	13·1	13·5	12·1	19·0	18·2	15	33
Paris (Eiffel Tower)†	22·8	22·0	20·8	19·6	18·7	17·0	16·6	17·9	18·4	20·8	20·7	22·0	19·8	15·2	16·2	Long period	1000
Moscow	10·8	10·6	10·3	9·6	9·7	8·9	8·9	8·6	9·1	9·9	10·3	10·1	9·7	11·3	11·3	10	—
Lindenberg—Berlin	13·2	12·3	12·1	11·4	10·8	10·1	9·6	10·3	9·9	10·5	11·7	12·3	11·3	16·8	15·0	30	30
Prague Karlov	6·7	6·2	5·7	10·0	7·2	7·4	6·5	6·4	7·2	6·4	6·2	5·1	6·7	49·3	23·9	—	—
Brindisi	10·8	12·2	13·5	9·8	9·6	10·2	11·1	11·1	9·8	10·4	11·2	12·6	11·0	22·7	12·7	3	Surface
Tripoli	11·2	9·4	9·4	10·3	10·2	9·2	8·0	8·9	9·5	8·4	10·6	10·5	9·6	16·7	16·7	3	Surface
Ramallah (Jordan)	16·2	13·2	15·3	12·6	11·3	11·9	12·6	10·7	10·0	9·5	10·0	11·3	12·1	36·0	21·5	4	70
Seistan (Persia)	4·2	5·2	5·7	7·5	9·6	11·7	14·2	13·7	11·9	6·9	5·2	4·0	8·3	71·0	51·8	Long period	22
Perim (Adan)	14·5	14·0	14·1	13·1	10·0	9·8	11·2	10·1	8·8	13·5	14·7	14·1	12·3	19·5	28·5	,,	—
Karachi	7·7	8·8	11·3	12·2	14·5	16·6	16·4	14·9	11·7	8·2	6·6	11·4	11·7	42·0	43·6	,,	44
U.S.A. (New England)	11·0	11·2	11·4	11·0	9·8	9·0	8·6	8·4	8·8	9·6	10·5	10·6	10·0	14·0	16·0	Long period	22 stations
Southern hemisphere																	
Wellington (N.Z.)	9·3	9·7	8·9	8·1	8·0	8·2	7·5	8·2	9·3	10·2	10·1	9·5	8·9	14·6	15·7	3	35
Port Elizabeth (S.A.)	12·2	11·8	11·5	10·8	9·6	10·0	10·2	11·0	12·6	13·3	13·3	12·9	11·6	14·6	17·2	8	50
Perth (W. Australia)	13·8	13·5	12·8	10·7	10·6	10·6	11·2	11·8	11·8	12·6	13·4	13·9	12·2	13·9	13·1	20	71
Sydney (N.S.W.)	8·9	8·1	7·5	7·0	6·8	7·1	7·2	7·4	8·0	8·2	8·5	8·9	7·8	14·1	12·8	26	58
Adelaide (S. Australia)	9·9	8·8	8·3	8·0	8·1	8·3	8·5	9·2	9·2	9·8	9·9	9·9	9·0	10	11·1	30	75
Ascuncion (Paraguay)	2·9	3·0	3·0	3·2	3·6	4·0	4·6	4·7	5·0	4·5	3·5	3·0	3·7	35	22	—	—
Rio Gallegos (Argentina)	9·3	10·5	9·9	9·5	7·8	7·3	5·9	5·8	8·4	10·1	11·7	10·5	8·8	33	34	—	—
Buenos Aires	10·1	10·5	8·9	9·9	8·4	8·8	9·8	10·0	11·0	10·1	10·3	10·2	9·8	12	14	—	—
Campos (Rio de Jan.) Brazil	9·2	10·3	7·4	5·6	6·0	6·9	6·3	7·4	9·2	7·4	9·6	8·5	7·8	32	28	—	—
Recife (Pernambuco) Brazil	13·6	13·0	11·6	12·5	15·9	14·5	13·6	13·6	14·5	13·6	13·0	12·5	13·6	17	15	—	—

* Average for 20 stations. † At 1000 ft.

are often great and are unpredictable much in advance. Although the five months from May to September are rather less windy than the remaining seven, during which the strongest gales most frequently occur, neither a calm spell nor a high wind can be depended upon at any date. In some other countries with less temperate climates there is rather more regularity in the daily wind during certain seasons of the year.

There are, however, quite strongly marked tendencies each year in the monthly mean wind speeds. These are indicated in Table VI which gives monthly mean speeds for a number of widely spaced places in both the northern and southern hemispheres. The largest percentage variations in the table happen to be for districts near to large mountain masses in hot climates but, even for especially windy sites on well exposed hilltops near the coast in Great Britain, the maximum and minimum percentage variations are of the order of 25 per cent.

Owing to the cube law relating wind power and wind speed the variations in energy production from month to month are even more marked as shown in Fig. 16 which refers to the Meteorological Office station at Butt of Lewis (NW. Scotland). Fig. 17 shows the energy which would be produced by different wind speeds at the same station, during the whole year and during January and August (the windiest and least windy months). This shows that, while in August wind speeds between 10 and 30 m.p.h. are most important for energy production, for the windy months, and for the year as a whole, the range 15–50 m.p.h. contributes most of the energy.

Monthly wind speed variations are important, (*a*) if the wind power is to be used in conjunction with hydro-electric power when the coincidence, or otherwise, of wind and water availability may affect the efficacy of the complete installation (this question, as it affects France, is discussed by P. Ailleret, Ref. 6) and (*b*) if a small or medium-sized wind-driven plant is to be used as the sole (or main) supply to remote premises: the energy requirements at different seasons of the year must then be related to the probable wind speeds during those seasons. The ratios between maximum and minimum monthly wind velocities are usually much smaller than corresponding ratios for the water flow in a steady river.

(*c*) *Annual wind régimes.* Measurements to determine the annual wind régime at a site include those of wind direction as well as of speed. The wind's direction is a matter which mainly affects the choice of a wind power site and is discussed in Ch. 5. Wind speed records may give mean hourly values throughout the year, or only the numbers of hours for which the speed lay within certain limits

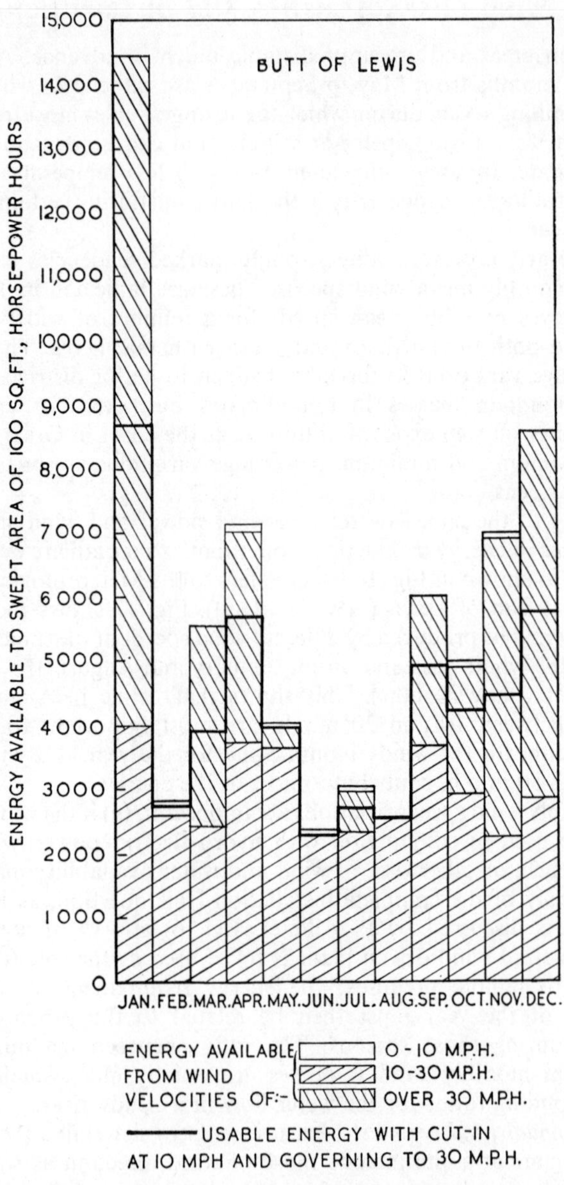

FIG. 16. *Energy available monthly at Butt of Lewis*

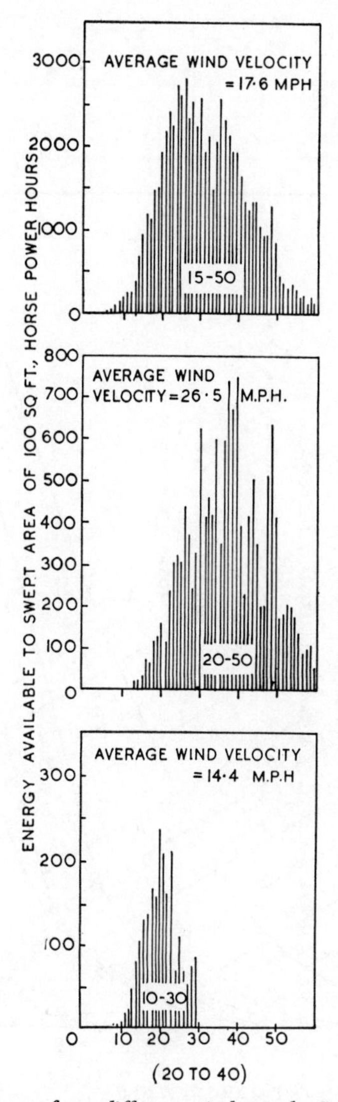

FIG. 17. *Energy from different wind speeds, Butt of Lewis*
Upper figure is for the year, middle figure for January, and bottom for August
The figures in brackets give the most appropriate 20 m.p.h. range of operation for this station

5

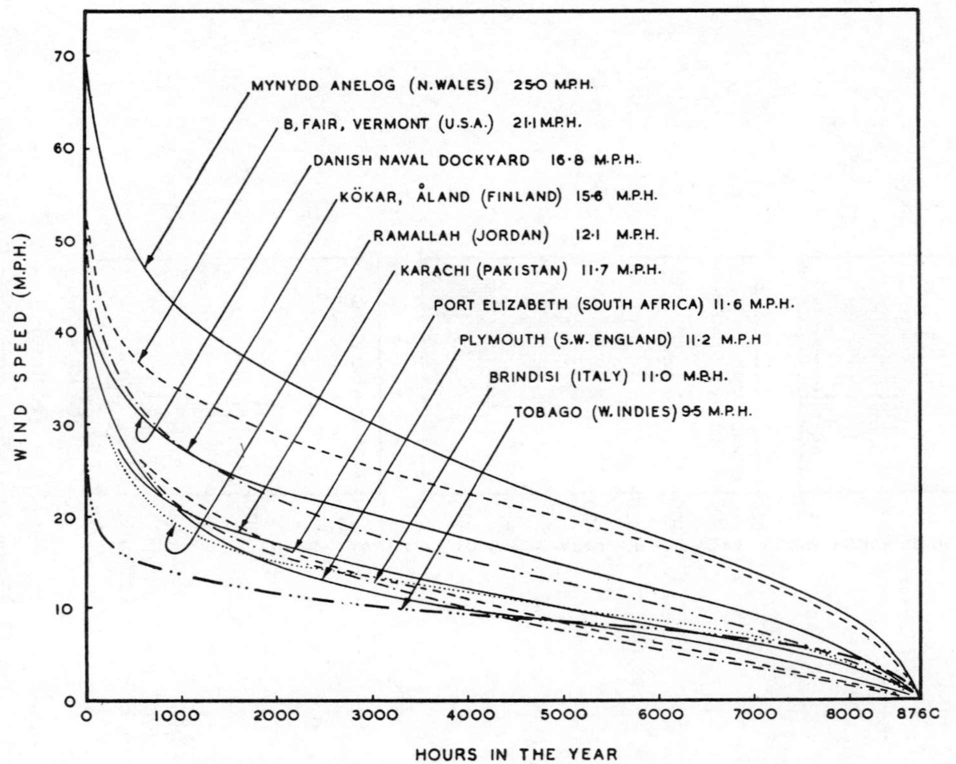

FIG. 18. *Velocity duration curves for widely separated sites*

such as those recognized in the Beaufort scale (see p. 101). The most useful form of presentation for wind power purposes is the velocity-duration curve which has been introduced in Ch. 3.

In Fig. 18 velocity-duration curves are shown for a range of annual mean wind speeds while in Fig. 19 are shown power curves for three sites having annual mean speeds of 15, 20 and 25 m.p.h. The power curves emphasize the importance of a high wind speed. The velocity-duration curves have been drawn for sites, in different parts of the

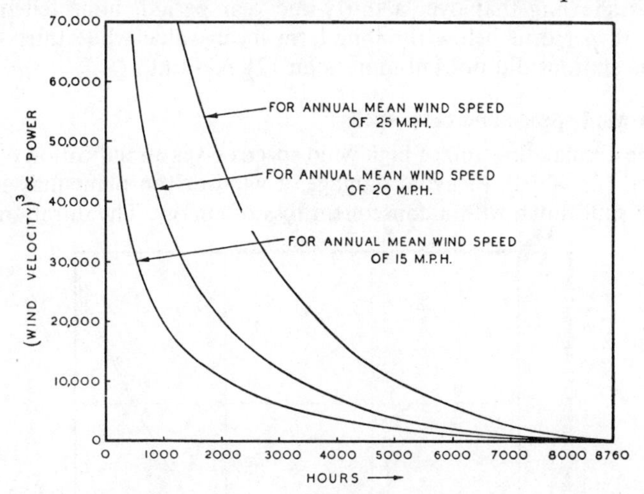

FIG. 19. *Power duration curves for three annual mean wind speeds*

world, with different annual mean wind speeds. It will be noted that the curves for approximately the same annual mean wind speed are so similar in shape that they almost coincide.

As might be expected, annual wind speeds vary widely for different regions and altitudes. Because of the variations from point to point, even in any particular region, according to the exposure, altitude and surrounding terrain, it would be wasteful of space to attempt a list of these annual means. They range, however, from under 1 m.p.h. to 40 m.p.h. or even more. Thus, Dibrugarh (Assam), which must be one of the calmest places for which wind records exist, has an annual mean speed of 1·0 m.p.h. (Ref. 2) while Putnam (Ref. 1) gives a value of 34 m.p.h. for a five year period at the summit of Mt. Washington (6300 ft). As shown in Table VI fairly common values lie in the range 10–15 m.p.h. A number of specially chosen sites in Great Britain and Ireland at altitudes up to 2000 ft have annual speeds between 24 and 28 m.p.h.

(*d*) *Long-term mean wind speeds.* A natural question which arises, when the régimes are considered, is whether these are fairly constant over a long term. Fortunately, for wind power purposes, they are. Thus, for example, forty-two years of continuous records for Southport (England) show that the annual mean wind speed varied between 84 and 118 per cent of the long-term average while for thirty-seven of these years the variation was only between 90 and 110 per cent.

P. H. Thomas (Ref. 7) gives figures for fifty stations in the United States showing that over a thirty-one year period, none fell more than 18 per cent below the long-term mean value while thirty-one of the stations did not fall more than 12½ per cent.

High wind speeds and calm spells

The annual duration of high wind speeds gives an indication of the period for which, to avoid damage, a wind-driven plant may have to be shut down with a consequent loss of energy. The duration for

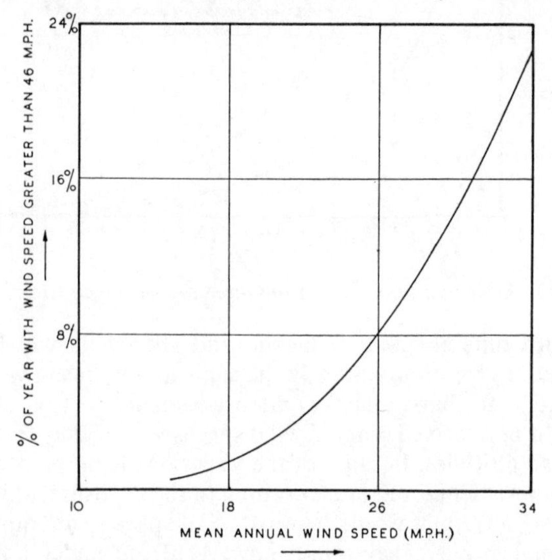

FIG. 20. *Annual duration of wind speeds exceeding* 46 *m.p.h.*

wind speeds above (say) 60 m.p.h. is a very doubtful quantity but it is more easily possible to give the annual duration of speeds above some reasonably high value such as 46 m.p.h.—the upper limit of Beaufort force 8. This increases with the annual mean wind speed for the site as shown in Fig. 20 which has been drawn from records for a number of widely separated sites. Thus, at good sites in Great

Britain where the annual mean wind speed is around 26 m.p.h., hourly wind speeds above 46 m.p.h. will persist for some 8 per cent of the year. At these sites, wind speeds exceeding 50 m.p.h. for more than 24 hrs continuously are not uncommon while speeds above 70 m.p.h. are observed for a few hours each year, as shown in Table VII.

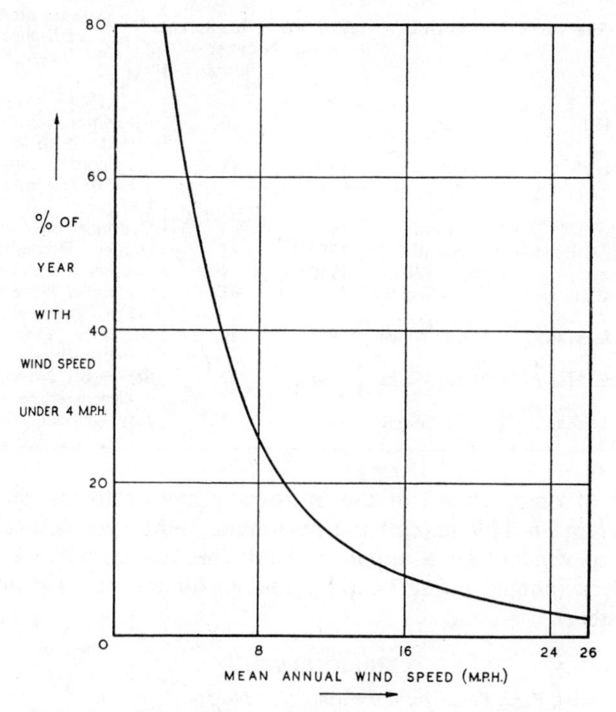

FIG. 21. *Annual duration of wind speeds under* 4 *m.p.h.*

The annual duration of calm spells is important because it indicates the period which must be covered by storage when small or medium-sized wind-driven plants are used autonomously. Fig. 21 plotted—as a mean—from records for a large number of stations in different parts of the world, shows how the annual percentage of winds under 4 m.p.h. increases as the annual mean wind speed falls. This percentage is seen to range from 30 per cent when the mean wind speed is 7 m.p.h. (below which wind power can scarcely be economic) down to perhaps 3 per cent or less when the mean wind speed is around 25 m.p.h.

The number of consecutive hours of calm weather bears especially on storage requirements. At even moderately windy places lengthy periods of low wind speeds can occur. For example, in a ten year

TABLE VII

Site	District	Year	Mean hourly wind speed over 70 m.p.h. Number of hours	Remarks on months of occurrence
Costa Hill	Orkney	1949	30	All winter months (October to March)
Costa Hill	Orkney	1950	11	No record for November or December 1 hr June
Carn Brea	Cornwall	1950	3	December only
St. Agnes Beacon	Cornwall	1950	11	October, December
Foel Eryr	South Wales	1950	4	October, December
Rhossili Down	South Wales	1951	32	September, November, December, February
Mynydd Mawr	North Wales	1950	19	November, December, February
Mynydd Mawr	North Wales	1951	32	September, November, December, February
Slieve Leahan	Co. Donegal	1951	2	July, October

period of observations, at the Stornoway meteorological station, (annual mean 14·9 m.p.h.) the maximum number of consecutive hours of wind under 8 m.p.h. was 110 (in August) while that at Aberdeen (annual mean 9 m.p.h.) the maximum was 196 hrs (in November).

BIBLIOGRAPHY

(1) PUTNAM, P. C. *Power from the wind.* Van Nostrand (1948).

(2) IYER, V. DORAISWAMY. Wind data for windmills. *Scientific Notes*, Vol. VI, No. 63, India Meteorological Department (1948).

(3) KIDSON, E. and EWART, M. E. A year's wind records. Meteorological Office Note No. 14, *The New Zealand Journal of Science and Technology*, Vol. XV, No. 3, p. 208 (November 1933).

(4) *Surface winds of South Africa.* Weather Bureau, Department of Transport, Pretoria, W.B. (June 1949).

(5) BURGESS, C. R. Local weather. *Weather*, Vol. 6, Nos. 5 and 6, pp. 144 and 163 (May and June, 1951).

(6) AILLERET, P. L'énergie éolienne: sa valeur et la prospection des sites. *Revue générale d'électricité* (Mars 1946).

(7) THOMAS, PERCY H. *Electric power from the wind.* Federal Power Commission (1945).

(8) BERRY, F. A., BOLLAY, E. and BEERS, NORMAN R. *Handbook of Meteorology.* McGraw-Hill Book Company Inc. (New York, 1945).

(9) *Climatological atlas of the British Isles*. Air Ministry, Meteorological Office, H.M. Stationery Office (1952).

(10) PICK, W. H. *A short course in elementary meteorology*. Air Ministry, Meteorological Office, H.M. Stationery Office (1938).

(11) CAMBILARGIU, E. La energía del viento en el Uruguay. Instituto de Máquinas, *Publicacion No.* 12, Montevideo (Uruguay, 1953).

(12) Studiengesellschaft Windkraft e.V., Stuttgart. *Windkraft Mitteilungen*, No. 3 (1st April 1954).

(13) WORLD METEOROLOGICAL ORGANIZATION. Energy from the Wind. Assessment of suitable winds and sites. *Technical Note No.* 4. WMO—32.TP.10 (1954).

(14) SNOWDEN, D. FLORA. *Tornadoes of the United States*. University of Oklahoma Press, at Norman, Oklahoma, U.S.A., 1953.

(15) PRIVETT, D. W. Some new wind charts for the south Atlantic Ocean. *Weather*, Vol. IX, No. 1, pp. 9–13, January, 1954.

(16) RAMDAS, L. A. and RAMAKRISHNAN, K. P. Wind energy in India. *Symposium on Wind and Solar Energy* (New Delhi, 1954). UNESCO/NS/AZ/191/Ann. 18.

(17) BHATT, U. J. Possibility of wind power generation in Saurashtra. *Symposium on Wind and Solar Energy* (New Delhi, 1954). UNESCO/NS/AZ/191/Ann. 17.

(18) RAMIAH, R. V. Some problems in the utilization of wind power in India. *Symposium on Wind and Solar Energy* (New Delhi, 1954). UNESCO/NS/AZ/191/Ann.4.

(19) FRENKIEL, J. Wind power research in Israel. *Symposium on Wind and Solar Energy* (New Delhi, 1954). UNESCO/NS/AZ/191/Ann.22.

(20) GOLDING, E. W. *The economic and practical aspects of utilizing wind energy in arid areas*. United Nations Educational, Scientific and Cultural Organization (Paris, 10th August 1953). UNESCO/NS/AZ/139.

(21) SERRA, L. Le vent en France et ses possibilités d'utilisation. *La Météorologie* (October–December 1953), pp. 273–292.

See also meteorological references listed in Appendix I.

CHAPTER 5

WIND POWER SITES

It is almost unnecessary to say that a favourable site for wind-driven plant must have a high average wind speed, although the actual value of this speed for economic operation will vary greatly with circumstances—particularly with the generating costs for alternative means of power production. This will be considered more fully in a later chapter. We are concerned here rather with the features which lead to an exceptionally high wind speed and with other factors influencing the choice of sites.

Average wind speed

The annual average wind speed at any site depends upon:

(*a*) Its geographical position.

(*b*) Its more detailed location as, for example, its altitude and distance from the sea.

(*c*) Its exposure; in particular, its distance from higher ground likely to give screening, especially in the direction of the prevailing wind.

Under the same heading should be included the nature of the surrounding ground, even though it is not high; very broken ground with rocks, woods or groups of buildings, can have a serious retarding effect upon winds near the surface.

(*d*) The shape of the land in the immediate vicinity. Since, to gain altitude and so obtain an increased wind speed, sites are chosen on hill tops, this means, in fact, the shape of the hill. A steep, though smooth, hill may accelerate the wind over the summit, through compression of the lower layers of air by those above, and so produce a wind speed which may be 20 per cent or more higher than that which would be expected at the altitude of the summit (see Ch. 7).

On the other hand, the hill should be without precipitous faces which produce turbulence and, for the same reason, the peak, or summit ridge, should not be too sharp.

Other factors affecting the choice of site

A low cost of construction for wind-driven plant being essential for economy, factors such as the distances from railways and main roads, the accessibility of the summit by road or track capable of carrying heavy loads, and the nature of the ground itself—as affecting the cost of foundations—are all important. The probable purchase price of the land on the hill-top also needs some consideration. Fortunately, very windy hill-tops are usually so devoid of vegetation or, indeed, of any covering other than peat and low heather or coarse grass (in such countries as Great Britain) that their agricultural value is very low.

The distances from an electricity supply line, and from load centres, also influence the choice of a site. The importance of this point depends upon the capacity of the plant which may be installed on the hill and on its distance from other probable sites. If a large total capacity can be installed, the cost of connection to a network at some distance may be a negligible part of the total cost of construction, though long transmission distances to load centres must involve undesirable energy losses.

Typical good sites

Plates VII, VIII and IX show views of three hills which have already been proved, from measurements over periods varying from one to three years, to have the characteristics required for excellent wind power sites (Ref. 1). Maps are given also to show their geographical positions and main contours. Plate VII shows Costa Hill (500′) on the Mainland of Orkney on which a 100 kW experimental wind-driven generator has been built. It has a long-term annual average wind-speed of 25 m.p.h. at 10 ft above ground.

Plate VIII shows Mynydd Anelog (628′) on the Lleyn peninsula in Caernarvonshire, where another 100 kW experimental wind power unit may be built. Its long-term annual average wind speed is 26 m.p.h. at 10 ft above ground.

Plate IX shows Bloody Foreland (1038′) in Co. Donegal which has a long-term annual average wind speed of 27 m.p.h. at 30 ft above ground.

All three hills lie very close to the sea, are almost equally well exposed and have smooth, steep slopes. Costa Hill is a short ridge running roughly north to south but the others are more conical in shape. Although this may be coincidental, it is interesting to note that the magnitudes of the average wind speeds are in the same order as the heights of the hills. This might be expected, but it is perhaps more significant that the average wind speeds are so similar for sites situated so far apart.

The grouping of wind power units

Some very windy sites are to be found on the tops of hills with a summit area of only one or two acres. Unless very small units were to be considered, there is obviously no space on such a hill for more than one. On the other hand, hills in the form of a 'mound', with fairly steep sides but having a much greater area at the top, may be satisfactory, provided that the summit area is not so flat that the wind breaks away from the surface and passes over the flat area at a height greater than that of a windmill located there. Ridges whose directions runs athwart the prevailing wind are likely to give good results and often have a considerable total area along their summits.

On mounds or ridges, it may be possible to locate more than one generating unit with consequent economy both in construction and maintenance costs. Usually a service road will have to be laid up the hillside and its cost will be almost the same whether a single unit, or several, are to be installed. The number of units will depend (i) upon what is eventually proved to be the most economic size in capital cost per kilowatt of capacity, (ii) upon the total capacity which it is intended to install at the site and (iii) on the extent to which one wind turbine will affect the performance of another located down wind from it.

Although there is not yet conclusive evidence from which the most economic size of unit can be derived it is suggested in Ch. 16 that this may have a capacity of around 3000 kW. Thus an intended total capacity of up to this value could be perhaps supplied from one unit but for a larger installation several might be needed.

Precise information on the influence of one wind turbine upon a second one, when directly down wind from the first, is lacking but, energy having been taken out of the wind by the first machine which reduces the wind velocity as it passes through its disc, the second will be screened. The amount of the screening will depend upon the distance apart of the two machines. At a great distance the original velocity of the retarded air will be recovered through its contact with the surrounding unretarded air. It is probable that the air velocity will be sufficiently recovered for the effect to be negligible after a distance equal to some eight diameters of the disc swept by the wind turbine. A 3000 kW unit would have a disc diameter of about 200 ft when designed for wind régimes in British coastal districts so that a spacing of 1600 ft might be needed to avoid interference completely. It is, however, doubtful whether such a large spacing will prove necessary. The wind direction changes, by at least a few degrees, so frequently that any effect experienced would not be continuous.

In some estimations which have been made of the potentialities of certain hills a horizontal spacing of a quarter of a mile (i.e. 1320 ft) has been assumed and this may be taken as a reasonable spacing of any large sets.

Experimental data from a wind turbine installation must be obtained before any more accurate estimates can be made. For this purpose high speed measurements of wind distribution over an imaginary disc area should be included because it is possible that turbulent effects in the wake of a machine may be more serious than the actual diminution of wind speed due to its screening action.

The uncertainty concerning this spacing, however, only affects the grouping of wind power units on single, flat-topped hills or mounds. Long ridges, on the summits of which a number of units could be located, are much less likely to be influenced by it because, in the first place, their summit lines are seldom dead straight and, secondly, on the occasions when the wind blows along the general line of the ridge it is unlikely that it will follow this line throughout without being deflected.

Wind-power areas

In some parts of Great Britain and Ireland—and probably also in other countries—there exist clusters of hills on each of which one or more wind turbines could be installed to provide, by local interconnection, a significant total capacity.

The development of such hill clusters which would, in effect, constitute a 'wind-power area' comparable with a hydro-electric scheme, appears to have some advantages over scattered installations although the latter, if strategically placed relative to a supply network, might be of operational benefit to the system. Some of these advantages are:

(i) Reduction of transmission costs which would improve the economic potentialities of a group of sites at some considerable distance from the supply network.

(ii) Reduction of the cost of transporting the materials of construction: a central depot, storehouse and assembly workshop could be established.

(iii) Reduction of maintenance costs through the possibility of a small staff being able to concentrate on attention to a close group rather than having to travel round an extensive district.

(iv) Clusters of hills suitable for such development are of little value for agriculture or any other purpose; they are bare of trees and other vegetation, while their smoothly-rounded shapes do not render them specially attractive scenically. As an energy source they constitute, however, an important asset which could be used without any obvious disadvantage.

The same needs for good exposure and for the absence of screening by higher ground, especially in the direction of the prevailing wind, influence the choice of such hill clusters as they do that of more isolated hills. It frequently happens that the individual hills in the group are of approximately the same height so that 'internal' screening may not be very serious.

Only by wind measurements over a period of a year or more can the question of suitability be settled with any certainty but, from inspection of various windy districts and from the results of wind surveys in some of them (Ref. 1) possible 'wind power areas', in Great Britain and the Republic of Ireland, can be suggested. These include part of the Orkney Mainland, in particular the central ridge of hills running from Erie to Finstown, the range of hills on the west coast of Sutherland, the Mull of Kintyre, the hills on South Uist and those in the west and south west of Harris, the Antrim plateau (although the average wind speed there is not very high), the Sperrin mountains in Co. Tyrone, the Derryveagh group of hills in Co. Donegal and the peninsula lying west of the line joining Ardara and Killybegs, also in Co. Donegal. The situation of the last of these—a good example of what is in mind—is shown in Fig. 22 which gives the general position of the hill cluster and also an enlargement to show the main contours of the hills. The peninsula consists, in fact, of a ring, about twelve miles diameter, of smoothly-shaped hills varying in height between 1200 and 1700 ft with a relatively shallow depression forming a plain, or tableland, within them. On the lower ground there is some agriculture but the hills themselves are used only for sheep grazing. The exposure to the sea is good with no screening from south-east to north and only at a considerable distance from the east.

Effect of wind direction: prevailing winds

The exposure of a site in the direction of the prevailing wind is an important criterion affecting its choice for wind power purposes but this importance should not be exaggerated. The term 'prevailing wind' means that the wind blows from one direction more commonly than from any other direction. But often the percentage of the year during which it blows is not outstandingly higher than the percentages from some other directions. Taking the eight main points of the compass, a site with no prevailing wind would have winds from each of these directions for $12\frac{1}{2}$ per cent of the year. The prevailing wind direction may thus mean simply that the percentage from that direction is greater than $12\frac{1}{2}$ and, in fact, it commonly lies between 15 and 20 per cent in Great Britain. In some parts of the world this percentage is much higher.

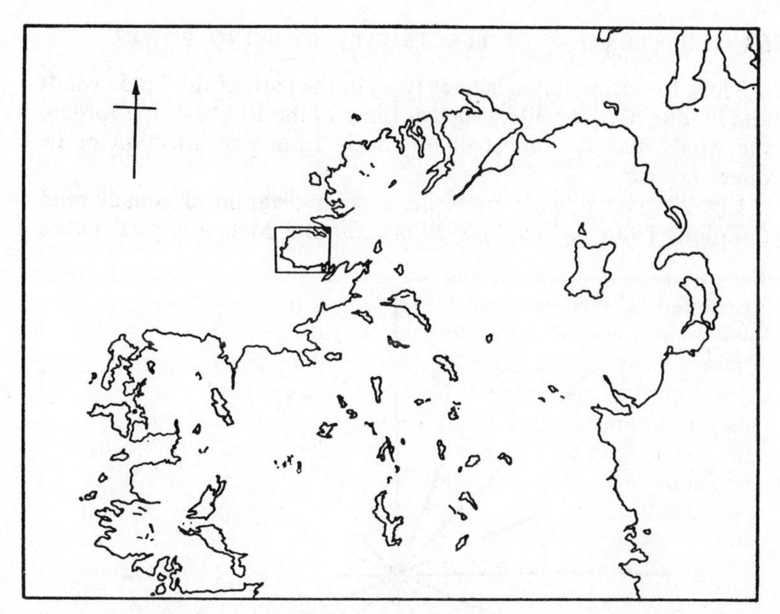

FIG. 22 (a). *Location of Killybegs Peninsula, Co. Donegal*

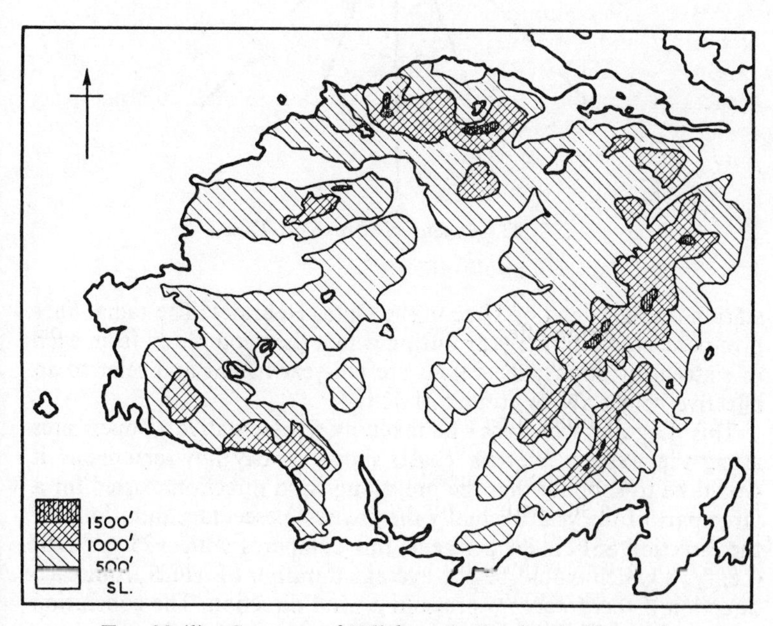

FIG. 22 (*b*). *Contours of Killybegs Peninsula, Co.Donegal*

Thus, for example, on islands lying in the path of the Trade winds and in long narrow valleys such as those of the Rhône or the Jordan, the winds may be almost continuously from one direction or its direct reverse.

Fig. 23 gives a 'wind rose', i.e. a polar diagram of annual wind durations from different directions, for the Meteorological Office

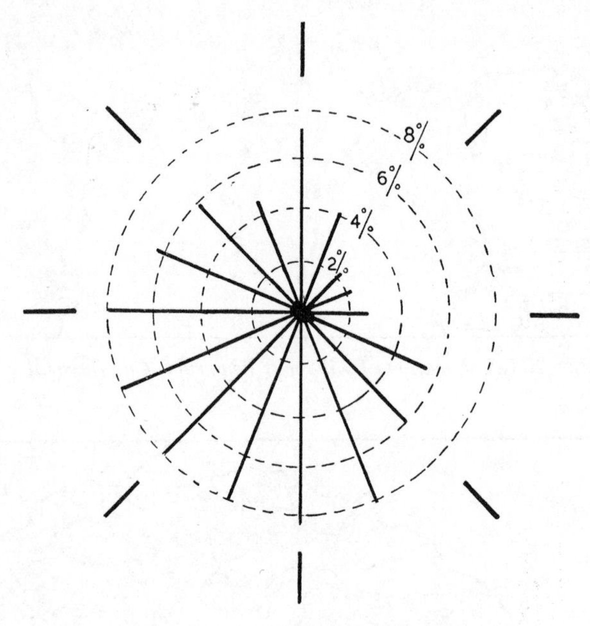

FIG. 23. *Annual wind rose for Tiree*

station on the island of Tiree in the Outer Hebrides. The radial lines give percentages of the year during which the wind blows from each of sixteen directions; the values are ten-year means and refer to an effective height above ground of 42 ft.

This diagram, which may be taken as representative of open sites along the western Scottish coast, shows clearly how erroneous it would be to suppose that the prevailing wind direction existed for a large part of the year. Actually the greatest percentage duration, for the direction SSE. is 8·2 per cent; this compares with 6·25 per cent (i.e. $\frac{100}{16}$) which would be the average duration of winds from each direction if there were no prevailing wind direction. The conclusion to be drawn from the diagram is that, except for the directions between ESE. and NNE., the wind through the year is fairly uniformly distributed between all directions.

Fig. 24, taken from a recent E.R.A. report (Ref. 1), is for a much shorter duration of time but is included because the wind roses shown have a direct bearing upon the influence of wind direction on wind energy. The diagrams, which relate to Costa Hill, Orkney, in the month of September 1949, show:

(*a*) a wind rose giving, for the eight main directions, the numbers of hours of wind of mean hourly velocities between 17 m.p.h. and 60 m.p.h. (probable limits of operation for a wind-turbine);

(*b*) a rose of wind-run, in miles, within the same velocity limits;

(*c*) an energy rose; this gives the energy, per unit swept area, available from different directions assuming the same limits of operating wind speed and, further, that the power generated will be constant, at its full value, for speeds between 30 m.p.h. and 60 m.p.h.

The most striking feature of the three rose diagrams is their similarity; from this the conclusion can be drawn that, at least for the short period considered (one month), no direction of wind is responsible for more than its due proportion of energy; otherwise expressed, a wind rose giving durations of wind directions is also a good indication of the energy contributed by each of these directions. It is probable that this conclusion applies also to longer periods of time so that there would appear to be little point in undertaking the laborious task of drawing energy roses for wind power sites.

As an example of a locality with a very definite prevailing wind direction Aqaba Bay, on the northern tip of the Red Sea, can be chosen. Its wind rose is shown in Fig. 25 from which the effect of the long narrow valleys of the Jordan and the Wadi Araba, in inducing northerly winds, can be seen.

The annual percentage durations of wind at this site for the period 1945–47 from the eight main directions are given in Table VIII from which it is seen that for 78 per cent of the year the wind is from the north or north-east.

TABLE VIII

Aqaba Bay

From data supplied by the Israel Meteorological service

Direction	N.	NE.	E.	SE.	S.	SW.	W.	NW.	Calm
Percentage of year	59	19	4	1	3	5	1	8	—

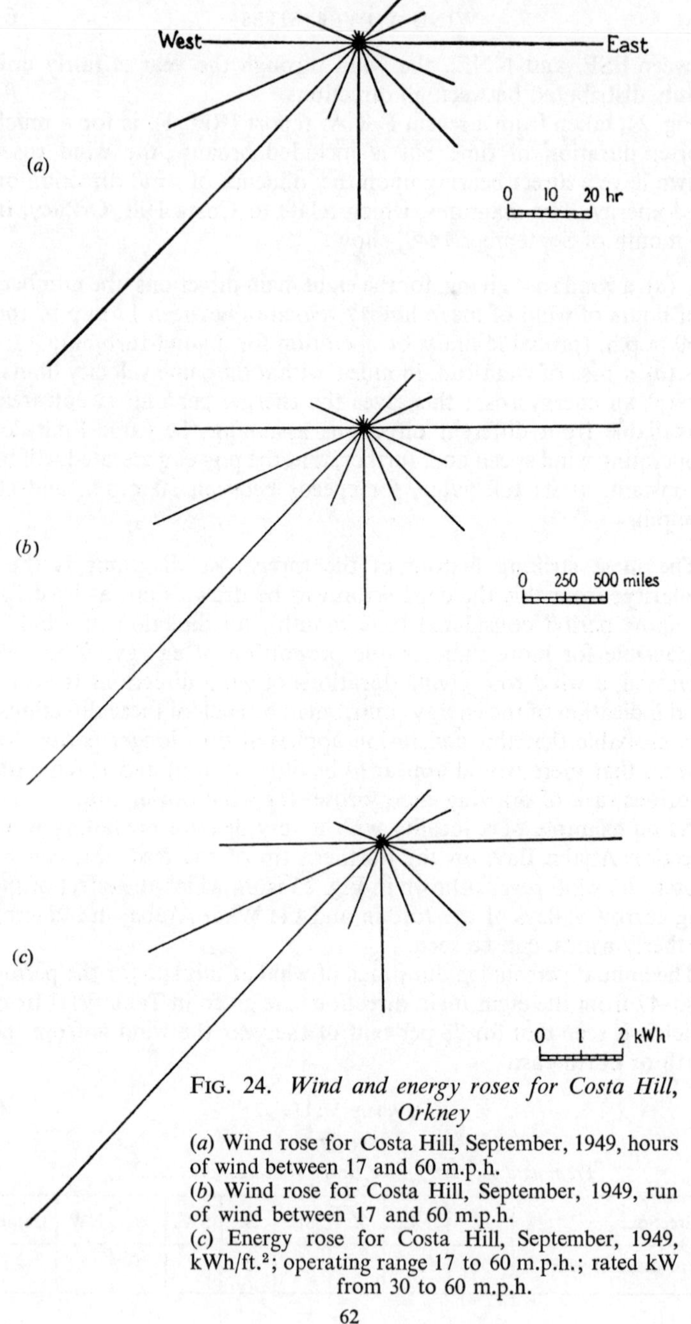

FIG. 24. *Wind and energy roses for Costa Hill, Orkney*

(*a*) Wind rose for Costa Hill, September, 1949, hours of wind between 17 and 60 m.p.h.
(*b*) Wind rose for Costa Hill, September, 1949, run of wind between 17 and 60 m.p.h.
(*c*) Energy rose for Costa Hill, September, 1949, kWh/ft.^{2}; operating range 17 to 60 m.p.h.; rated kW from 30 to 60 m.p.h.

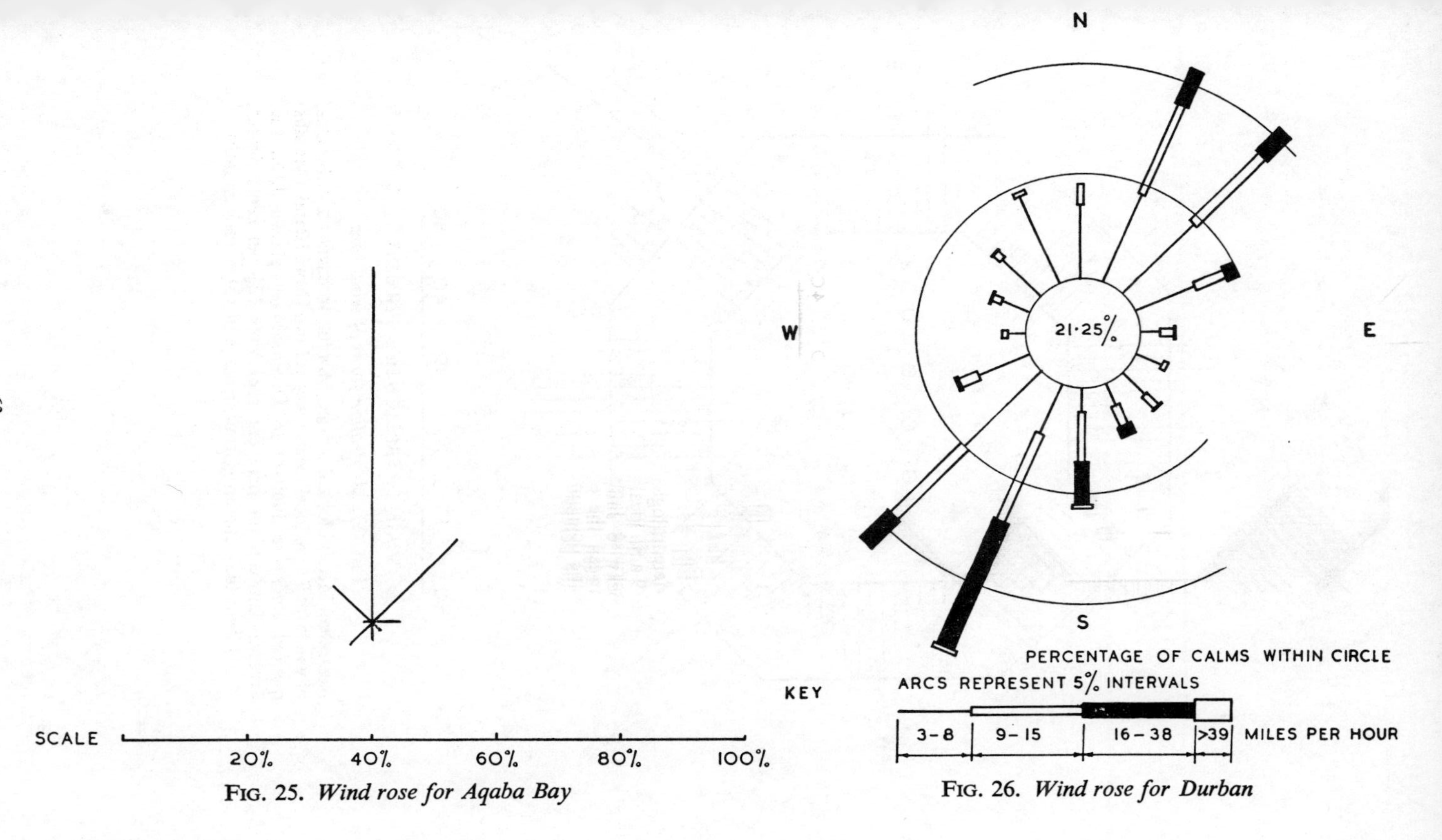

FIG. 25. *Wind rose for Aqaba Bay*

FIG. 26. *Wind rose for Durban*

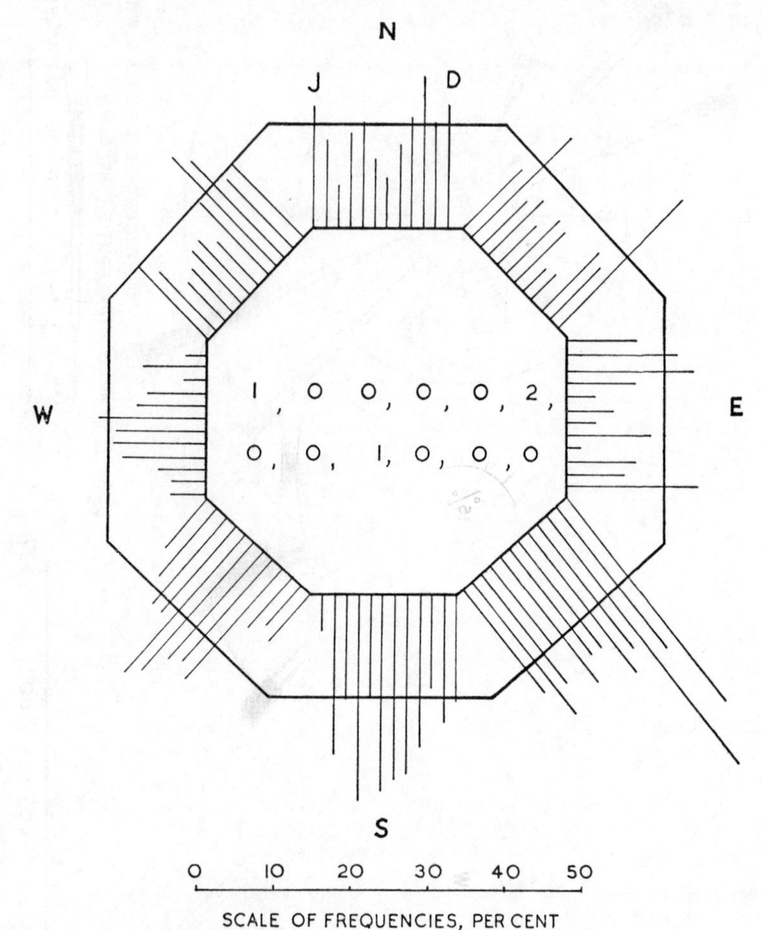

FIG. 27. *Australian form of wind rose*

Australian Meteorological Service. Monthly frequency distributions of winds (at 9 a.m. local mean time) at Lord Howe Island. (Months marked clockwise January to December—as shown J.D.: the distance between the octagons represents $12\frac{1}{2}$ per cent; figures within the octagon give percentage of calms each month)

As a contrast, the annual wind directions at Saskatoon (Saskatchewan) compiled for the period 1922–47 (Ref. 3) are given in Table IX.

TABLE IX

Saskatoon

Direction	N.	NE.	E.	SE.	S.	SW.	W.	NW.	Calm
Percentage of year	13	8	9	15	13	13	16	12	1

Here there is no marked tendency towards a 'prevailing' wind.

Fig. 26 gives, in more elaborate form, the wind rose for Durban (Ref. 5) where the commonest winds, from either the south-west or the north-east, are also the strongest.

Fig. 27, included to illustrate a different method of drawing a wind rose, used by the Australian Meteorological service, shows the percentage durations of winds from eight directions for each of the twelve months. The persistence of a wind direction beyond the average percentage of $12\frac{1}{2}$ (for eight directions) is indicated by the lines which project beyond the outer octagon. Although the prevailing winds are from the quadrant SW. to SE. the directions are here uniformly distributed.

BIBLIOGRAPHY

(1) GOLDING, E. W. and STODHART, A. H. The selection and characteristics of wind power sites. Electrical Research Association, *Technical Report*, Ref. C/T108 (1952).

(2) PUTNAM, P. C. *Power from the wind.* Van Nostrand (1948)

(3) BOUGHNER, C. C. and THOMAS, M. K. *Climatic summaries for selected meteorological stations in Canada, Newfoundland and Labrador*, Vol. II. Meteorological Division, Ministry of Transport, Toronto, Ontario (May 1948).

(4) Studiengesellschaft Windkraft e.V., Stuttgart. *Windkraft Mitteilungen*, No. 3 (1st April 1954).

(5) *Surface winds of South Africa.* Weather Bureau, Department of Transport, Pretoria, W.B. (June 1949).

(6) WORLD METEOROLOGICAL ORGANIZATION. Energy from the Wind. Assessment of suitable winds and sites. *Technical Note No.* 4. WMO—No. 32. TP.10 (1954).

(7) AILLERET, P. L'énergie éolienne: sa valeur et la prospection des sites. *Revue générale d'électricité* (March 1946).

(8) GOLDING, E. W. *The economic and practical aspects of utilizing wind energy in arid areas.* United Nations Educational, Scientific and Cultural Organization (Paris, 10th August 1953). UNESCO/NS/AZ/139.

(9) SERRA, L. Le vent en France et ses possibilités d'utilisation. *La Météorologie* (October–December 1953), pp. 273–292.

See also meteorological references listed in Appendix I.

CHAPTER 6

WIND SURVEYS

When installing a small wind-driven generator for electricity supply to isolated premises there is often only a very limited choice of site. The generator—usually of low voltage—must be placed close to the premises to avoid excessive voltage drop. The energy afforded is thus only that which may be obtained within this limitation. A very high output, though desirable, is not an economic necessity. The amenity value of any reasonably high output is often great enough to justify the installation and there may be no really feasible alternative source of power.

On the other hand large plant, supplying its output directly to a network, is in competition with the generating plant in the power stations. It is essential for economy to ensure an adequate annual output. This demands an exceptionally windy site. Hence, a wind velocity survey is the first, and most obvious, step in investigating large-scale wind power potentialities in any country or region.

Such a survey will consist of several stages and, although these may vary in detail according to the ideas of the investigators and to geographical and other conditions, they are likely to take the same general form. Thus, there will be a first-stage, or preliminary survey to provide an indication of the relative windiness of a large number of widely separated sites and to enable the best to be chosen. This may be followed by a second-stage survey to determine the wind régimes at the chosen sites, and finally—at least at certain sites—more comprehensive measurements may be made for the study of the 'structure' of the wind at them. By 'structure' is meant the detailed behaviour of the wind, including the variations of instantaneous and mean wind speeds with altitude, and of the horizontal distribution of the wind. Measurements of these components of the wind structure give an indication of its distribution over the area which might be swept by a wind turbine. The horizontal and vertical components of the wind at various heights and spacings, and the rate of change of magnitude and direction of gusts, may also be included in the measurements to be made in the final stage.

Meteorological records

The problem of site selection is, of course, greatly simplified if adequate information on mean wind speeds, obtained from long-term records, is available for the region under consideration. Even though the observation stations may be widely separated—to the extent of fifty miles or so—and may be located on low ground, long-term records for well-exposed sites which can be taken as representative of a district are invaluable as a basis. They make it possible, at the outset, to eliminate some districts and to concentrate on those which have proved windier throughout a period of years. Fairly short-term measurements of wind speed on potential wind power sites can then be given a long-term significance by comparing them with similar measurements made at the nearest meteorological station for the identical period of time. The effects of any differences in the methods of measurement used can be checked by installing at the station an anemometer of the same type as that used on the potential site.

Such intercomparison over the same period of time is necessary because, as shown in Ch. 4, there may be considerable differences in the wind speed, in any locality, from year to year.

Unrelated measurements at a selected site may, therefore, be misleading; a high speed over a short period may be due, in part, to the fact that the wind speed for the surrounding district was higher than usual during the period of measurement.

Annual mean wind speeds, computed from continuous records, are the most reliable, and easily applied, data as a basis for site selection. Readings of instantaneous wind speed taken at certain specified times during each day may indicate, in a general way, the relative windiness of different districts but they are not sufficiently reliable for wind survey purposes.

As an example of the usefulness of meteorological information, consider the map given in Fig. 28. This gives 'isovents' (i.e. lines joining places having equal average wind velocities) for Great Britain and Ireland, the average wind speeds in this case being annual ones. This map was prepared by the Meteorological Office from its station records over the period 1926—40. The wind speeds refer to a height of 33 ft (10 m) above ground in open situations. Although, naturally, such an isovent map cannot take into account small local variations in average wind speed it serves the purpose as a first guide to the relative windiness of different districts and thus points to those which are most worthy of fuller investigation.

Contour maps

As already mentioned in Ch. 5, hills of certain favourable shapes accelerate the wind over them so that its average velocity at the summit is greater than may be accounted for by altitude effects alone.

The theory of wind flow over hills is complex and, although some useful conclusions may be drawn from elementary considerations such as those given in the next chapter, evidence is still lacking on the precise shape of hill for maximum accelerating effect. Nevertheless, it has proved possible, from experience, to choose hills, such as

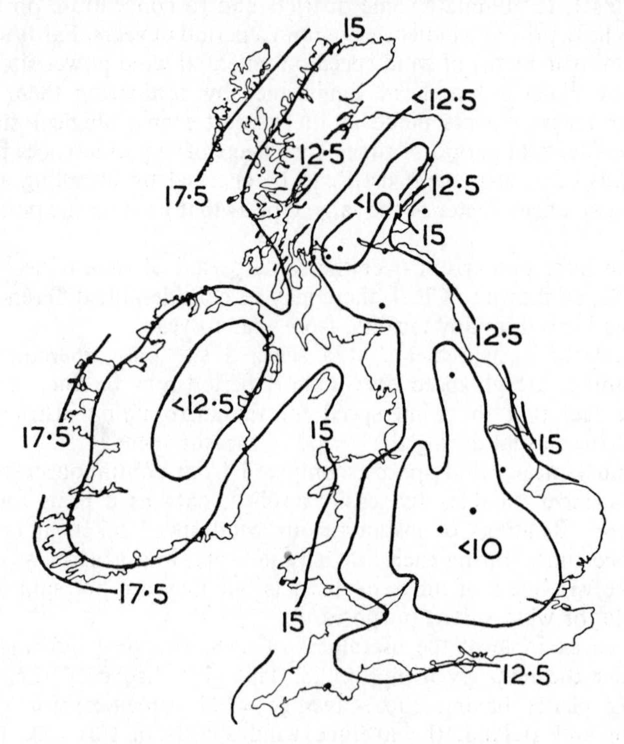

FIG. 28. *Annual isovent map for Great Britain*
(*Air Ministry, Meteorological Office*)

those shown in Plates VII, VIII and IX, giving good—although not, perhaps, optimum—results.

Following examination of meteorological data to select windy districts the next step is, therefore, to study contour maps of those districts and so to make a preliminary selection of their most promising hills. This is best done by first using a $\frac{1}{2}$ in. to the mile map on which the high ground is distinctively coloured and then examining a 1 in. to the mile map giving contours at 50 or 100 ft intervals. The former is easier to read but the latter gives more precise information.

Since the time and effort required to make such surveys on the actual sites are sufficiently great without adding to them unnecessarily, it is wise to use the map fully by observing other details. The most important of these are the nature of the terrain—looking in particular for woods, or rocky ground near the summit, both of which are to be avoided—and also the best approach route to the summit. If no roads or footpaths are shown on the hill, search must be made for a point on the nearest road from which the easiest climb can be made, avoiding if possible long ascents, very steep gradients and marshy ground. It must be remembered that, if the hill is eventually chosen as a potential site, fairly heavy measuring equipment, followed later, perhaps, by the wind power machine itself and constructing gear, may have to be carried over the route chosen.

Larger scale maps such as 6 in. to the mile, or even larger, are not generally useful in the early stages of a survey.

Inspection of sites

However carefully maps are studied, site selection by this means alone should not be relied upon: complementary inspection of the site itself is well worth while. In addition to assisting the investigator to assess the probable screening effects of the nearest high ground and of the disturbing effects, upon surface winds, of minor irregularities in the shape of the hill itself, ascent to the summit affords an opportunity of judging the accelerating effect of the hill through the difference which will be noticed between the wind speeds at the top and bottom of the hill. Frequently it happens that hills chosen from preliminary map studies are discarded after inspection and are replaced by others which did not, from the map, appear so promising. It may well be that with greater knowledge of the subject such anomolies will disappear but they must be accepted, and allowed for, until that knowledge is acquired.

Even in the present "state of the art" it is possible for an experienced observer to use such inspections to place the individual hills in a scattered group in order of windiness with considerable accuracy. This is very helpful when a decision is to be made on the distribution of measuring equipment.

Measurements for the selection of sites

Information has been published on systematic wind-velocity surveys forming part of wind-power investigations in the United States, France and Great Britain. In Denmark, Holland, Germany, Italy, Russia and India, wind data have been collected for this purpose but until quite recently the information has been mainly from established meteorological stations which are not often located at sites which would be suitable for wind power installations. It may be that in some countries there has been insufficient appreciation of the fundamental importance of site selection though, doubtless, the lack of especially favourable hill sites—in flat countries—and the fact, in others, that the intention has been to consider only the possibilities of relatively small installations, supplying isolated communities, have influenced the methods followed.

United States survey

For the Smith-Putnam 1250 kW project at Grandpa's Knob (Ref. 1) a wind survey was undertaken between the years 1940 and 1945 principally in the Green Mountains in Vermont though a few observation stations in adjacent states were included.

Twenty sites were selected. They were at altitudes up to around 4000 ft and were chosen, not so much with the intention of discovering a number of suitable sites, but to obtain information on the probable long-term wind régimes at a few of them from measurements made both on those sites themselves and on others up-wind or down-wind of them (relative to the prevailing westerly winds).

Anemometers (see Ch. 8, p. 125) were erected, usually in groups of three or four, at different heights above ground up to 110 ft, or sometimes more. These were kept under observations for fairly short periods, not often exceeding six months.

It is interesting to note also that ecological evidence, in the form of 'brushing', 'flagging' or 'carpeting' of trees on the hills, was collected in an attempt, thereby, to educe a scale of wind speeds. In the report of this work it is stated that above a critical value of mean wind speed of about 27 m.p.h., trees cannot survive. This is confirmed—though under perhaps rather different climatic conditions—in Great Britain where neither trees nor any vegetation growing to heights greater than a few inches are ever found on hills with annual mean wind speeds above about 23 m.p.h. Plate X shows the permanent effect of strong winds upon trees. This photograph of a row of beech trees was taken at about 1000 ft on a hill slope in the Snaefell range in the Isle of Man but there are no trees on the summits themselves.

The American investigators paid considerable attention to the 'speed-up' factors (see Ch. 5) of the hills selected and they attempted also to formulate rules governing these factors in relation to the ratio of height to base-width of ridges.

It seems a little surprising that the periods of observations were frequently so short—only a few weeks in some instances—though more lengthy periods may not have been necessary for the special purpose which they had in mind. Otherwise, a measuring period of at least one year would appear to be desirable to include all the climatic conditions likely to be experienced. From the results of some of the surveys in Great Britain it is clear that false conclusions on the relative merits of a group of sites could be drawn from observation periods of merely two or three months. There have been several instances of a particular hill showing up very well for the first few weeks and thereafter proving much less satisfactory. This has doubtless been due to a special phase of weather causing the apparently good site to be subjected to abnormal winds in the initial period.

P. H. Thomas (Ref. 2), in discussing the development of wind power, recommends that, in addition to the analysis of Weather Bureau records for diversity and duration-curve data, it is essential to obtain information on wind velocities at the summits of steep ridges. He stresses the need for long-term average speeds to be determined and also emphasizes the importance of obtaining complete duration-curve data from which the available energy can be found.

Thomas suggests that 'while actual and continuous readings on all likely sites would be desirable, as in the case of water power sites, a great deal would be accomplished by readings on a few selected sites, from which the value of other sites could be estimated by means of pattern data'.

French survey

P. Ailleret (Ref. 3) described in March, 1946, a programme, for the selection of wind power sites, which had been approved by the French committee on wind energy. The survey involved was to be based on a specially developed anemometer and equipment (see Ch. 8) recording the energy, in kilowatt-hours, which could (theoretically) be extracted per square metre of swept area by a wind turbine installed on the site.

Starting in October, 1946, most of the 150 instruments then ordered have since been installed on sites distributed over all parts of France. They have been placed mainly on existing structures—such as lighthouses when near the coast—at heights above ground varying from 6 to 300 m. This last height is that for the instrument at the top of the Eiffel Tower the records from which are used as the basis of comparison for all the rest. This is, no doubt, sound because wind velocity records for many years past are available for the Eiffel Tower.

The sites chosen in France are very varied in character and their altitude above sea level ranges from 1 to 1912 m. In some instances two anemometers have been installed at different heights above ground at the same sites to obtain an indication of the increase in wind energy with height. Other instruments have been located very near to the edge of cliffs to obtain information on their influence upon the wind.

While, as mentioned in Ch. 8, there are some objections to measurements integrating wind energy if attempts are made to interpret the results in terms of the energy which might, in fact, be extracted by a wind turbine, they have advantages for a preliminary survey. These are, particularly, the simplicity of the record and the fact that the method is not dependent upon readings being taken at regular intervals.

Surveys in Great Britain and Ireland

These surveys, made by the Electrical Research Association in conjunction with the electricity supply authorities in the areas concerned, are divided into successive stages rather more definitely than those described in the preceding paragraphs. Thus, the procedure has been followed:

(*a*) A first-stage survey, using single, cup anemometers mounted on 10, 15 or 30 ft poles (or masts) with or without a recorder.

(*b*) A second stage in which masts about 70 ft high are used to carry cup-contact anemometers at two or three heights above ground, these instruments operating recorders on the ground.

(*c*) A final stage which is limited to sites actually chosen for the installation of pilot wind power plants. This stage involves a tower —somewhat similar to the 'Christmas tree' used for the Grandpa's Knob measurements in Vermont—carrying a variety of meteorological instruments for the very comprehensive measurements needed to determine the detailed structure of the wind up to at least the height of the hub of the proposed wind-turbine.

The measurements made in this final stage will be discussed in Ch. 11 since they are rather too specialized to be included in a description of more general wind surveys.

First-stage survey. In the E.R.A. Technical Report Ref. C/T108 the sites selected, up to July 1951,* for wind velocity measurements in the first-stage survey are shown in the map reproduced in Fig. 29. A comparison with the isovent map of Fig. 28 shows that most of

* In Fig. 29 sites used up to the end of 1953 have also been included.

FIG. 29. *E.R.A. wind survey sites, Great Britain and Ireland*

the sites investigated have been in the windier western coastal districts.

The instruments used (described in Ch. 8) and the methods followed in their application, have been as follows:

(*a*) Counter-type cup anemometers. These are mounted on 10 ft poles so that the instrument can be read easily from the ground when the visibility is good. Three steps, made from light metal strip were added after some difficulty had been experienced in taking readings during bad weather. This low height of 10 ft is sufficient to avoid disturbances from irregularities of the ground on the bare and smooth hilltops selected. Readings—of run-of-wind, in miles—are taken at weekly intervals by local observers who send them to E.R.A. headquarters where the records are kept.

(*b*) Long-term counter-type, cup. anemometers. Normally the counters return to zero after 10,000 miles of wind have been registered. On high, or very remote, hills whose summits cannot be visited weekly, it is necessary to install counters which will cover a longer period than one week between readings so that a modified instrument reading up to 100,000 miles is used mounted on 10 ft poles.

(*c*) Contact-type cup anemometers. These are usually mounted on 30 ft masts and an impulse recorder (see p. 117) is used with them. This is housed in a box or other form of portable shelter at the foot of the mast unless the summit of the hills is high or inconvenient for local observers to climb weekly. Then, twin cables buried in a shallow trench just sufficient to protect it against damage from animals which may be grazing on the hill, is run down from the instrument to the recorder placed in a shelter on lower ground.

The recorder charts have to be changed weekly and sent back to headquarters for analysis. Mean hourly wind speeds are extracted, classified, and used to plot the velocity-duration curve for the site.

(*d*) Contact-type, cup anemometers with counters. Again, if the hill is high or inconvenient for the local observer to climb frequently, a contact type of anemometer may be used to operate, through a twin cable descending the hill, a telephone message register located on low ground. This form of measuring equipment gives only the run-of-wind, in miles, and is, in fact, only an alternative to (*b*) above.

Method (*c*), which provides data for drawing the annual velocity-duration curve for a height of 30 ft at the site, gives the fullest information but the impulse recorders are rather expensive. They are installed, therefore, only at what are judged, at the outset, to be good sites representative of the best in the district. It has been found by experience (Ref. 4) that the wind régimes and, hence, annual velocity-duration curves, for sites in the same district are so similar that the annual mean wind speed for any site can be used as a measure of the annual wind energy which might be extracted by an aerogenerator with a given range of operation. This energy can be estimated closely, for the site at which the recorder is located, by drawing its power-duration curve—derived from its velocity-duration curve. By comparing the annual mean wind speed at this site with that at another which has a counter instrument, the annual energy for the latter can be obtained with sufficient accuracy for the purposes of a first-stage survey. In this way the work of carrying out the survey can be simplified considerably, and its cost reduced.

It should not be overlooked, however, that the E.R.A. surveys depend for their success upon the assistance received from local observers. In the main, these observers are not technically trained people; they are farm workers, shepherds, postmen, mountaineering club members, or others living near to the sites. Often, however, they have considerable natural aptitude for the semi-technical work of attending to the wind measuring instruments. Somewhat different methods may prove necessary in regions where the local inhabitants cannot be relied upon for such assistance.

In Plates XI and XII are shown some details of installations such as those described. The ancillary equipment used should be robust and simple in construction, light and convenient to carry, and easy to erect.

Second stage survey. Sites for this stage are among those which have already been included in the earlier stage survey and which have shown sufficient promise for them to be considered seriously as sites for wind-driven generators acting as pilot plants.

Measurements are made of run-of-wind at different heights above ground up to about 70 ft. Contact-type cup anemometers are used in conjunction with a photographic recorder (see p. 120) which records the readings of the anemometers and the indications of a wind-direction indicator, half hourly. Thus, hourly mean wind speeds at the different heights, related to wind direction, are easily obtained by analysis and these are used both to confirm the suitability of the site for wind power and to determine the characteristics of the hill in accelerating winds from different directions. The recording paper has to be removed and the camera reloaded weekly by someone having a little experience in photography. To carry the anemometers, tubular steel masts supported by steel guys are used. Unless these are strong enough to be climbed, the anemometers must be duplicated to avoid lengthy interruption of the record when an instrument fails.

Plates XIII and XIV show an installation of this kind together with a specimen of the photographic record.

BIBLIOGRAPHY

(1) Putnam, P. C. *Power from the wind.* Van Nostrand (1948).

(2) Thomas, Percy H. *Electric power from the wind.* Federal Power Commission (1945).

(3) Ailleret, P. L'énergie éolienne: sa valeur et la prospection des sites. *Revue générale d'électricité* (March 1946).

(4) Golding, E. W. and Stodhart, A. H. The selection and characteristics of wind power sites. Electrical Research Association, *Technical Report*, Ref. C/T108 (1952).

(5) Nilakantan, P. Some considerations in the choice of areas for preliminary wind power surveys in India. *Symposium on Wind and Solar Energy* (New Delhi, 1954). UNESCO/NS/AZ/191/Ann.8.

(6) Frenkiel, J. Wind power research in Israel. *Symposium on Wind and Solar Energy* (New Delhi, 1954). UNESCO/NS/AZ/191/Ann.22.

(7) Serra, L. Le vent en France et ses possibilités d'utilisation. *La Météorologie* (October–December 1953), pp. 273–292.

See also meteorological references in Appendix I.

Chapter 7

WIND FLOW OVER HILLS

It is common knowledge that the speed of the wind over a hilltop is generally greater than that over lower-lying flat ground near to it. We find a number of hills named 'windy hill' dotted over Great Britain and the equivalent in other countries such as, in Denmark, Vejrhøj (Weather hill or Stormy hill) or Blaesebjerg (Blowy hill) and, in France, Mont Ventoux.

Previous chapters have discussed both the desirability and the technique of choosing hill sites for windmills whenever the choice of site is not strictly limited by the need to locate the machine close to the place where its power is to be used. It is probable, as suggested by Putnam (Ref. 1) in describing the investigations, already referred to in previous chapters, connected with the Grandpa's Knob aerogenerator, that certain hills, if suitably shaped, cause an acceleration of the wind over their summit to give a speed which is somewhat greater than that due to altitude alone. His conclusion at the end of these studies was, however, that "We have found no criteria by which to make an economically useful quantitative prediction of the effects of topography upon wind flow".

This remains generally true in spite of the considerable amount of both experimental work and theoretical study of the subject which has been undertaken during recent years. It would be very convenient if it were possible to use a precise formula relating the wind speed at a certain height over the summit of a hill of given altitude and shape to the undisturbed wind speed at the same altitude but at a distance up-wind from the hill. This would facilitate the choice of sites; one might almost choose them from close study of contoured maps, if sufficiently detailed, without being familiar with the actual localities. But no such method can be followed: it is not possible to work from a formula. Experience in selecting sites through inspection of the hills, followed by wind measurements over a fairly lengthy period, and attempts to advise others on the technique of site selection, have tended rather to emphasize the value of this experience, or 'knowhow', or whatever other term may be applied to it. This is not to disparage the studies which have been made, and which are continuing at several research centres, with the object of throwing more light on the subject but there are so many factors which occur in nature, influencing the wind flow, that it is a matter of great complexity. The results of these studies must, of course, aid the choice of a site of especial windiness but wind measurements on the spot will probably prove to be the quickest and surest way to find what is needed for wind power production.

Some of the results already obtained will be discussed, in so far as they bear on this particular need or on the effects of wind phenomena upon a windmill subjected to them, and the difficulties encountered in experimental studies on the subject will be mentioned.

The effect of altitude

G. Hellman (Ref. 2 and 2*a*) gave the formula

$$V_h = V_{10}[0{\cdot}2337 + 0{\cdot}656\ log_{10}(h + 4{\cdot}75)]$$

relating the wind speed V_h at an altitude of h metres above flat open ground to the speed V_{10} at a height of 10 m. A more recent formula given by N. Carruthers (Ref. 3) is $V_h = kh^{0{\cdot}17}$ where k is a proportionality constant.

(To illustrate the difference between the two formulae, the wind speed at a height of 200 m, corresponding to a speed of 10 m.p.h. at a height of 10 m, is 17·5 m.p.h. from the Hellman formula, and 16·6 m.p.h. from the Carruthers formula.)

At coastal sites the effect of altitude is reduced, so that, for example, when 70 per cent of the area surrounding a site is open sea, a correction factor of 0·7 must be applied.

One cannot, however, apply formulae of this kind to long-period average wind speeds over hill summits with any precision and the fact that their altitude is not alone responsible for high average speeds is indicated by the figures given in Table X from measurements made on hills in Great Britain and Ireland.

There appears to be some correlation between altitude and mean wind speed at the hills in group (*c*) but for other hills, in this group, at which the anemometer height was 30 ft, altitudes of 1574 ft and 1260 ft gave mean speeds of only 21 m.p.h. and 18 m.p.h. respectively.

The effect of the greater wind speed in the upper layers of the air flowing over a hill is to compress the stream lines in the lower layers; the faster moving upper air needs greater forces to deflect it from its straight line flow.

There is a wealth of literature on the subject of the increase of wind speed with height above level ground, above the sea and over mountains. L. W. Bryant (Ref. 4) reported the results of four sets of observations of wind velocity made at sea, during the period 1920–23 and he analysed them on the basis of the 'eddy conductivity' theory of G. I. Taylor (Refs. 5 and 6). He showed that the vertical

PLATE I. (*Top*). Chinese vertical-axis windmill pumping brine. (From *Farmers of Forty Centuries* by F. H. King, Jonathan Cape Ltd.)

PLATE II. (*Left*). One of the brine-pumping windmills at the Aden saltworks

PLATE III. (*Right*). A typical English post mill at Outwood, Surrey. Built in 1665, it is the oldest windmill still working in England. (From *Windmills in England* by Rex Wailes, Architectural Press)

PLATE IV. (*Above left*). A typical English tower mill—Pakenham Mill, Suffolk. (From *Windmills in England* by Rex Wailes, Architectural Press)

PLATE V. (*Above right*). Experimental windmill at La Cour's windmill research station, Askov, Denmark

PLATE VI. (*Right*). 30 kW Lykkegaard windmill with sails furled

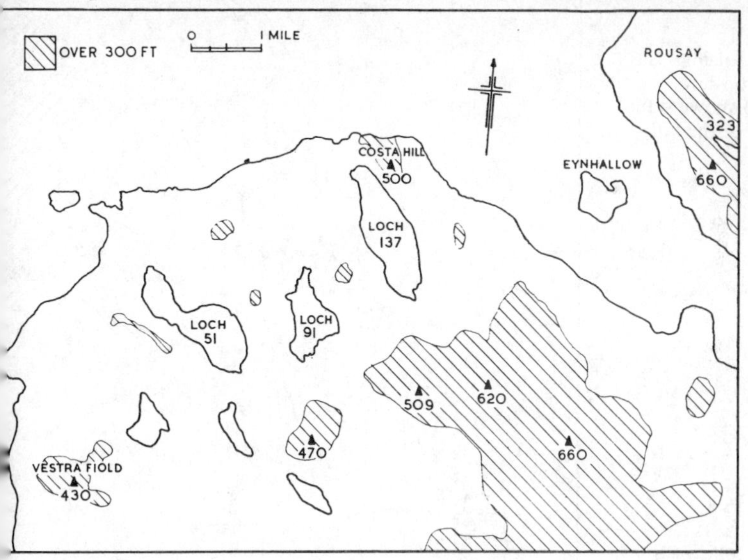

PLATE VII. Costa Hill, Orkney. The map shows the position of the hill, in relation to the surrounding terrain

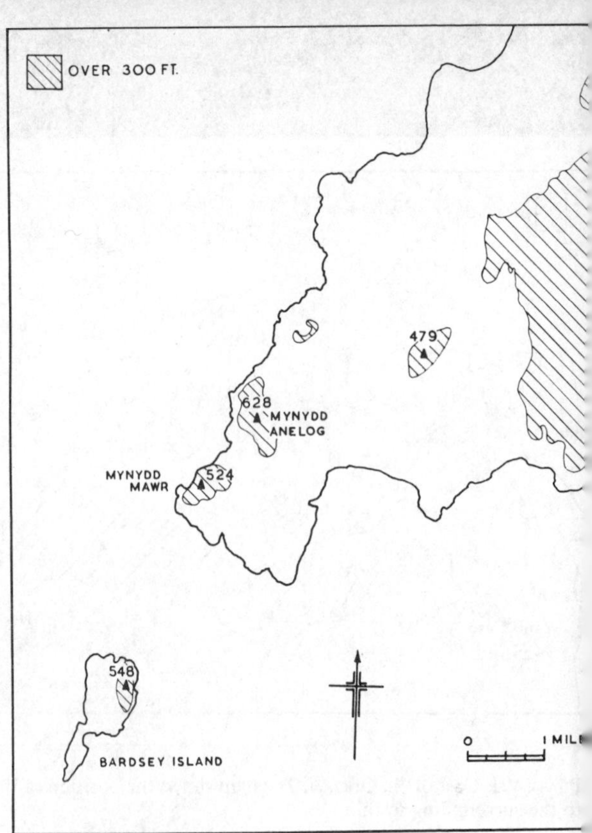

PLATE VIII.
Mynydd Anelog,
Caernarvonshire,
with map showing the
exposure of the hill

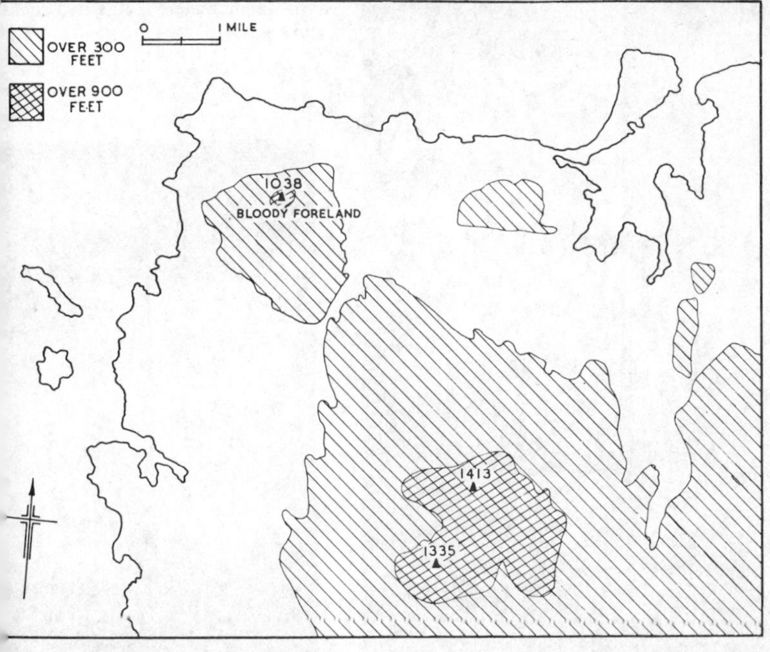

PLATE IX. Bloody Foreland, Co. Donegal, with map showing the surrounding terrain

PLATE X. A beech hedge, at 1000 ft in the Isle of Man, showing the effect of wind shear

PLATE XI. Typical wind survey ir stallation; 10 ft pole with counte type anemometer. It has two sets c three guys

PLATE XII. 70 ft mast carrying contact-type anemometers at 35 ft and 70 ft

PLATE XIII. 30 ft mast used in wi surveys. It carries a contact-ty anemometer

PLATE XIV. Electrolytic tank and associated equipment used for tests on hill models

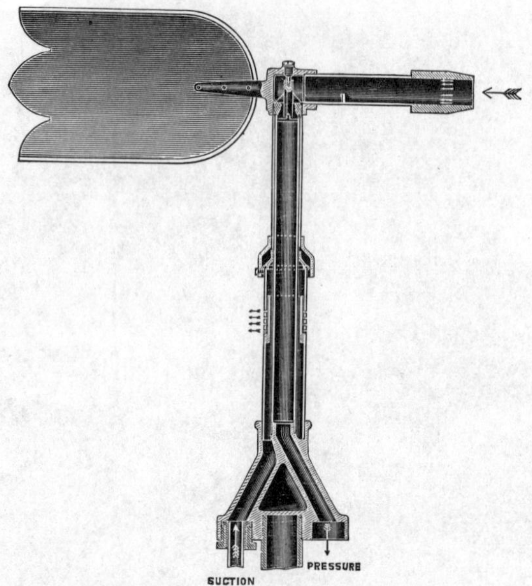

Plate XV (a). Section through the vane of a Dines anemometer (*R. W. Munro Ltd.*)

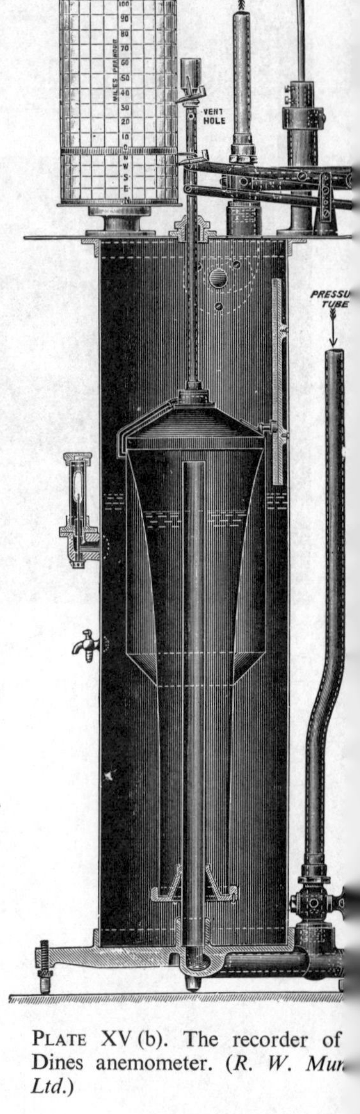

Plate XV (b). The recorder of Dines anemometer. (*R. W. Mun* *Ltd.*)

Plate XVI. Balsa wood windmill-type anemometer

PLATE XVII. (*Above*). Counter-type anemometer. (*R. W. Munro Ltd.*)

PLATE XVIII. (*Right*). Portable hand anemometer (Kelvin Hughes) (*Kelvin and Hughes Ltd.*)

PLATE XIX. Photographic recorder for five contact-type anemometers and wind-direction indicator

9382	7171	0602	4848	5216
9381	7017	0468	4697	5765
9380	6870	0342	4549	5620
9379	6718	0210	4399	5470

PLATE XX. Portion of record from photographic recorder. The figures are from five contact-type anemometers and a wind-direction indicator

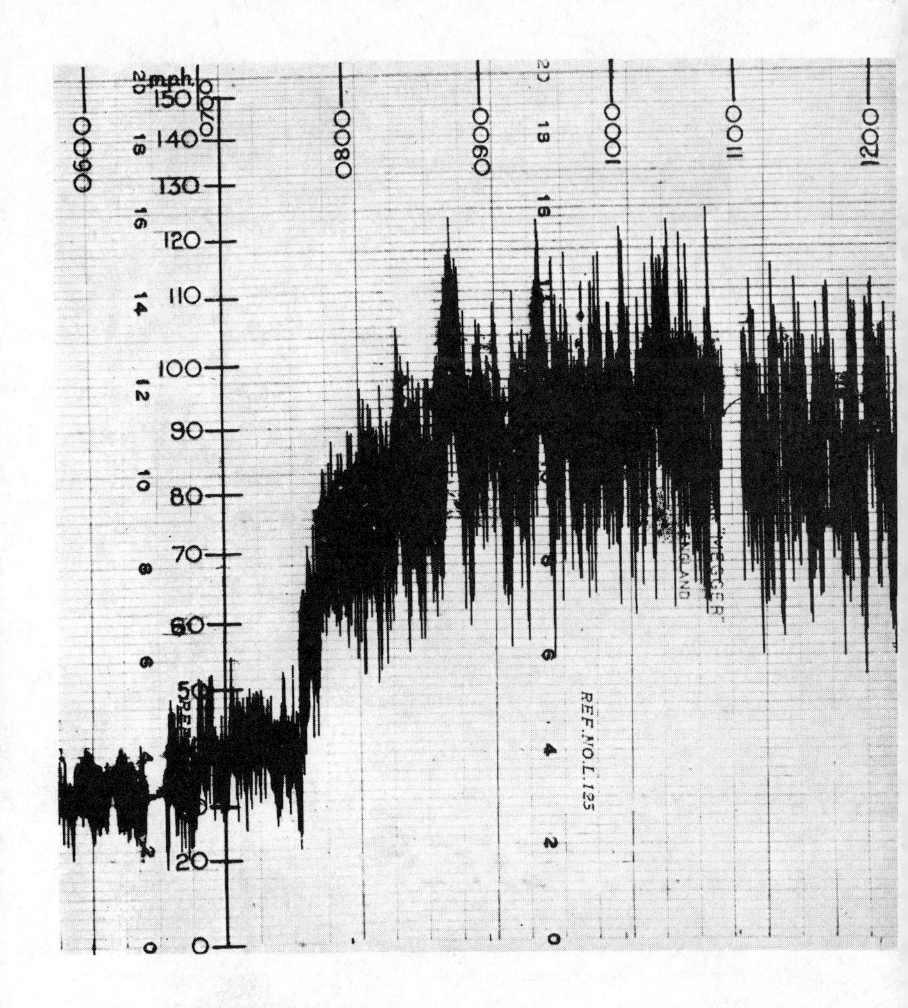

Plate XXI. Portion of record from special cup generator anemometer installed on Costa Hill, Orkney. The record shows gusts up to 125 m.p.h. during the storm of 31st January 1953

PLATE XXII. (*Above*). E.R.A. gust anemometer for vertical components

PLATE XXIII. (*Right*). E.R.A. gust anemometer for horizontal components

PLATE XXIV. (*Left*). 100 kW (John Brown) wind-driven generator on Costa Hill, Orkney

PLATE XXV. (*Right*). 45 kW (SEAS) wind-driven generator at Bogø, Denmark

PLATE XXVI. Showing the 100 kW Enfield-Andreau wind-driven generator during its installation on St. Albans test site. (*Enfield Cables Ltd.*)

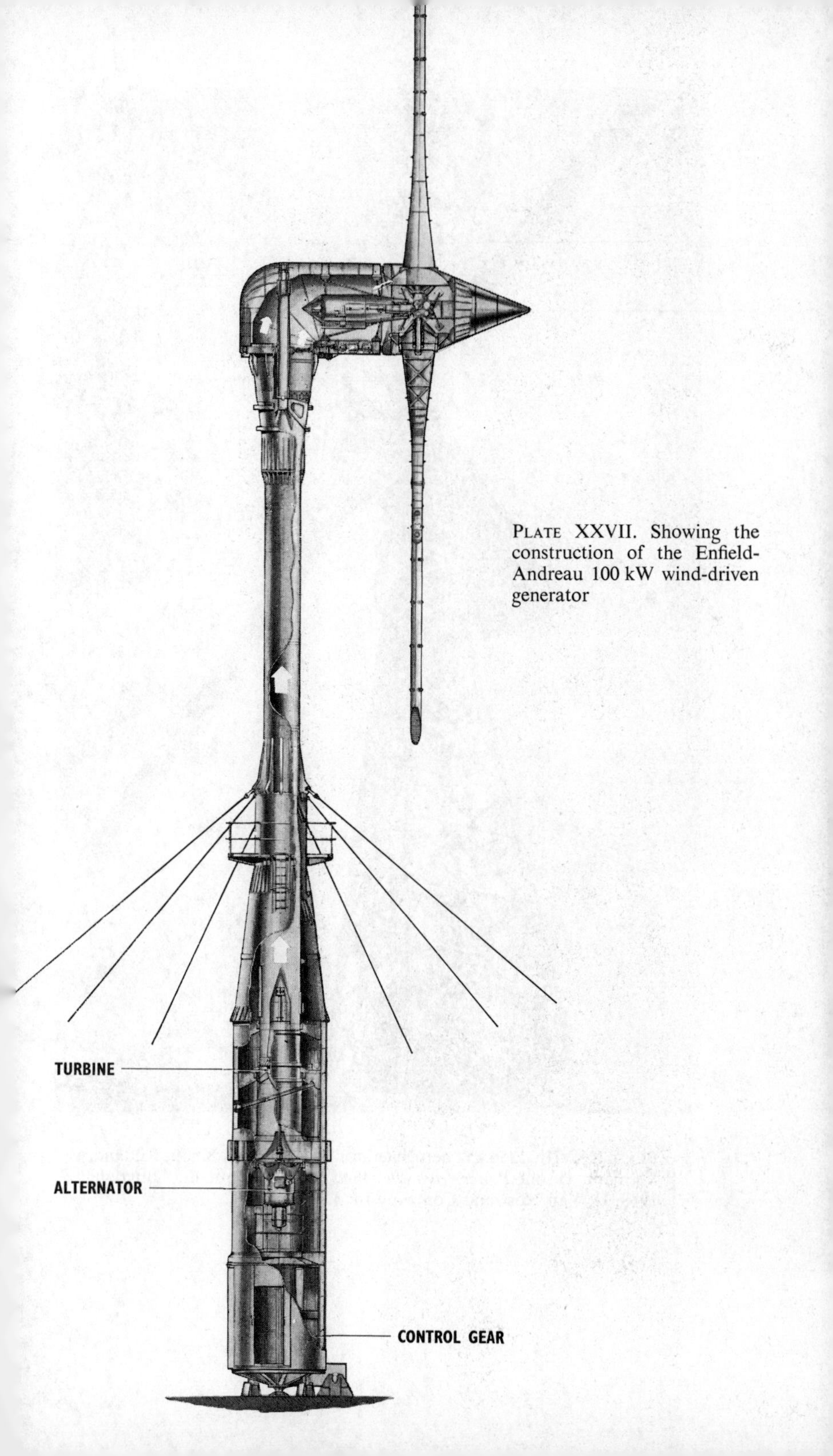

PLATE XXVII. Showing the construction of the Enfield-Andreau 100 kW wind-driven generator

PLATE XXVIII. 1250 kW aerogenerator at Grandpa's Knob, Rutland, Vermont. (From *Power from the Wind* by P. C. Putnam. Copyright 1948. D. Van Nostrand Company Inc.)

wind gradient was highest when the eddy conductivity was low, under overcast conditions with a light wind. There was no evidence that the wind gradients over the sea differed significantly from those over flat country.

Sir David Brunt (Ref. 27) and R. Geiger (Ref. 23) have discussed very thoroughly the question of the variation with height and they make frequent references to the work of G. Hellman (Ref. 2), O. G. Sutton (Refs. 45, 46 and 51), A. C. Best (Ref. 37), H. U.

TABLE X

Annual mean wind speed for the district in which the hills are situated (m.p.h.)	Altitude of hill summit (feet)	Annual mean wind speed at the summit (m.p.h.)
(*a*) 17·5	2034	29
	863	28
	1038	27
	1515	25
(*b*) 15	500	25
	663	24
	1253	23
	430	23
	348	22
	704	21
	523	20
	1137	19
(*c*) 12·5	2152	26
	1894	24
	827	20
	665	17

Height of anemometer above ground: 30 ft in group (*a*); 10 ft in groups (*b*) and (*c*).

Sverdrup (Refs. 47 and 48), L. Prandtl (Refs. 32, 43 and 52), C. G. Rossby and R. B. Montgomery (Ref. 53) and Barkat Ali (Ref. 54). These investigators have made observations on the vertical wind gradient under different climatic conditions and at different heights above ground. Some have been concerned with wind speed very close to the ground, as part of microclimatic studies, while others have been interested in the winds at much greater heights. C. W. Thornthwaite (Refs. 24, 49 and 50) and Maurice Halstead (Ref. 25) have worked mainly on wind speeds at low altitudes, particularly from the point of view of their possible effects upon growing plants, and they describe the construction of special types of anemometer used in their measurements.

Much of the work mentioned has centred round the formula for vertical wind gradient, $V_h = V_0 h^\alpha$ where V_0 is the wind speed at some standard height and α is an exponent which varies with the height, the time of day and season of the year, the nature of the ground, the mean wind speed and, above all, with the temperature gradient.

As Geiger emphasizes, the law of wind variation with height is only a statistical law which holds good in a long series of observations but not necessarily in individual instances.

W. R. Morgans (Ref. 7), in his valuable summary of the knowledge (in 1931) of wind flow over various kinds of country, deals with both the height and horizontal distance of the influence of hills on the wind. He describes the experimental and theoretical work of a group of German meteorologists (Refs. 8, 9, 10, 11, 12, 13 and 14) who made some important contributions to knowledge of the subject. In this work balloons and smoke-producing rockets were used with a ciné camera. W. Georgii (Refs. 9, 10 and 11) worked on the turbulence produced by the increase in friction when winds pass over the coast from sea to land. He showed also that with a wind velocity at the summit less than 10 m/sec, stationary eddies can occur on both the windward and leeward sides. Above this velocity, which he regarded as critical, turbulence sets in. Under these conditions 'any theory depending on streamline flow can have no application to the flow of air over obstacles' (Ref. 7).

The flow of the wind on the leeward side is often less regular than on the windward side. Experiments by H. Koschmieder (Ref. 13) and K. O. Lange (Refs. 15 and 16) established the occurrence of progressive leeward eddies and gave wave lengths, periods and diameters of such eddies.

More recently, R. S. Scorer (Refs. 17, 18, 19 and 20) has worked on the theory of airflow over mountains, with particular reference to the formation of lee waves, and on winds in mountain gaps. G. Manley (Ref. 21), in his experimental studies of the Helm Wind of Crossfell, has drawn attention to a group of phenomena found in many different parts of the world when winds blow across mountain ridges. Under favourable conditions 'standing waves' occur to leeward of the ridge and these are accompanied by strong downward and upward currents which may endanger aircraft. From the viewpoint of wind power research such waves are mainly important in so far as they cause changes in wind speed and direction at a distance of some miles from the hill which initiates them (Ref. 22) and these effects may introduce difficulties if a windmill were located near the hill.

Frictional Drag

Air has viscosity, so that it is not possible to calculate accurately the windspeed which will be found at very low altitudes; the nature of the ground surface will have an effect upon the air flow, especially at the surface itself where frictional drag may reduce the speed considerably. This effect will be increased, of course, by the existence of trees or bushes, buildings, rocks or other similar sources of unevenness (see Refs. 23, 24, 25 and 27).

Although this drag, due to ground friction, may be over-ridden by much greater accelerating forces at any appreciable heights over a hill summit so that, for a smooth hill without rocky outcrops, it may be neglected, it is still true that the lower layers of the wind approaching the hill may have been slowed down by passage over a long length of broken ground. Hence the superiority of coastal sites when the prevailing winds are from the sea. There is also a frictional effect due to eddies, to express which the term 'eddy viscosity' has been introduced. The temperature gradient in the air has an important influence upon the wind speed at different altitudes because temperature differences cause 'interlocking' between the layers of air with the result that the upper layers are retarded by the slower-moving lower layers (Refs. 4, 7, 27 and 28).

Experimental work on the vertical wind gradient over hills

M. A. Giblett (Ref. 26) has given some interesting results obtained from a large number of measurements of mean wind speed at heights of 150 ft (V_{150}) and 50 ft (V_{50}) over level country. Classified as the ratios of these speeds are in Table XI under the headings 'Day' and

TABLE XI

Mean wind speed at 150 ft (V_{150})	Ratios of mean wind speeds $\frac{V_{150}}{V_{50}}$	
	Day	*Night*
(m.p.h.)		
10–14	1·02	1·08
15–19	1·07	1·17
20–24	1·16	1·19
25–29	1·17	1·20
30–34	1·20	1·20

'Night', and also according to the values of the wind speeds at the greater height, they illustrate the effect of temperature differences, in the daytime, in reducing the vertical gradient of speed by the interlocking which has just been mentioned. Clearly this effect is greatest at low wind speeds. When the speed exceeds 30 m.p.h. the vertical gradient is the same for day and night and gives the ratio $\frac{V_{150}}{V_{50}} = 1{\cdot}20$. This value is in exact agreement with Carruthers' formula.

Experimental work by J. Juul of the SEAS Company in Denmark, by the research department of Electricité de France, by P. C. Putnam and his associates in the United States of America and by the Electrical Research Association in Great Britain has been aimed at measurement of the vertical wind gradient over hilltops as well as over level ground. This work has had wind power development in view and the results obtained are, therefore, of particular interest.

Juul (Ref. 29) mounted anemometers, giving direct readings of wind speed, at each 5 m of height on towers 25 m high and also on the arches of bridges to give a height of 60 m. A large number of simultaneous readings was taken and average wind speeds were determined. In Fig. 30 curves for wind speed over flat ground and over a low hill are reproduced from Juul's report. The curve for flat ground refers to a site at about 100 m from trees and buildings and these obstructions probably account for low wind speeds nearer the ground but at heights of 25 m and over there is little difference between the speeds over the hill and over level ground. The curve for the hill shows an increase in wind speed up to a height of some 15 m, then a decline to about 25 m followed by a gradual increase as the height increases. One cannot draw any general conclusion from these results; in this instance the wind speed at a low height over the summit of the hill is the same as that at a height of 50 m above level ground.

The experiments of Electricité de France (Refs. 30 and 61) were on increase of wind speed with height over flat ground (using smoke-producing rockets with photographs of the track of the smoke taken at 1 sec or $\frac{1}{2}$ sec intervals) and on the wind flow over obstacles of different shapes (determined from wind-tunnel tests on models). The rocket tests confirm, for heights up to 150 m, and wind speeds of 5 m/sec and 10 m/sec, the formulae already given for the effect of altitude. The wind tunnel tests provided some interesting qualitative information on the effects of widely different shapes of obstruction—representing a cliff face and hills of varying steepness—upon the wind flow and giving an indication of the optimum location of a wind power plant on the hilltop.

P. C. Putnam (Ref. 1) has given in some detail the assumptions, concerning wind behaviour in its flow over hills, which were made at the outset of the Grandpa's Knob wind power project. Much attention was paid to the question of 'speed-up' over hills, i.e. the

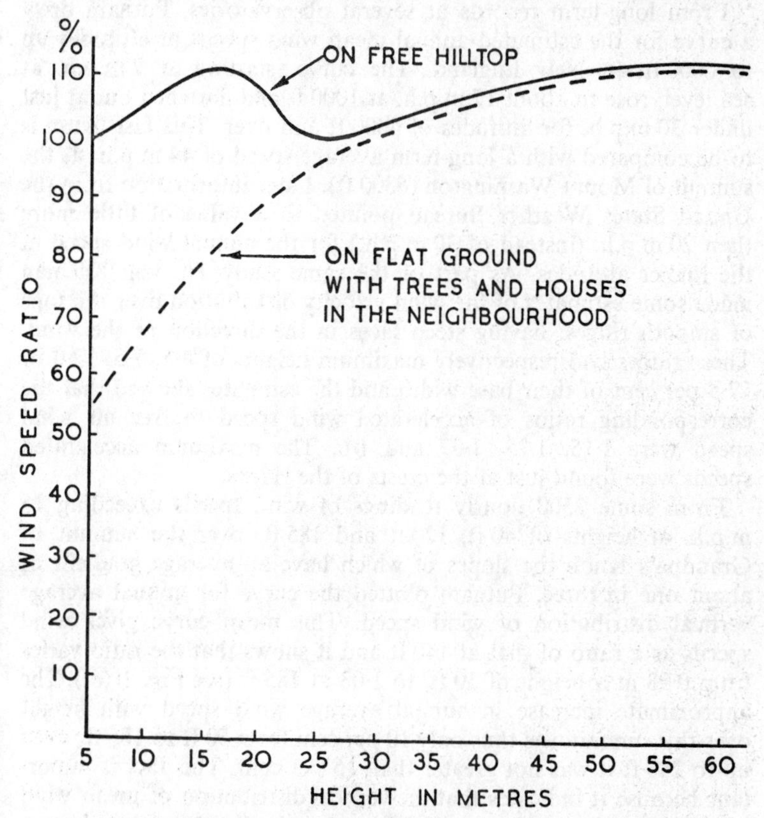

FIG. 30. *Variation of wind speed with altitude over flat ground and over hill tops*

ratio of the wind speed over the summit to the wind speed in free air at the same altitude. He emphasizes the difficulty in stating a value for the latter or in measuring it. Computations were made from data obtained from pilot balloon observations and pressure maps. When dealing with altitudes of several thousand feet, at which the effect of ground friction can be neglected, the free-air wind speed or 'gradient wind' is calculable from measurements of barometric pressure (Refs. 27 and 28) but for lower altitudes it is not possible to calculate wind speeds with precision because of the number of variables which may be introduced. Some of these are the nature of the ground surface, the temperature gradient in the atmosphere and the order of magnitude of the wind speed itself.

From long-term records at several observatories, Putnam drew a curve for the estimated annual mean wind speeds at altitudes up to 6000 ft. in New England. The curve, starting at 7 m.p.h. at sea level, rose to about 17 m.p.h. at 1000 ft and flattened out at just under 30 m.p.h. for altitudes of 4000 ft and over. This last figure is to be compared with a long-term average speed of 44 m.p.h. at the summit of Mount Washington (6300 ft). Later information from the United States Weather Bureau pointed to a value of little more than 20 m.p.h. (instead of 30 m.p.h.) for the annual wind speed at the higher altitudes. As part of the same study Th. von Kármán made some estimates of the wind velocity distribution over the tops of smooth ridges, having steep faces in the direction of the wind. These ridges had respectively maximum heights of 4·5, 7·5, 12·0 or 17·5 per cent of their base width and the estimates showed that the corresponding ratios of accelerated wind speed to free air wind speed were 1·15, 1·23, 1·37 and 1·6. The maximum accelerated speeds were found just at the crests of the ridges.

From some 2500 hourly readings of wind speeds exceeding 15 m.p.h. at heights of 40 ft, 120 ft and 185 ft, over the summit of Grandpa's Knob the slopes of which have an average gradient of about one in three, Putnam plotted the curve for annual average vertical distribution of wind speed. This mean curve gives wind speeds as a ratio of that at 140 ft and it shows that the ratio varies from 0·93 at a height of 50 ft, to 1·03 at 185 ft (see Fig. 31(*a*)). The approximate increase in annual average wind speed with height over this summit was thus only 10 per cent from 50 ft to 185 ft; even up to 240 ft it was not greater than 15 per cent. This fact is important because it indicates that the vertical distribution of mean wind speed over the swept area of a large windmill rotor located on a hilltop may be fairly uniform. The variations of the hourly readings from the mean lie within $\pm$ 3 per cent at a height of 40 ft.

In the wind surveys made in Great Britain by the Electrical Research Association hourly measurements of wind speed at various heights over three hilltops have been made continuously (except for instrument failures) during periods of one to four years (see also Ch. 6). The results for two of these hills are shown in the curves of Fig. 31(*b*) and (*c*)

The curves are for mean values of the wind speed ratio $\frac{V_h}{V_{66}}$ because

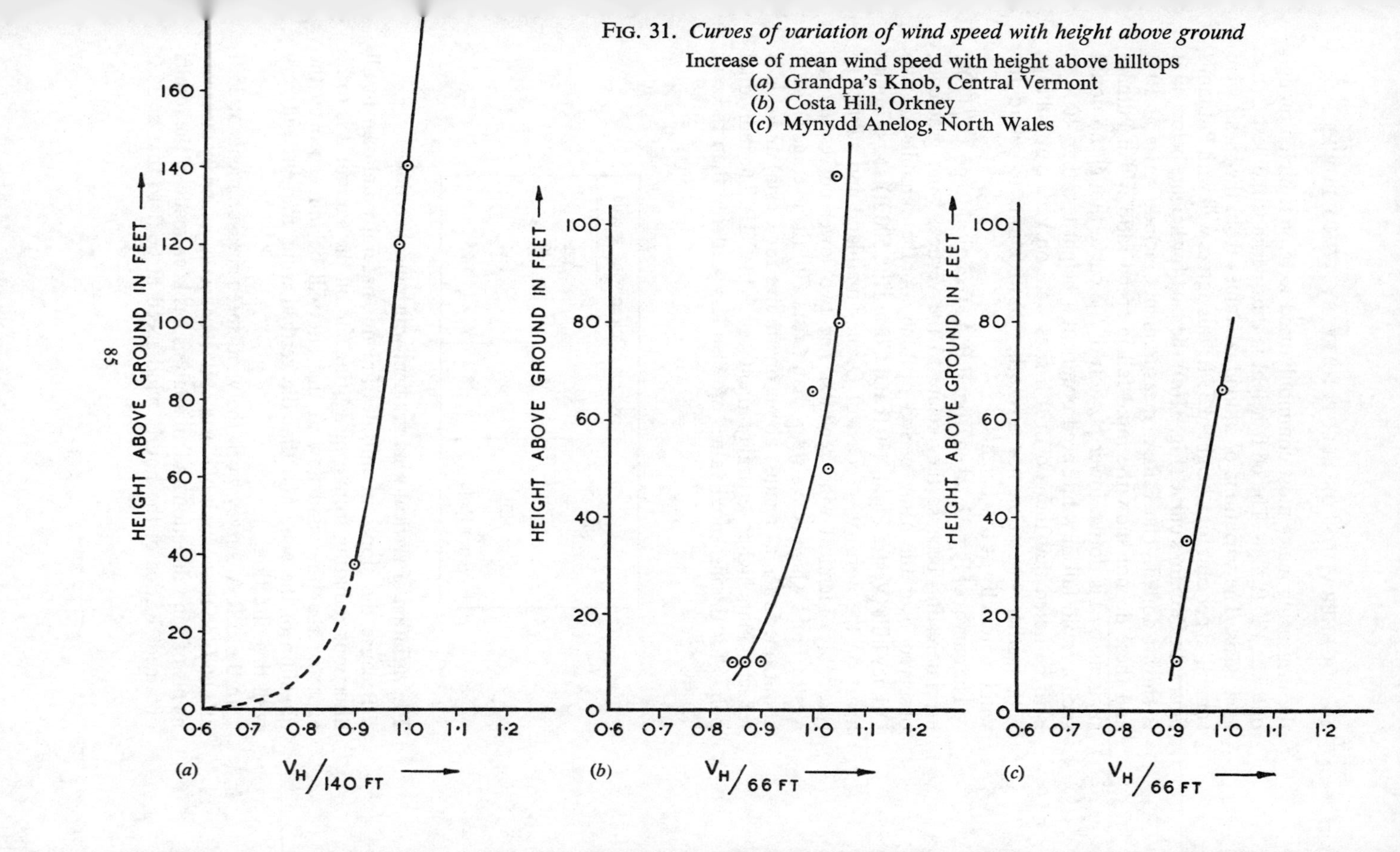

FIG. 31. *Curves of variation of wind speed with height above ground*

Increase of mean wind speed with height above hilltops
(*a*) Grandpa's Knob, Central Vermont
(*b*) Costa Hill, Orkney
(*c*) Mynydd Anelog, North Wales

the measuring masts most commonly used were 66 ft high although one was 120 ft high. The 66 ft height is thus perhaps the best to use as a basis. The variations of individual values from the mean were some ± 3 per cent for a height of 35 ft; this agrees well with Putnam's figure. In Putnam's curve (Fig. 31(*a*)) the part for heights below 40 ft is shown dotted. The E.R.A. measurements include series at 10 ft and these do not show the rapid fall in speed ratio which Putnam suggests by his dotted lower portion of the curve though no doubt such a rapid fall in wind speed occurs at a height less than 10 ft. It is to be noted that the two British sites were on very bare hilltops with neither vegetation nor any other obstructions to retard the wind materially at low heights.

The curves of Fig. 32, taken from E.R.A. Technical Report C/T108, are interesting because they relate to the vertical wind gradients, observed over the same period of time, on the summits of two Orkney hills (Vestra Fiold, 430 ft and Costa Hill, 500 ft) some 7 miles apart on the north west coast of Orkney mainland and subject to (probably) identical wind régimes. The two lower curves are for Vestra Fiold which is less steep than Costa Hill (its average slope is about one in ten as against about one in five for Costa) and this is reflected in its higher vertical gradient (see Table XII). The individual points for the measured values are also less scattered than those at Costa.

TABLE XII

Site	Ratios of mean wind speeds	
	$\frac{V_{35}}{V_{10}}$	$\frac{V_{66}}{V_{10}}$
Costa Hill	1·06	1·11
Vestra Fiold	1·18	1·25

The variation of vertical wind gradient with wind direction

Because the slopes of hillsides are not naturally uniform in all directions radiating from a measuring site on the summit it is to be expected that the wind flow at this site will be influenced by the direction of the wind and that the vertical wind gradient will vary with this direction.

In the E.R.A. survey work the wind speed measurements at 35 ft and 66 ft on the summit of Costa Hill were accompanied by measurements of wind direction and an analysis has been made of the results obtained. Ratios of the hourly wind speed at 66 ft to that at 35 ft

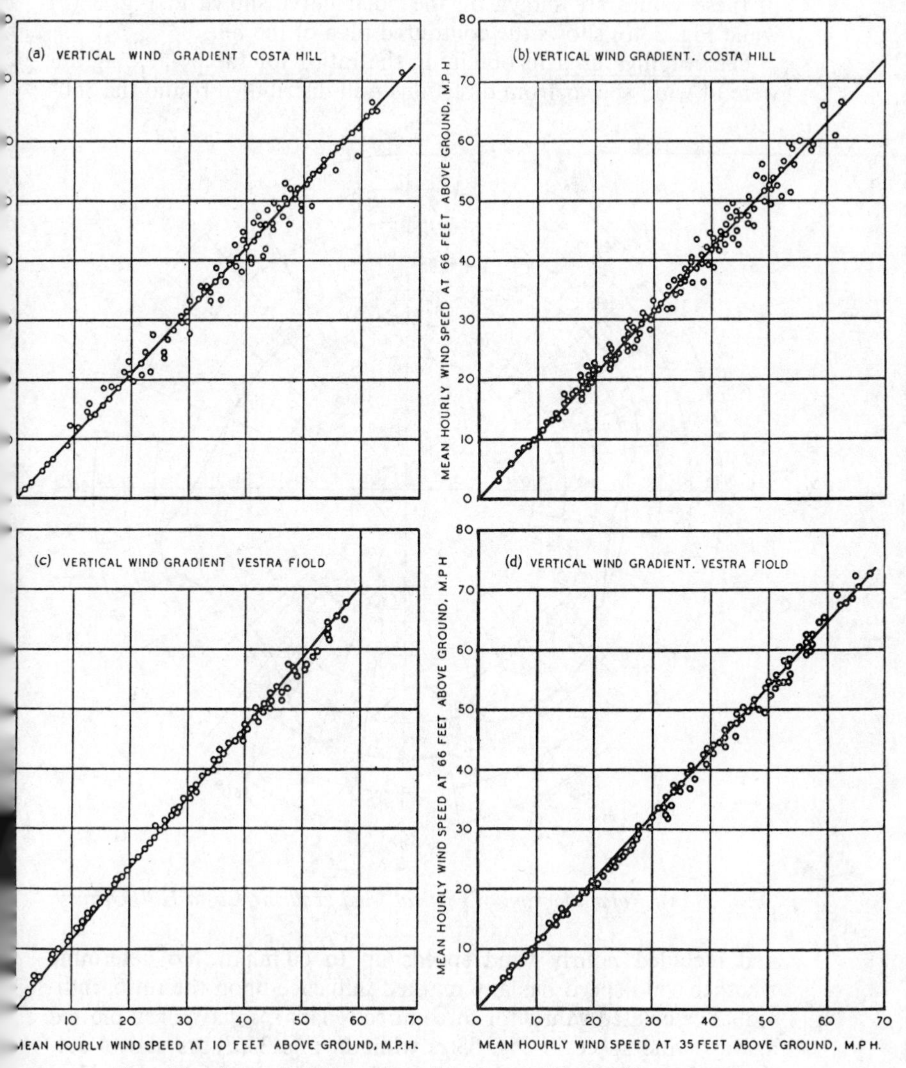

FIG. 32. *Vertical wind gradient for Costa Hill and Vestra Fiold*

were calculated for a large number of pairs of readings and the means of these values are shown on the polar curve shown in Fig. 33(*a*) while Fig. 33(*b*) shows the contoured plan of the hill.

The readings used in obtaining the ratios for the polar diagram were of wind speeds from directions well distributed round the 360°

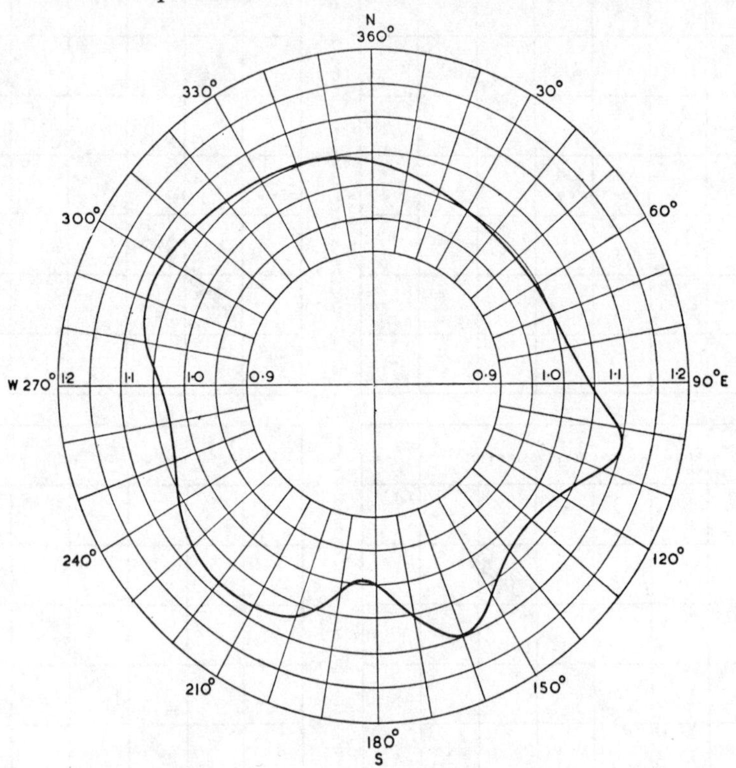

FIG. 33 (a). *Polar diagram of vertical wind gradient, Costa Hill, Orkney*

and included hourly wind speeds up to 60 m.p.h. To determine whether wind speed has any marked influence upon the ratio, individual calculated values for three directional axes, have been plotted against wind speed in Fig. 33(*c*) while in Fig. 33(*d*) are shown sections of the hill along these axes. It can be seen from Fig 33(*c*) that in the direction N. to S. (000° to 010°) the ratio $\frac{V_{66}}{V_{35}}$ is greater than unity over the whole range of wind speeds 0 to 40 m.p.h. In this direction the wind is passing from the sea, over a 400 ft cliff, and then over almost level ground for some 1700 ft. In the reverse direction the wind is ascending a slope and the ratio is only greater than unity at low wind speeds. For directions 120° and 300° the

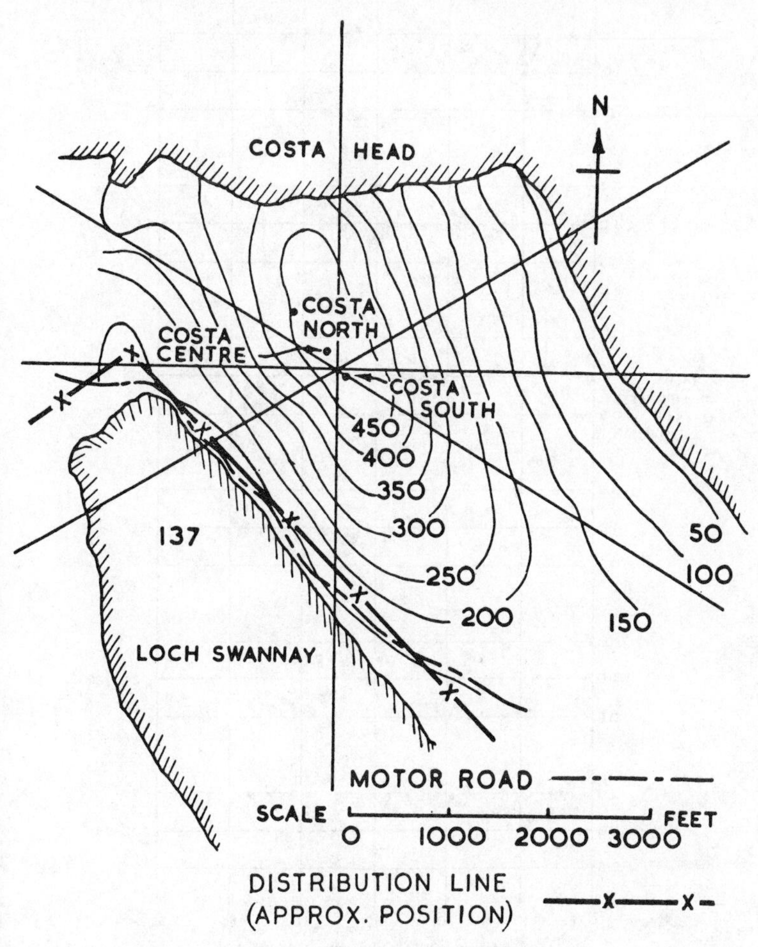

FIG. 33 (b). *Contour map of Costa Hill, Orkney*

wind is passing over only gentle slopes and the ratio is again greater than unity. But for directions 60° and 240°, for which the hill section is more steeply sloped, the mean value of the ratio is probably close to unity. It is noticeable that there is much more scatter in the calculated values of the ratio when the wind approaches the

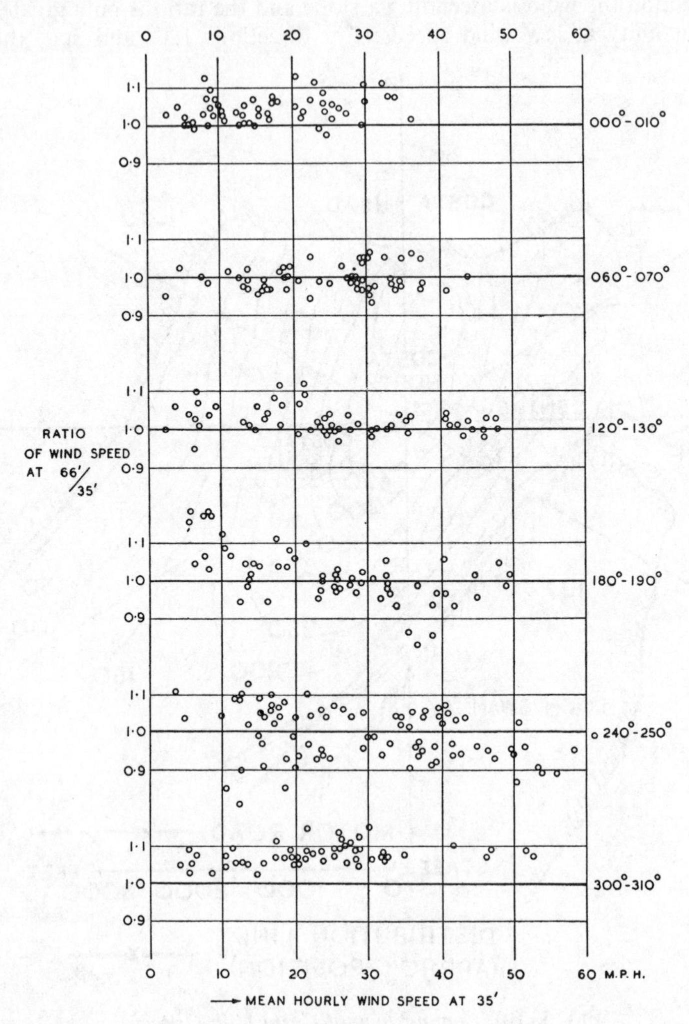

FIG. 33 (c). *Vertical wind gradient for six directions across Costa Hill, Orkney*

hill from over the land (directions 180° and 240°) than when it is coming from the sea (in the other directions shown).

J. Knudsen (Ref. 58) has reported measurements of hourly wind speed at heights, above ground, of 2·2 m, 6 m and 10 m, and with winds from all directions, on the summit of a rounded hill about 30m high on the island of Sula (63° 51′ N., 8° 27′ E.) off the Norwegian Coast near Trondheim. The measurements were at heights which are rather too low to be of major interest for wind

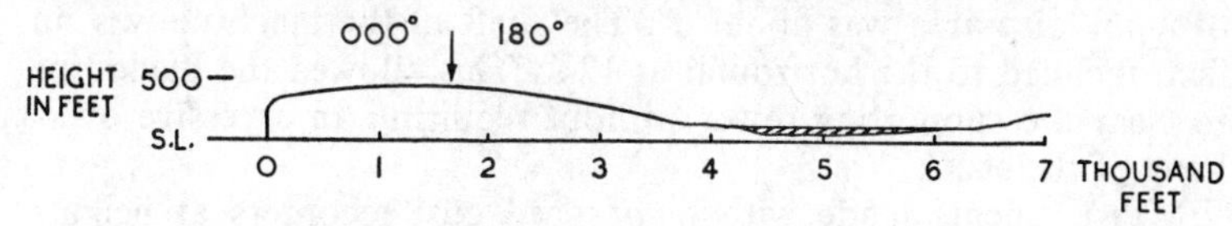

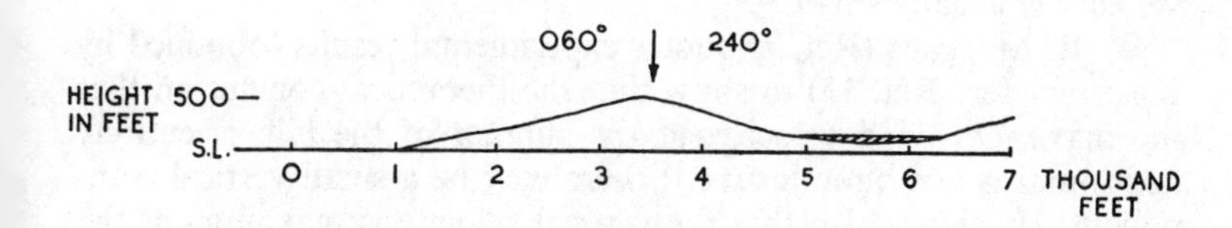

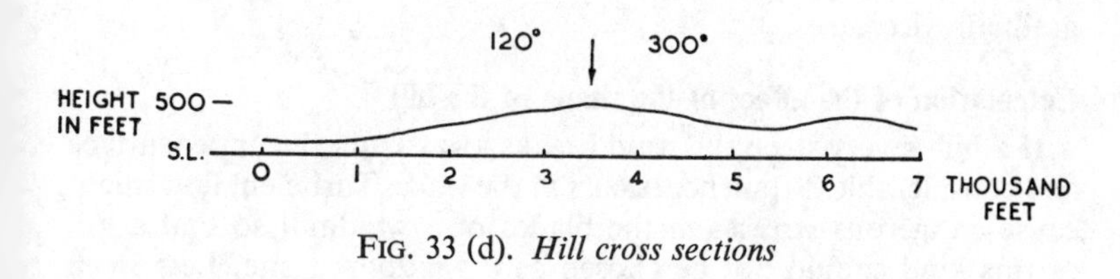

FIG. 33 (d). *Hill cross sections*

power but it is interesting to find that the vertical wind gradient showed a decrease of wind speed with height for all wind directions, with $\frac{V_{10}}{V_{2\cdot2}}$ varying from 0·68 to 0·96 and $\frac{V_{10}}{V_6}$ from 0·79 to 0·96. These ratios varied with wind speed (over the range 2·5 to 15 m/sec) except for the sector 124°—202° when it was constant at 0·94.

Vertical component of wind speed at a given height

When considering the location of a large wind-driven machine on a hilltop the inclination of the wind to the horizontal, as it passes through the rotor, is important. In the estimation of energy output from wind speed measurements it is assumed that these speeds are in the horizontal plane so that the winds always pass through the rotor disc normally. In fact, the inclination of the wind may vary with height above the summit of the hill, with its own general direction (which will affect the contours over which it flows in approaching the summit) and with its mean velocity and the weather conditions at any particular time.

Putnam observed the inclination at Grandpa's Knob to vary from 10° at 20 ft above ground to zero at 185 ft. The average inclination over the disc area was about 5°. The shaft of the machine was, in fact, inclined to the horizontal at 12½°. This allowed the blade tips to clear the supporting tower without requiring an excessive overhang of the shaft.

Measurements made with high speed gust recorders at heights between 50 and 110 ft on Costa Hill, Orkney, though by no means exhaustive, have confirmed that the vertical component of wind velocity is usually small.

W. R. Morgans (Ref. 7) quotes experimental results (obtained by Koschmieder: Ref. 13) to show that the theoretical conclusion that the maximum wind velocity, at the summit of the hill, is entirely horizontal is not quite justified; there may be a small vertical component. He shows also that the vertical velocity is maximum at the middle of a uniform slope and that it usually decreases with height but sometimes it first gradually increases with height and then gradually decreases.

Calculation of the effect of the shape of the hill

If a hill is very steep the wind breaks away from the upper surface and considerable turbulence occurs in the wake. Turbulent flow might cause dangerous stresses in the blades of a windmill so that a hill of this kind should not be chosen as a windpower site. Less steep hills can often be found, having an aerofoil contour, and calculation of the wind speed at low altitudes over the summit can be made assuming that the flow is aerodynamic, the fluid being non-viscous and incompressible. Such calculations cannot, of course, take into account either the variation with altitude of the wind speed approaching the hill or the effects due to the viscosity of the air. Nevertheless, they can give some indication of the influence of the shape of the hill upon the wind speed at the summit.

H. H. Rosenbrock (in an unpublished note) has investigated this question for (*a*) long ridges and (*b*) circular hills.

(*a*) *Long ridge*. Consider a cylinder, of radius a, lying with its axis in a horizontal plane, and a stream, of uniform speed U, approaching it in a direction perpendicular to its axis (see Fig. 34).

The streamlines are defined (Sir H. Lamb, Ref. 31) by the equation

$$U\left(r - \frac{a^2}{r}\right) \sin \theta = \psi$$

ψ is a constant. Different values of ψ indicate different stream lines. The length r and angle θ are the polar co-ordinates of a point on the streamline (e.g. when $\psi = o$ the equation is satisfied by $r = \pm\, a$ and $\theta = o$ which means that the streamline for this value of ψ lies along the positive x axis up to the cylinder, passes round the half cylinder and then extends along the negative x axis).

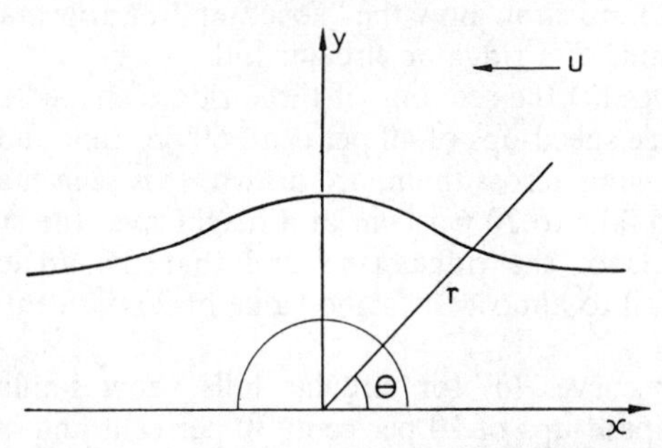

FIG. 34. *Diagram of flow over a cylinder*

Now, any streamline can be replaced by a solid surface of the same shape without influencing the stream flow above it so that, if the ridge considered has a section which coincides with a streamline (say) $\psi = P$ the wind flow over it can be calculated by obtaining the shapes of streamlines for values of ψ greater than P.

The wind speed at any point on the vertical, or y, axis is given by

$$U\left(1 + \frac{a^2}{y^2}\right)$$

from which it is possible to calculate the speed at different heights above the ridge just mentioned.

(Incidentally, if $y = a$ the speed is obviously $2U$ which means that there is a speed-up factor of 2 over the top of the cylinder.)

(*b*) *Circular hill.* The streamlines near a sphere are defined by Lamb as

$$\psi = \tfrac{1}{2}\, U\left(r^2 - \frac{a^3}{r}\right) \sin^2 \theta$$

and the speed at any point on the y axis is $U\left(1 + \dfrac{a^3}{2y^3}\right)$.

It can be shown that the surfaces $\psi = constant$ represent sections through hills of circular plan so that the streamline equations for a sphere can be applied to calculations on wind flow over hills of this kind (by putting $y = a$ it is seen that the speed-up factor for a sphere is

$$\frac{U\left(1+\frac{a^3}{2a^3}\right)}{U} = 1{\cdot}5)$$

Following these methods, Rosenbrock has drawn the curves given in Fig. 35 to show how the "speed-up" disappears with height over the summit of a ridge or circular hill.

In the curves (*a*) the sections of three ridges shaped so as to give summit surface speed-ups of 40 per cent, 60 per cent and 80 per cent to a wind blowing across them, are drawn. It is seen that the 40 per cent speed-up falls to 20 per cent at a height over the summit equal to the altitude of the ridge itself and that the greater speed-up percentages fall to almost the same value of 20 per cent at the same height.

The upper curves (*b*) for circular hills show similar effects in the cases of speed-ups of 20 per cent, 30 per cent and 40 per cent.

The curves show that, when a wind is blowing across a ridge, the speed-up is greater than that given by a circular hill and it falls away more slowly with height above the summit. It is found from wind measurements, however, that, even when a ridge lies athwart the direction of the prevailing wind, the wind direction does not generally prevail long enough in the year for this high speed-up to counterbalance the much lower one for other wind directions moving towards that of the line of the ridge itself. Circular hills, having the same—if rather smaller—speed-up for all wind directions, usually give better results in annual mean wind speed.

Another point to be made from Fig. 35 is that the greater speed-up from a steep hill disappears more quickly with increase of height above the summit than does that produced by a rather less steep hill so that, if the height of a windmill to be placed on a hill were comparable with the altitude of the hill itself, it would be better to choose a hill having a surface speed-up of only perhaps 20–30 per cent.

Model tests

It would greatly aid the selection of windpower sites, and throw light on the wind structure over different terrains, if scale models could be used to provide reliable data from laboratory tests. The most obvious method would appear to be that of placing a model in

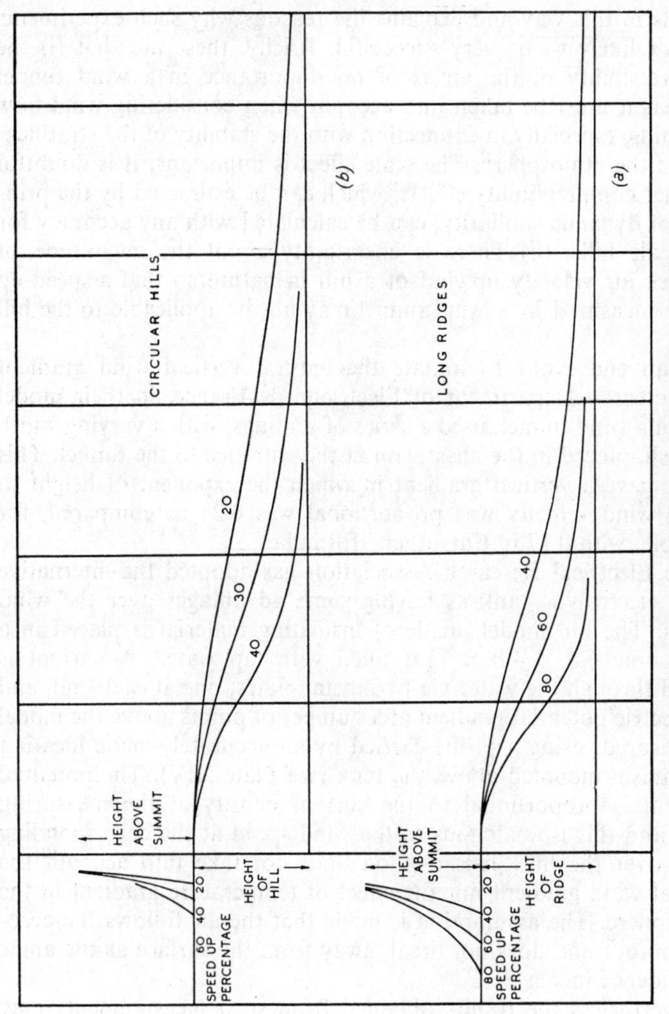

FIG. 35. *Speed-up curves for circular hills and ridges*

a wind tunnel and measuring the wind speed at various points close to its surface with different orientations of the model in relation to the wind stream. P. C. Putnam (Ref. 1) reports an attempt to use models in this way and explains the reasons why such experiments are not likely to be very successful. Briefly, these are that (i) the compressibility of the air is of no importance in a wind tunnel whereas it must be taken into account when considering wind flow over hills, especially in connection with the stability of the stratification in the atmosphere. The scale effect is important; it is doubtful whether compressibility effects, which can be estimated by the principle of dynamic similarity, can be calculated with any accuracy for full scale hills. (ii) There is uncertainty about the magnitude of the free air velocity upwind of a hill in nature so that a speed-up factor measured in a wind tunnel may not be applicable to the hill itself.

In an endeavour to imitate the natural vertical wind gradient the Research Department of Electricité de France, in their model tests in a wind tunnel, used a series of gratings, with a varying width of mesh, placed in the air stream at the entrance to the tunnel. This device gave a vertical gradient in which the exponent of height to which wind velocity was proportional was 0·23 as compared, for example, with 0·17 in Carruthers' formula.

The Electrical Research Association has adopted the alternative of an electrolytic tank as having some advantages over the wind tunnel. The hill model, made of insulating material is placed in a tank some 6 ft × 4 ft × $1\frac{1}{2}$ ft filled with tap water. A current is passed through the water via two metal plates, one at each end, and the electric potential gradient at a number of points above the model is measured, using a probe carried by an accurately-made locating mechanism mounted above the tank (see Plate XIV). The potential gradient is proportional to the current density at the measuring point and this is analogous to the wind speed at the corresponding point over the hill. This method does not take into account the vertical wind gradient nor the effect of temperature gradient in the atmosphere. The assumption is made that the air follows the aerofoil contour and does not break away from the surface as the angle of incidence increases.

Nevertheless the results obtained from such measurements may give some guidance in site selection, at least, by indicating the relative merits of widely different shapes of hills and hill groups. Both the technique of measurement of potential gradient in the tank, and the relationship between the results obtained from it and from measurements of wind speed on the actual hill whose model is tested are, however, still being investigated. It is therefore, not yet possible to say how successful the tank method will be.

BIBLIOGRAPHY

(1) PUTNAM, P. C. *Power from the wind.* Van Nostrand (1948).

(2) HELLMAN, G. *Uber die Bewegung der Luft in den untersten Schichten der Atmosphäre.* Königlich Akademie der Wissenschaften, Sitzungsberichte, Part I, Vol. 32, pp. 174–197 (Prussia, 1917).

(2*a*) HELLMAN, G. Über die Bewegung der Luft in den untersten Schichten der Atmosphäre. *Meteorologie Zeitschrift*, Vol. 32, pp. 1–16. Braunschweig (January 1915).

(3) CARRUTHERS, N. Variation in wind velocity near the ground. *Quarterly Journal*, Royal Meteorological Society (London) (October 1943).

(4) BRYANT, L. W. Note on change of wind with height. *Aeronautical Research Committee Reports and Memoranda*, No. 1407. A.R.C.R. & M. (1931).

(5) TAYLOR, G. I. Eddy motion in the atmosphere. *Phil. Trans. A.*, Vol. CCXV (1915).

(6) TAYLOR, G. I. Phenomena connected with turbulence in the lower atmosphere. *A.R.C.R. & M.*, 304 (1916).

(7) MORGANS, W. R. Relation between ground contours, atmospheric turbulence, wind speed and direction. *A.R.C.R. & M.*, No. 1456 (December 1931).

(8) FICKER, H. VON. Die Wirkung der Berge auf Luftstromungen. *Met. Zeits* XXX, p. 608 (1913).

(9) GEORGII, H. Hangwindstudien. *Z.F.M.*, p. 1 (1923).

(10) GEORGII, H. Die obere Begreuzung der abgelenkten Luftstromungen an Hindernissen. *Z.F.M.*, p. 107 (1925).

(11) GEORGII, W. Beitrag zum Stromfeld der Luft im Luv von Gebirgen. Veroff des Forschungs-Inst. der R.R.G. No. 1 (1926–7).

(12) KOCH, H. Graphische Analyse der Strohmung im Luv einer Düne. Veroff des Forschungs-Inst. der R.R.G. (1928).

(13) KOSCHMIEDER, H. Die Arbeiten des Messtrupps wahrend des 2. Segel flugwettbewerbes in Rossitten. *Zeitschrift fur Met.*, p. 235 (1924).

(14) RAETHJEN, P. Kinematographische Flugvermessung. Veroff des Forschungs-Inst. der R.R.G. (1928).

(15) LANGE, K. O. Windstromungen uber dem Gelande der Wasserkruppe Rhon. Veroff des Forschungs-Inst. der R.R.G. (1926–27).

(16) LANGE, K. O. Uber Windstromungen an Hugelhindernissen. Veroff des Forschungs-Inst. der R.R.G. (1929).

(17) SCORER, R. S. Theory of waves in the lee of mountains. *Quarterly Journal*, Royal Meteorological Society (London), Vol. 75, No. 323 (January 1949).

(18) SCORER, R. S. Forecasting the occurrence of lee waves. *Weather*, Vol. 6, No. 4 (April 1951).

(19) SCORER, R. S. Mountain-gap winds; a study of surface winds at Gibraltar. *Quarterly Journal*, Royal Meteorological Society (London), Vol. 78, No. 335 (January 1952).

(20) SCORER, R. S. Theory of airflow over mountains: II. The flow over a ridge. *Quarterly Journal*, Royal Meteorological Society (London), Vol. 79, No. 339 (January 1953).

(21) MANLEY, GORDON. The helm wind of Crossfell, 1937–39. *Quarterly Journal*, Royal Meteorological Society (London), Vol. LXXI, Nos. 309–310, July–October, 1945.

(22) WARD, F. W. Helm-wind effect at Ronaldsway, Isle of Man. *Meteorological Magazine*, No. 947, Vol. 92 (August 1953).
(23) GEIGER, R. *The climate near the ground.* Harvard University Press, 2nd Edition (1950).
(24) THORNTHWAITE, C. W. Micrometeorology of the surface layer of the atmosphere. The Johns Hopkins University, Laboratory of Climatology, Seabrook, New Jersey. *Interim Report*, No. 18 (April 1, 1952, to 30 June, 1952).
(25) HALSTEAD, MAURICE H. The relationship between wind structure and turbulence near the ground. The Johns Hopkins University, Laboratory of Climatology, Seabrook, New Jersey. *Supplement to Interim Report*, No. 14 (1 April, 1951 to 30 June, 1951).
(26) GIBLETT, M. A. The structure of wind over level country. *Meteorological Office Geophysical Memoir*, No. 54, Vol. VI (1932).
(27) BRUNT, SIR DAVID. *Physical and dynamical meteorology*, 2nd Edition. Cambridge University Press (1952).
(28) BRUNT, SIR DAVID. *Weather study*. Thomas Nelson & Sons Ltd. (1942).
(29) JUUL, J. Investigation of the possibilities of utilization of wind-power. *Elektroteknikeren*, Vol. 45, pp. 607–635, No. 20 (October 22 1949).
(30) Organization for European Economic Co-operation. Vol. II of the *Final Report of the proceedings of the working group*, No. 2 (*Wind power*) (1954).
(31) LAMB, SIR HORACE. *Hydrodynamics*. Cambridge University Press, 5th Edition (1924).
(32) PRANDTL, L. and TOLLMIEN, W. The distribution of wind over the surface of the Earth calculated from the laws of flow in pipes. *Zeitschrift fur Geophysik Jahrg*. 1. *Heftl/2*, pp. 47–55 (1924–25).
(33) IDRAE, P. Etude sur les conditions d'ascendance du vent favorable au vol à voile. *Mem. de l'Office Nat. Met. de France* (1923).
(34) POCKELS, F. Zur Theorie der Niederschlagsbildung an Gebirgen. *Ann. d. Physik* (3), Vol. IV, p. 459 (1901).
(35) SHAW, SIR NAPIER. *The air and its ways*. Cambridge University Press (1923).
(36) WARD, A. Wind direction at North Front, Gibraltar. *Meteorological Magazine*, No. 977, Vol. 82, p. 322 (November 1953).
(37) BEST, A. C. Transfer of heat and momentum in the lowest layers of the atmosphere. Great Britain, Meteorological Office, *Geophysical Memoir*, No. 65, Vol. VII, No. 8 (1935).
(38) CORWIN, LEONARD B. A photographical recorder for the determination of wind velocity gradients. American Geophysical Union, *Transactions, Part I*, pp. 142–144 (1943).
(39) DEACON, E. L. Vertical diffusion in the lowest layers of the atmosphere. *Quarterly Journal*, Royal Meteorological Society, Vol. 75, No. 323, pp. 89–103 (January 1949).
(40) DEFANT, A. Die Zirkulation der Atmosphäre in den Gemassigten Breiten der Erde. *Geografiska Annaler*, Vol. 3 (1921).
(41) HALSTEAD, M. H. A stability term in the wind gradient equation. American Geophysical Union, *Transactions, Part II*, pp. 204–208 (October 1943).
(42) LAIKHTMAN, D. L. O profile vetra v prizemnom sloe statskonarnykh Usloviiakh (On the profile of wind in the terrestrial layer of the atmosphere under steady-state conditions.) U.S.S.R., *Akad. Nauk., Geographical and Geophysical Series*, Vol. 25, pp. 58–77 (1947).
(43) PRANDTL, L. Über Flussigkeitsbewegung bei sehr kleiner Reibung. *Intern. Mathematiker-Kongresses III, Verhandlungen* (Heidelberg, 1904).

(44) SHEPPARD, P. A. The aerodynamic drag of the earth's surface and the value of von Karman's constant in the lower atmosphere. Royal Society (London), *Proceedings, Series A, Mathematical and Physical Sciences*, Vol. 188, No. 1013, pp. 208–222 (30 January 1947).

(45) SUTTON, O. G. The logarithmic law of wind structure near the ground. *Quarterly Journal*, Royal Meteorological Society (London), Vol. 62, pp. 124–127 (1936).

(46) SUTTON, O. G. The logarithmic law of wind structure near the ground. *Quarterly Journal*, Royal Meteorological Society (London), Vol. 63, pp. 105–107 (1937).

(47) SVERDRUP, H. U. Note on the logarithmic law of wind structure near the ground. *Quarterly Journal*, Royal Meteorological Society (London), Vol. 62, pp. 461–462 (1936).

(48) SVERDRUP, H. U. Second note on the logarithmic law of wind structure near the ground. *Quarterly Journal*, Royal Meteorological Society (London), Vol. 65, pp. 57–60 (1939).

(49) THORNTHWAITE, C. W. and HALSTEAD, MAURICE. Note on the variation of wind with height in the layer near the ground. American Geophysical Union, *Transactions, Part II*, pp. 249–255 (1942).

(50) THORNTHWAITE, C. W. and KASER, PAUL. Wind gradient observations. American Geophysical Union, *Transactions, Part I*, pp. 166–182 (October 1943).

(51) SUTTON, O. G. Note on the variation of the wind with height. *Quarterly Journal*, Royal Meteorological Society, Vol. 58, pp. 74–76 (1932).

(52) PRANDTL, L. *Beitrage zur Physik der freien Atmosphäre*. Vol. 19, p. 188 (1932). Akademische Verlagsgesellschaft, Leipzig.

(53) ROSSBY, G. C. and MONTGOMERY, R. B. Massachusetts Institute of Technology, *Physical and Meteorological Papers*, Vol. 3, No. 3 (1935) (also Vol. 7, No. 4 (1940), and Vol. 1, No. 4) (1932).

(54) ALI, BARKAT. Variation of wind with height. *Quarterly Journal*, Royal Meteorological Society (London), Vol. 58, pp. 285–288 (1932).

(55) DOBSON, G. M. B. *Quarterly Journal*, Royal Meteorological Society (London), Vol. 40, 123 (1914).

(56) CHAPMAN, E. H. The variation of wind velocity with height. *M.O. Professional Notes*, No. 6 (1919).

(57) BERRY, F. A., BOLLAY, E. and BEERS, NORMAN R. *Handbook of meteorology*. McGraw-Hill Book Co. Inc., (New York, 1945).

(58) KNUDSEN, JON. Über Windmessungen an Hindernissen. Bergens Museums Årbok 1941, *Naturvitenskapelig rekke* Nr. 10.

(59) BURGESS, C. R. Local weather. *Weather*, Vol. 6, Nos. 5 and 6, pp. 144 and 163 (May and June, 1951).

(60) Studiengesellschaft Windkraft e.V., Stuttgart. *Windkraft Mitteilungen*, No. 3 (1st April 1954).

(61) CORBY, G. C. The airflow over mountains; a review of the state of current knowledge. *Quarterly Journal*, Royal Meteorological Society (London), Vol. 80, p. 491 (1954).

(62) SERRA, L. Le vent en France et ses possibilités d'utilisation. *La Météorologie* (October–December 1953), pp. 273–292.

See also meteorological references listed in Appendix I.

CHAPTER 8

THE MEASUREMENT OF WIND VELOCITY

It is perhaps not surprising that a natural phenomenon such as the speed of the wind has been the subject of measurement for many centuries. The earliest methods were by personal estimates from observation of its effects upon stationary or moving objects. Although somewhat rough and imprecise, these methods were easily applied; the only difficulty was in formulating a scale by which relative wind strengths could be distinguished and recorded.

Interest in the subject is evidenced by the fact that instruments for accurate measurement of wind speed and direction have been in existence for some 300 years. During this period, of course, instruments of many different types, often very ingenious, have been introduced, but they are far too numerous to be described here. An excellent review by Cleveland Abbe (Ref. 1) traces their development from the earliest times to the year 1888 while several of the works mentioned in the Bibliography at the end of this chapter deal fully with more recent developments. Another useful review of wind-measuring instruments has been given by Louvan E. Wood (Ref. 2). This includes a table comparing the different types and giving notes on transmitting mechanisms, indicators and recorders which may be used with them.

The instruments and methods described in the following pages are thus confined mainly to those especially applicable to wind measurements bearing upon the question of power production.

Personal estimation without instruments: wind scales

These apparently rough-and-ready methods are still in use by coastguards, lighthouse keepers and other observers who regularly report their estimates to meteorological services in many parts of the world. The general direction of the wind at an open site, clear of obstructions, can be judged to within a few degrees with little difficulty, and even estimates of its velocity are remarkably accurate—probably within $\pm$ 10 per cent, or even better—when made by experienced observers. Of the many scales used to express wind strength—Abbe made a collection of two hundred of them—that devised by Admiral Sir Francis Beaufort, Hydrographer to the British Navy, in 1805 and now known as the 'Beaufort Scale' has been most widely adopted internationally.

TABLE XIII

The Beaufort scale of wind forces

Beaufort number (B)	Description of wind	Equivalent mean velocity in knots	Limits of mean speed at 33 ft above flat ground in an open situation*			Mean wind force in lb/ft² at standard density ($P = 0{\cdot}0105B^3$)
			Knots	*Statute miles per hour*	*Metres per second*	
0	Calm	0	Less than 1	Less than 1	Less than 0·3	0
1	Light air	2	1– 3	1– 3	0·3– 1·5	0·01
2	Light breeze	5	4– 6	4– 7	1·6– 3·3	0·08
3	Gentle breeze	9	7– 10	8– 12	3·4– 5·4	0·28
4	Moderate breeze	13	11– 16	13– 18	5·5– 7·9	0·67
5	Fresh breeze	19	17– 21	19– 24	8·0–10·7	1·31
6	Strong breeze	24	22– 27	25– 31	10·8–13·8	2·3
7	Moderate gale	30	28– 33	32– 38	13·9–17·1	3·6
8	Fresh gale	37	34– 40	39– 46	17·2–20·7	5·4
9	Strong gale	44	41– 47	47– 54	20·8–24·4	7·7
10	Whole gale	52	48– 55	55– 63	24·5–28·4	10·5
11	Storm	60	56– 63	64– 72	28·5–32·6	14·0
12	Hurricane	68	64– 71	73– 82	32·7–36·9	18
13	—	76	72– 80	83– 92	37·0–41·4	23
14	—	85	81– 89	93–103	41·5–46·1	29
15	—	94	90– 99	104–114	46·2–50·9	35
16	—	104	100–108	115–125	51·0–56·0	43
17	—	114	109–118	125–136	56·1–61·2	52

The figures in this table have been abstracted from a specification published by the Air Ministry Meteorological Office which gives also descriptions of the effects of the various wind forces to enable them to be estimated. *Approximate corrections: for 50 ft add 10 per cent, for 100 ft add 25 per cent, for 20 ft subtract 10 per cent, for 10 ft subtract 20 per cent.

This scale, which divides wind speeds into 17 strengths or 'forces' is given, in its latest, revised form in Table XIII.

As for the wind direction, this is usually expressed in terms of the points of the compass but it is often convenient to express it in degrees (0 to 360°) reckoning north as zero and measuring the angle in a clockwise direction.

Instruments for the measurement of wind velocity

Abbe (loc. cit.) states clearly the possible means of measuring wind velocity by its effects as follows:

"A current of air sets in motion a system susceptible of rotation about an axis; produces an augmentation or diminution of pressure on different portions of fixed or constrained bodies; cools surfaces of higher temperature by convection; accelerates evaporation; evokes musical sounds of different pitch or intensity from properly constructed apparatus susceptible of vibration." He also describes methods of direct measurement of wind speed by observations on floating objects, by smoke trails, kites and balloons and by the effect of the wind on the velocity of sound waves.

All of these methods have been used though some have been much more common than the rest, and it is interesting to note the variants of them which are now pressed into service in new designs of instruments for special purposes.

In wind survey work instruments based on the first of Abbe's effects—rotation—used in conjunction with various forms of recorder are most important though the pressure tube anemometer, utilizing his second effect, is still the basic instrument for general meteorological purposes. An instrument acting on the same principle (see Ref. 4) is used also for the measurement of wind speed in wind tunnels in which rotational anemometers are calibrated.

DINES PRESSURE-TUBE ANEMOMETER. This instrument, devised by W. H. Dines, must claim first place in description, not because it is suitable for use on potential wind-power sites, when the measuring equipment is usually only a temporary installation, but because it has been widely used as the standard anemometer in permanently established meteorological stations the records from which serve as a basis in the search for especially windy districts.

In stream line flow, Bernouilli's equation $\frac{1}{2}\rho v^2 + p = C$ expresses the fact that the total head, i.e. the velocity head plus the static pressure, is constant. The terms on the left-hand side of the equation represent these two components of pressure, ρ being the density of the moving fluid and v its velocity while p is the static pressure. When an open-ended tube is placed in, and facing, the stream the constant

C is the total pressure acting at its mouth. Experimental evidence shows that such tubes, over a wide range of size and shape, can be used for the accurate measurement of total head even in turbulent flow (see Ref. 4).

Clearly, if the difference between the total head and the static pressure is measured, the result is proportional to v^2. This is the underlying principle of pressure tube anemometers of which the Dines instrument is a modification.

A correction for air density is necessary with all types of pressure anemometers when they are located at heights which are appreciably different from that of the place of standardization.

W. Ferrel (Ref. 5) has given a formula, for the conversion of velocity to pressure, which is theoretically correct if air viscosity is neglected. This is,

$$p = \frac{0{\cdot}002698}{1 + {\cdot}004t} v^2 \cdot \frac{P}{P_0}$$

p = pressure in pounds per square foot.
P_0 = standard barometric pressure (760 mm).
P = barometric pressure at the observation station.
t = temperature in °C.
v = air velocity in feet per second.

At a temperature of 15° C and $P = 760$, the formula becomes

$$p = 0{\cdot}00255\, v^2$$

This compares with the Smeaton-Rouse formula

$$p = 0{\cdot}00492\, v^2 \text{ (see p. 15)}$$

Since Smeaton's time a number of values of the constant have been proposed. These are all in sufficiently close agreement for the value 0·003 to be accepted as sufficiently accurate for most purposes. Such pressure-velocity formulae hold strictly only for small surfaces (up to about 1 sq metre).

The construction of the Dines anemometer is illustrated in Plate XV in which drawing (*a*) shows the head in section and (*b*) shows the recording apparatus.

The head, which is usually mounted on a rigid mast forty or more feet high, consists of a horizontal tube, open at one end and free to rotate, which is turned to face the wind by a vane. This head is supported by a vertical tube into which it leads. A little lower down, the vertical tube is surrounded by another, concentric, tube which is pierced by four rings of holes placed close together round its circumference.

The wind blowing into the horizontal tube creates a pressure which is communicated through a flexible metal tube to the recorder. Again, wind blowing through the small holes in the concentric tube creates a diminution of pressure, or suction, which is also transferred to the recorder by a second flexible metal tube.

In the recorder, a copper vessel, closed at one end, floats inverted in a cylindrical metal container partly filled with water and sealed from the outside air. The wind pressure, from the head, is transmitted to the space inside the float causing it to rise in the water and this is assisted by the suction which is applied to the space above the float. Thus, as the wind speed rises and falls, so does the float, and its motions are transferred, by means of a tubular rod passing through an air-tight collar in the top of the cylinder, to a pen tracing a record on the chart shown. The float is so shaped, in accordance with the law relating pressure to wind velocity, that the velocity scale on the chart is linear. A guide prevents the float from rotating and two stop cocks are fitted so that the spaces above and below it can be open to the outside atmospheric pressure for adjustment purposes.

The Dines instrument usually has a wind direction recorder combined with it so that both the speed and direction of the wind can be recorded on the same chart. The direction recorder must be located vertically underneath the vane in the anemometer head because it is operated from the direction shaft of the vane; this rotates a spindle to which a plain helix, operating a recording pen, is attached.

When the wind direction moves through north the roller, on the pen arm, moves along a short connecting groove between the ends of the helix to avoid breaks in the record or duplication of the direction chart.

Another form of the instrument, for use as a distant-reading anemometer incorporates a self-synchronous Selsyn system for transmission of both wind speed and direction to the recorder (see also p. 124).

A specimen of the continuous record of wind speed afforded by the anemometer is shown in Fig. 38 (on p. 109) and a wind direction record is shown in Fig. 36. Values of hourly mean wind speed are obtained from the chart simply by visual judgment of the mean of the oscillations in the record.

There is no question about the value of the Dines instrument for a permanent installation, but the need for a very rigidly-mounted mast to support the head and its pressure tubes, a building to house the recorder, provision for preventing the water in the float cylinder from freezing, and its general construction, make it rather expensive and unsuitable for wind surveys. Again, while hourly mean wind speeds can be obtained by observation of the mean height of the very irregular record during the hour, the variations in wind speed shown in the record are not very valuable for wind power investigations owing to the relatively slow response of the instrument to gusts—varying from about 1·7 sec at 60 m.p.h. to 2·3 sec at 21 m.p.h.

CUP ANEMOMETERS. By far the most extensively used instrument for the measurement of wind speed is the rotating cup anemometer. As first developed by T. R. Robinson (Ref. 6) in 1846 (and thereafter called the Robinson cup anemometer) its rotor consisted of a vertical

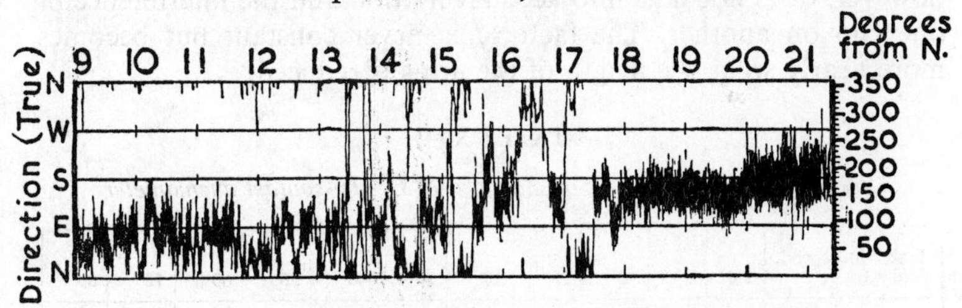

FIG. 36. *Wind direction record from Dines anemometer* (*R. W. Munro Ltd.*)

spindle carrying four horizontal arms, at right angles to one another, at the ends of which were hollow, hemispherical cups of thin sheet metal. The circular rims of the cups were in vertical planes passing through the common axis of rotation and their convex sides faced the direction of rotation. The action of the instrument depends upon the fact that the wind pressure on the concave sides of the cups is greater than on their convex sides; it thus rotates independently of wind direction.

Since its first introduction, many theoretical and experimental studies of this type of anemometer have been made (Refs. 1, 7, 8, 9, 10, 29 and 30). Its design has been improved by using three cups, which give a more uniform torque than four; by giving the cups beaded edges to reduce the effect of small-scale turbulence and improve the constancy of rate of revolution; by making them conical, instead of hemispherical, to reduce over-estimation of wind flow in fluctuating winds, and also by changing the relative dimensions of the cups and arms.

Theory and performance in steady winds. In spite of the apparent simplicity of the instrument its aerodynamic theory is complex and formulae for the relationship between its speed of rotation and the wind speed, under varying conditions, remain empirical.

It is clearly very desirable that rotational speed should be proportional to wind speed over a wide range; in other words, the anemometer factor $f = \frac{V}{v}$ where V = wind speed and v the circumferential speed of the cup centres, should remain constant.

J. Patterson (Ref. 8) developed a formula for f which, if the arms are very short so that corrections for their effect upon the torque can be neglected, reduces to $f = \frac{3 \cdot 717}{(Vr)^{0 \cdot 113}}$ where, V = wind speed in metres per second, and r = radius of cups in centimetres. This, however, does not take into account friction and the interference of one cup on another. The factor f is never constant but becomes more nearly so as the length of the arms is reduced.

TABLE XIV

Results of wind-tunnel test on standard M.O. cup-contact anemometer

True wind speed, m.p.h.	1·8	5	10	15	20	30	40	50	60	70	80
Factor, $\frac{V}{v}$	4·58	3·27	3·10	3·04	3·01	2·97	2·96	2·96	2·96	2·96	2·96
Indicated speed, m.p.h. for factor 2·98	1·2	4·55	9·61	14·7	19·8	30	40·2	50·4	60·45	70·5	80·5

Table XIV (Ref. 11) shows the variation of f and of indicated wind speed for varying true wind speeds for the latest pattern of the standard Meteorological Office cup-contact anemometer (the term "cup-contact" is explained later) which is designed to have a factor 2·98 and reads correctly at about 30 m.p.h.

These test results show that *in steady winds* up to 80 m.p.h. the error of the instrument does not exceed 0·6 m.p.h.

Incidentally, C. E. Brazier (Ref. 12) found, from wind tunnel tests, that there was practically no change in the factor f for cup anemometers when the plane of rotation of the cups was inclined to the wind at 30° which suggests that up to this inclination they measure the total wind velocity rather than merely the horizontal component.

Behaviour in fluctuating winds. When used in practice, cup anemometers are very frequently subjected to winds of fluctuating speed and their performance under such conditions is thus important.

F. J. Scrase and P. A. Sheppard (Ref. 10), following O. Schrenk (Ref. 13), have studied the question with results which are summarized below.

The equation of motion of the anemometer rotor is

$$\frac{I}{R}\frac{dv}{dt} = M - F$$

where I = moment of inertia, R = length of arm, v = circumferential cup speed, t = time, M = aerodynamic couple, F = friction moment.

Schrenk assumed that

$$\frac{M}{\frac{1}{2}\rho v^2 r^2 \pi R^2} = \lambda\left(\frac{V}{v}\right)^2 + \mu\left(\frac{V}{v}\right) + \alpha$$

where ρ = air density, V = mean wind speed, r = cup radius (for hemispherical cups) and λ, μ and α are constants determined from torque measurements in steady air flow.

The curves in Fig. 37 are those derived by Schrenk to show the

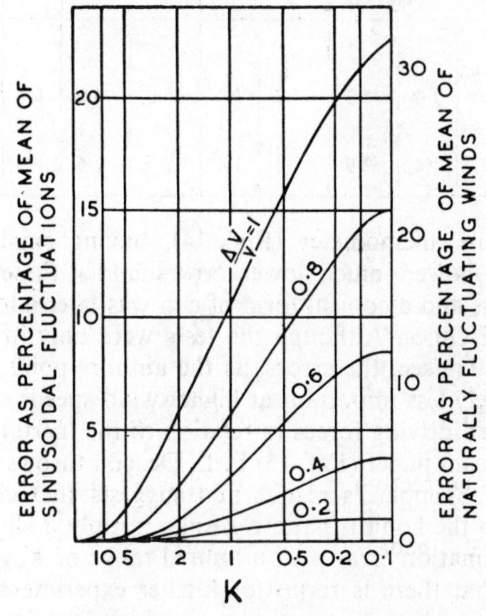

FIG. 37. *Anemometer error curves for fluctuating winds*

probable errors, both as a percentage of the mean velocity of sinusoidal fluctuations and of the mean in naturally fluctuating winds. The quantity $K = \dfrac{0{\cdot}55\rho R^2 r^2 TV}{I}$ where T = periodic time of the fluctuations, the units in the expression being in the M.K.S. system. ΔV is the amplitude of the velocity fluctuations.

While accepting Schrenk's analysis, Scrase and Sheppard point out that his constants will vary with changes in any of the parameters of the cup system. Their wind tunnel tests upon the standard M.O. cup anemometer (which then had hemispherical cups) showed that the over-estimation of wind speed in an irregularly fluctuating wind was reasonably in accordance with computations, from Schrenk's theory, given in Table XV.

TABLE XV

Errors of over-estimation in irregularly fluctuating Winds

V (m/sec)	2			5			10		
T (sec)	1	10	60	1	10	60	1	10	60
$\frac{\Delta V}{V}$	0·2	0·4	0·6	0·2	0·4	0·6	0·2	0·4	0·6
Over estimation (percentage of V) M.O. anemometer	1·4	5·7	7·8	1·4	5·0	4·0	1·4	3·8	1·5
Sheppard anemometer	1·4	3·9	1·8	1·4	2·3	0·6	1·3	1·2	0·1

The Sheppard anemometer (Ref. 14), having small and light conical cups, showed much lower errors under these fluctuating wind conditions and a conical form of cup was later adopted by the Meteorological Office. Although the tests were only in mean wind speeds up to 10 m/sec, the errors, as the authors point out, should become relatively less important at higher wind speeds owing to the greatly increased driving forces in relation to the inertia forces.

In a more recent paper (Ref. 15) E. L. Deacon throws some doubt on Scrase and Sheppard's results and suggests that changing the form of cup to the conical pattern results in only a slight decrease in the over-estimation error over a limited range of K.

It may be that there is room for further experimental work on the performance of such anemometers which, to date, has been mainly with fairly low wind speeds.

The following table (Ref. 16) shows the results obtained by inter-comparison of a Dines anemometer, a balsa wood windmill anemometer and a standard cup anemometer in a natural gusty wind with a moderately high mean value (see Fig. 38).

These results confirm Scrase and Sheppard's prediction of low errors at high wind speeds since they show that the readings from the three instruments agree to within 1 m.p.h. throughout a 24 hr period. Actually the agreement was better than appears from the table which gives measured wind speeds only to the nearest mile per hour.

TABLE XVI

*Intercomparison of anemometer results**

Costa Hill. 10.00 *hours* 18-8-51 *to* 10.00 *hours* 19-8-51

Type of anemometer	Mean hourly wind speeds (m.p.h.)																						
Dines	32	31	29	36	43	41	46	47	46	45	49	47	46	40	41	42	38	35	32	33	30	25	19	19
Cup-contact	33	31	30	36	43	42	46	48	46	45	49	47	46	40	42	41	38	35	32	33	30	24	19	19
Balsa wood windmill	33	32	29	37	42	42	46	48	47	45	49	48	45	40	42	41	38	35	32	33	30	23	20	19

* All anemometers sited at 80 ft above ground.

Thus, in spite of the apparent uncertainty indicated in the papers referred to above, there seems little doubt that measurements by

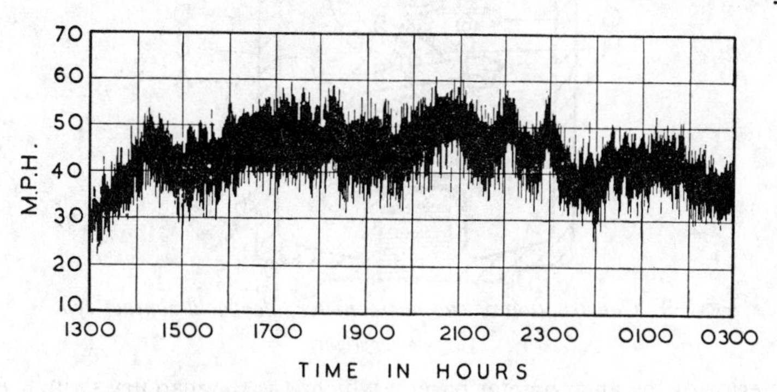

FIG. 38. *Wind speed record from Dines anemometer*

cup anemometers are sufficiently accurate for wind power purposes even when the wind speed is fluctuating.

Counter type cup anemometer. In this instrument the three cup rotor, of the standard M.O. pattern (discussed above) with 5 in. diameter beaded conical cups, drives a counter indicating the wind flow in miles (or kilometres), tenths and hundredths (see Plate XVII). The base has a tapped hole, $\frac{1}{2}$ in. B.S. pipe thread, so that the instrument can be screwed on to a corresponding length of pipe mounted on top of a pole or tower arm.

The mean wind speed is obtained by dividing the measured wind flow by the time taken.

Contact-type anemometer. This instrument has, below its standard three-cup rotor, a worm gear which tilts a glass tube containing mercury so that an electrical contact is made after the passage of each $\frac{1}{20}$th mile of wind. Fig. 39 shows the construction of the

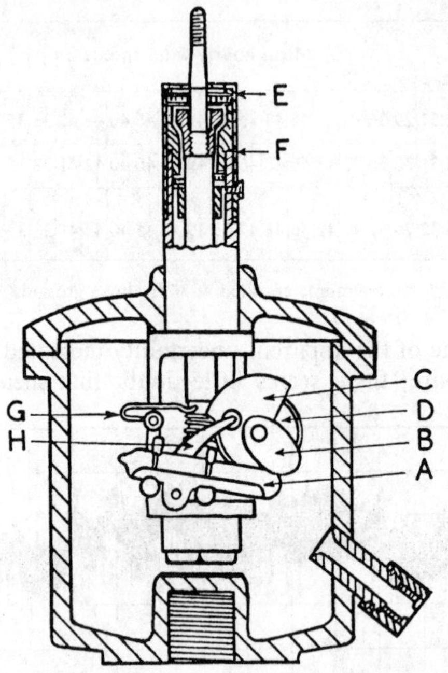

FIG. 39. *Construction of cup anemometer, electrical contact type* (*Crown copyright*)

interior of the anemometer base in which *A* is the mercury switch, *B* a cam which tilts it, and which is moved by a falling weight *C*. The weight is raised by a pin on the worm wheel *D* and falls after reaching its highest position. The mercury switch is included in a battery circuit to operate a telephone message register or some other type of electro-magnetic recorder. If, in service, the instrument is likely to be subjected to considerable vibration, or to very low temperatures, the mercury switch is replaced by a mechanical switch.

S. G. Crawford (Ref. 17) has described a sensitive cup-contact anemometer for accurate measurements of wind speed over the range 0·3 to 20 m/sec. This instrument is, however, not well suited to prolonged exposure under rigorous conditions.

L. E. Wood (Ref. 2) describes, with a diagram of connections, the use of a cup-contact instrument with a buzzer (or flashing electric lamp) to measure wind speed by counting the contacts made during one minute intervals.

For use in very severe weather conditions in the Antarctic (wind speed up to 100 knots, temperatures down to – 50° C, drifting snow and icing) a special design of cup-contact anemometer described

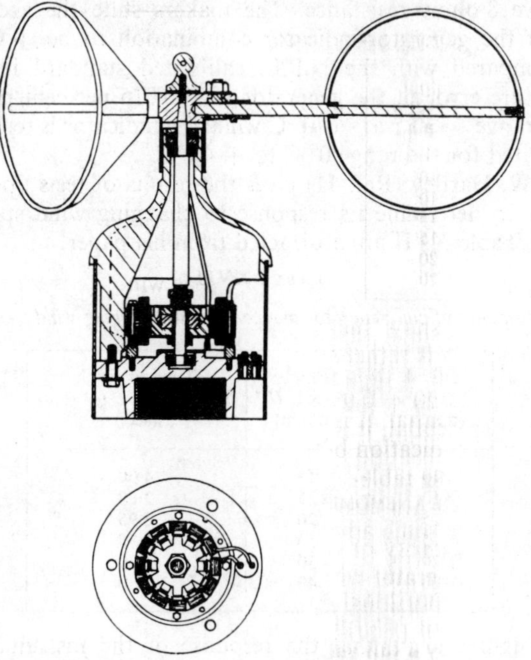

FIG. 40. *Construction of cup generator anemometer*
(*Crown copyright*)

by K. Langlo (Ref. 18) has been developed. This has no mercury but, instead, a robust wiping contact. The rotor has three unbeaded hemispherical cups, this shape of cup being used to minimize icing.

Cup generator anemometers. The function of these instruments is to measure instantaneous wind speeds in miles per hour, as distinct from wind flow, in miles, indicated by the two types described above. The latest Meteorological Office pattern (Ref. 11) has a three-cup rotor (conical cups 5 in. diameter with beaded edges) mounted on a short spindle running in ball bearings (Fig. 40). The spindle carries, at its lower end, a twelve pole permanent magnet system surrounded by a stator with low resistance windings; the whole forms a small a.c. generator giving 10 volts at 1000 r.p.m. The indicator takes the form of a moving-coil voltmeter with 240° scale, calibrated in m.p.h. from 5–100 m.p.h., supplied through a metal rectifier and taking 5 mA at full scale deflection.

As produced commercially, the generator is designed to supply two indicators calibrated on the assumption that the connecting leads have 8 ohms resistance. The makers state the accuracy (at 20° C) of the generator-indicator combination to be $\pm$ 0·5 m.p.h. when compared with the N.P.L. calibrated standard instrument. Temperature error of the generator is $-$ 0·015 per cent per °C rise over the range $-$ 30° to $+$ 70° C while the indicator is temperature-compensated for the range 0° C to $+$ 40° C.

G. E. W. Hartley (Ref. 11) gives the results of tests upon such a generator to determine its response to changing wind speeds. The figures in Table XVII are abstracted from his paper.

TABLE XVII

Response of cup generator anemometer to changing wind speed

Initial Speed, u (m.p.h.)	Final Speed, U (m.p.h.)	Time to reach Speed $u + 0{\cdot}9\,(U - u)$ (sec.)
10	15	4·00
10	25	2·60
15	20	2·95
15	30	1·95
20	30	1·80
20	40	1·30

These figures show that the response of the instrument, as at present designed, is rather slow. When the wind speed suddenly rises from 20–40 m.p.h. a time of 1·3 sec is taken for the indicated wind speed to rise to $20 + 0{\cdot}9\,(40–20) = 38$ m.p.h. The response curve not being exponential, it is not possible to give the time required to reach the final indication but it will, of course, exceed the time given in column 3 of the table.

WINDMILL TYPE ANEMOMETER. An alternative to the cup anemometer which may have some advantages in speed of response to changing winds and in linearity of scale, takes the form of a small windmill, driving a d.c. generator with permanent magnets, so that the voltage generated is proportional to wind speed. This voltage is supplied to an indicating, or recording, voltmeter. The windmill must be kept facing the wind by a tail vane which vane can also be used to operate a wind-direction recorder. Plate XVI shows the construction of a windmill-type anemometer made by the Electrical Research Association for their researches on the performance of wind-driven generators. It is used with an electronic counter and timer to obtain short-period mean wind speeds (down to a period of 2 sec). The propeller drives the permanent-magnet rotor of a small alternator which gives fifteen cycles per revolution. The number of cycles in the output during a given period is counted and is a measure of the mean wind speed during that period. A Bendix Friez anemometer of the windmill type is stated by the makers to have withstood satisfactory wind tunnel tests up to 200 m.p.h. They also give its error as within $\pm$ 0·5 per cent up to wind speeds of 80 m.p.h.

S. P. Fergusson (Ref. 7(*a*)) and J. H. Conover (Ref. 19) have described wind-tunnel, and open air, tests upon this type of instrument comparing it with a cup anemometer, with balsa-wood and duralumin windmill anemometers and with a pressure tube, to determine their relative performances under changing wind conditions. These tests indicate that the windmill anemometers give less overestimation in gusty winds than cup anemometers.

L. E. Wood (Ref. 2) describes an electronic wind-speed indicator containing a cup anemometer, of modified design, and an electronic circuit with a milliammeter calibrated in m.p.h. This circuit greatly reduces the load placed on the anemometer by the indicating instrument.

The French instrument manufacturers Jules Richard (formerly Richard Frères) make a rather different form of windmill anemometer. It has a rotor with six paddle-shaped blades. This gives an impulse, in an electrically-operated circuit, for every kilometre run of wind. The pen of the recorder is raised half a division at a time until 100 km have passed when the pen falls again to the bottom of the chart. On the same chart a second pen records the wind direction; a direction mark is made for each kilometre run of wind. The windmill transmitter and the corresponding recorder are made for either 1 km or 5 km intervals of wind-run and for four, eight or sixteen directions of wind.

The same manufacturers make another instrument, for use in combination with that just described, which records the vertical component of the wind.

COMPTEUR D'ÉNERGIE ÉOLIENNE. P. Ailleret has described (Ref. 20) a new form of cup anemometer which is used in the French wind power surveys. This instrument, which has been developed by Compagnie pour la Fabrication des Compteurs, of Montrouge, is illustrated in Fig. 41.

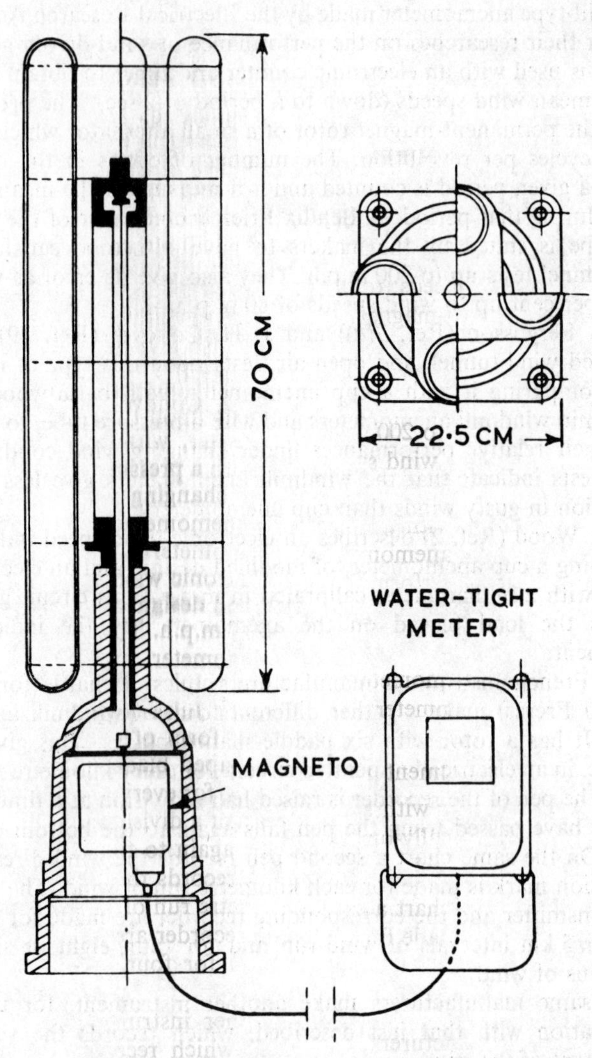

FIG. 41. *Anemometer by Cie de Compteurs*

It consists of two parts, the transmitter, which is a cup-generator anemometer with cylindrical cups with quarter-sphere ends, and the receiver, which is a modified form of kilowatt-hour meter.

J. Patterson (Ref. 8) tested anemometer rotors of this type, with short arms to obtain a constant anemometer factor, and found that they had high torque and high rotational speed, the factor being almost constant, at about two, for varying wind speed. Thus such a rotor can provide the small amount of power needed to operate the electric meter but bracing is necessary to withstand the large centrifugal forces. Ailleret reports that the Compteur has successfully withstood tests in a wind tunnel with wind speeds up to 50 m/sec.

There is, however, some reason to expect that the life of the rotor may be somewhat short on exceptionally windy sites. The spindle of the cup rotor carries the permanent magnet rotor of a small alternator housed in the base of the transmitter. The alternator, whose voltage and frequency are thus proportional to the wind speed within narrow limits, supplies both operating coils of the meter. The electrical constants of the circuit are adjusted so that the meter integrates the cube of the wind speed.

On the assumption that the Betz coefficient $\frac{16}{27}$, for the maximum possible fraction of available wind power which can be extracted by a windmill (see p. 192), is correct, the expression used for the power (in kilowatts) extracted can be written

$$P = \tfrac{16}{27} \times \tfrac{1}{2}\rho A V^3$$

where ρ = air density in kilogrammes per cubic metre, A = swept area in square metres, and V = wind speed in metres per second. Thus for one square metre, and taking $\rho = 1{\cdot}25$, $P = 0{\cdot}00037V^3$ and on this basis the meter dials are marked in kilowatt-hours per square metre.

The instrument starts to operate in a wind of about 3 m/sec (6·7 m.p.h.) and it follows the cubic law very closely up to about 13 m/sec (29 m.p.h.) after which speed the error increases until, at 25 m/sec, the reading is about 30 per cent low.

The advantages of the Compteur are its convenience in reading and its ease of maintenance: the meter is located near ground level so that it can be read without climbing and the absence of pens and charts avoids the need for frequent attention. Readings can be taken at lengthy intervals without any loss of the record which is, of course, purely an integration of the energy which theoretically could be extracted by a wind turbine at the site, It is thus very convenient for comparing the wind power potentialities of sites in a preliminary-stage survey.

Apart from the large errors at high wind speeds, the most serious disadvantage of the instrument is that it provides no information on the wind régime at the site. It is impossible to determine, at the end of a lengthy period of time, how the integrated total of kilowatt-hours per square metre has been made up; it could have been obtained by some hours of very high wind speeds—unusable by a wind turbine—followed by calms or, on the other hand, by many hours of moderate winds the energy of which would be usable.

HAND ANEMOMETER. An instrument of the cup-generator type which has proved very useful in wind surveys is the hand anemometer, one form of which is shown in Plate XVIII. It has a small four-cup rotor whose spindle, running in ball bearings, carries a four-pole magnet and keeper. A copper alloy drum, which is restrained by a hairspring, is located in the air gap between the magnet and keeper and the operating force is the magnetic drag between the magnet and the drum in which eddy currents are induced by the magnetic field of the rotating magnet. The drum carries a pointer moving over a semi-circular scale which is nearly uniform over the full range 0–60 knots. The makers give the errors as within $\pm$ 1 per cent of full scale deflection up to half scale, and within $\pm$ 2 per cent of the indicated reading over the remainder of the range.

Recorders used with cup anemometers

If, in wind measurements, anything more than total wind flow (in miles)—which can be obtained from a counter type cup anemometer—is required, some form of recorder is necessary for use with either a contact- or generator-type cup anemometer.

Several types of recorder—too many for detailed description here—have been developed for different purposes. It is proposed to concentrate on those which may be most useful in wind power investigations giving others, which are doubtless quite satisfactory for other purposes, merely a brief mention.

The recorders may be classified according to whether they record (i) approximately instantaneous, (ii) short period or (iii) long period wind speeds.

Approximately instantaneous recorders. These can take the form of a recording voltmeter used with a cup-generator anemometer. Wind speeds, changing with time, are recorded on a clockwork- or electrically-driven chart. As already mentioned, the response of the cup-generator instrument is rather slow—too slow to provide useful information on the magnitude and rate of rise of gusts. This defect, as well as the cost and maintenance requirements of the instrument, have been against its adoption in wind power investigations.

Short period recorders. The cup-contact anemometer, which can also be used with a battery circuit and buzzer to indicate wind speeds over very short periods of time, is widely used for recording wind flow from which mean wind speeds for short or long periods can be obtained.

The Bibby recorder (Ref. 21) uses the impulses received from the anemometer, whose mercury switch is included in a 6 V d.c. circuit, to drive forward a 120 tooth wheel, one tooth at a time. Through gearing, each advance raises a pen marking a chart. Thus, the passage of each $\frac{1}{20}$th mile of wind over the anemometer raises the pen by the same distance. At the end of three minutes a time switch, operated by a standard synchronous clock motor, makes a contact energizing an electromagnet and allowing the pen to fall to the bottom of the chart, the chart moves forward by a short distance at the same time. The record on the chart is a series of vertical lines, one for each three minutes, the height of each of which corresponds to the number of one twentieths of a mile of wind in three minutes ($\frac{1}{20}$th hour) so that the heights give, directly in miles per hour, the mean wind speeds during these successive time intervals.

I. Long (Ref. 22) gives details of a somewhat similar recorder constructed at Rothamsted Experimental Station. It uses a standard seven-day clock drum and records, on a chart on the drum, each fifty mile run of wind.

Long period recorders and analysers. (*a*) G. E. W. Hartley (Ref. 11) has described a recorder which gives mean hourly wind speeds over a period of one week. It has a long vertical cylindrical drum carrying a chart, and driven, through bevel gearing, by an impulse clock movement. The impulses are supplied from the cup-contact anemometer and these cause the drum to rotate, its angular movement being thus proportional to the run of wind in miles. A thirty-day clock is used to move a pen down a spindle, parallel to the drum surface, at the rate of $1\frac{1}{2}$ in. a day; the pen is re-set weekly at the top of the spindle by hand. A time switch depresses the pen to make hourly marks on the spiral record on the chart and the mean hourly wind speed is given by the distances between these marks.

(*b*) For a similar purpose the Electrical Research Association (Ref. 31), in their wind velocity surveys, use an impulse recorder containing the clockwork mechanism of a recording voltmeter, of the ribbon and chopper-bar type. The chopper bar is replaced by marking solenoids energized from a 6-V battery through the switch contact on the anemometer. A summator mechanism is used to integrate the $\frac{1}{20}$th mile impulses so that a single dot is made on the recorder chart after the passage of each two miles of wind. The chart is driven by a standard eight-day clock and moves at 3 in./hr. Twice the number of spaces between dots, in a 3 in. run of chart, thus gives the hourly wind speed. Up to four anemometers can be used with one recorder by providing an equal number of marking solenoids.

In Fig. 42 is shown the summator mechanism used in the recorder. An insulated disc is driven by a contactor coil through a ratchet wheel so that each $\frac{1}{20}$th mile anemometer impulse advances the

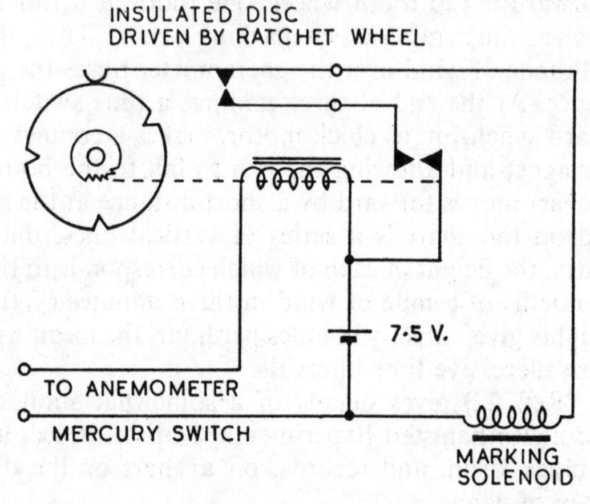

FIG. 42. *Summator mechanism of E.R.A. impulse recorder*

wheel $\frac{1}{120}$th of a revolution. The impulse also closes a subsidiary contact whose function is to prevent the marking solenoid from remaining energized between impulses, and so draining the 6-V dry battery used. The slots in the insulated disc allow the main contacts to close, and hence the marking solenoid to operate, three times during one revolution, which thus gives one mark for each two miles of wind.

Fig. 43 shows the construction of a later form of impulse recorder used by the E.R.A. and introduced because it requires less maintenance and skilled attention than the recorder just described.

It has a spindle, with 120 ratchet teeth, which rotates once for every six miles of wind passing the anemometer. This drives a paper tape, through gearing, at either 1 in. for sixteen miles of wind or at ten times that rate. A time switch, driven by clockwork, and a battery-operated solenoid make marks on the tape at any required time intervals such as hourly. Distances between time marks are proportional to the wind speed. A slight disadvantage of this instrument, which may however be important under some circumstances is that when there is no wind, the timing marks fall at the same place on the tape and it is thus difficult to determine the duration of the calm spell.

(*c*) A wind-run analyser developed at the Building Research Station (Ref. 23) also gives mean hourly wind speeds, or shorter period means if required. It uses a pen recorder, with a chart renewable monthly, and gives a record somewhat similar to that of the

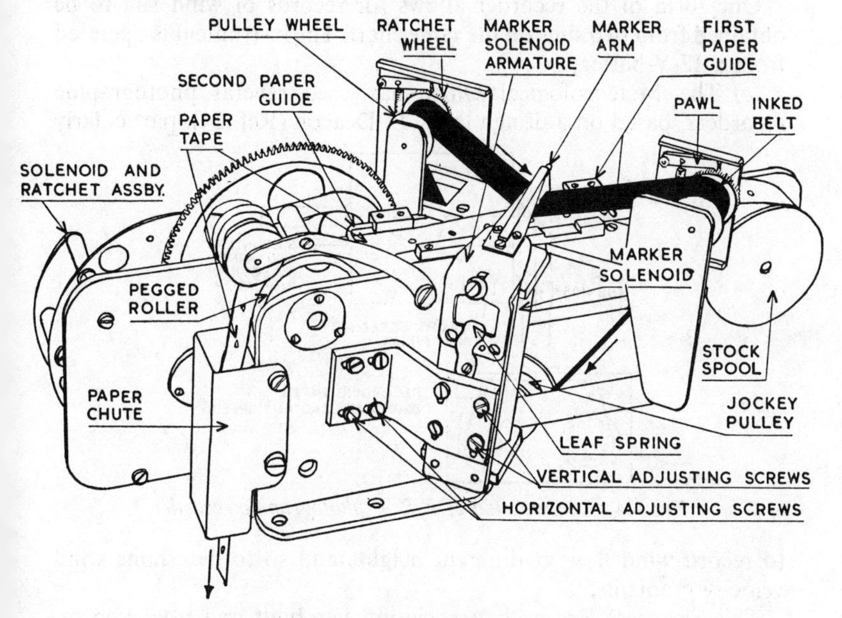

FIG. 43. *Construction of E.R.A. impulse recorder (Series II)*

Bibby recorder except that the line giving the hourly flow is not vertical but has a horizontal component corresponding to the hourly chart movement. While it is, no doubt, very suitable for an established research station, having a mains-voltage supply and facilities for maintenance, the complexity and probable cost of this instrument render it unsuitable for survey work.

(*d*) Another recorder giving wind run in the form of vertical line records is the Robitzsch 'Summenschreiber' (Fuess). Its operation is very similar to that of the Bibby recorder.

This instrument is arranged for use with a cup-contact anemometer operating two contact systems used alternatively. The first makes a contact for every 500 m of wind and the second for every $8\frac{1}{3}$m.

An interconnected change-over switch alters both the contact system in use and, at the same time, the chart speed which is governed by a two-rate, clockwork-driven, time switch. In this way either mean hourly wind runs during one month, or mean wind runs during each minute in a day, may be recorded, both in kilometres per hour. The corresponding chart speeds are 1 mm/hr or 0·5 mm/min.

One form of the recorder allows for records of wind run to be obtained from two sites on the same chart. The instrument is operated from a 12-V battery.

(*e*) The Meteorological Office has used several photographic recorders, based on a design by E. L. Deacon (Ref. 24), particularly

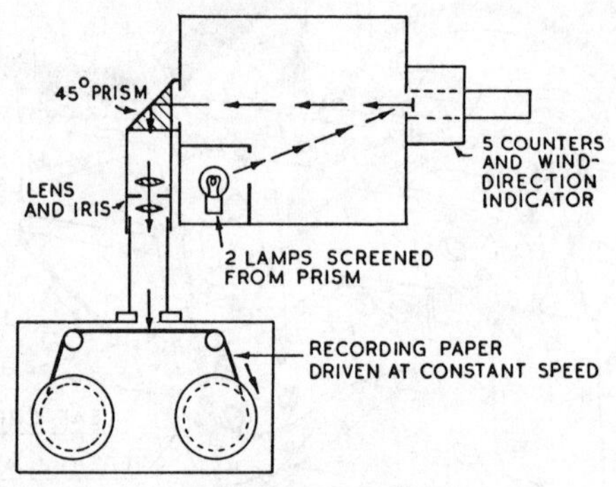

FIG. 44. *Construction of E.R.A. photographic recorder*

to record wind flow at different heights and so to determine wind velocity gradients.

The Electrical Research Association has built and used two recorders based on the same general design but adapted to the requirements of wind surveys on hill top sites. The essential features of their design are outlined in Fig. 44 while Plate XIX shows the appearance of the recorder. They have been used to record photographically, at thirty minute intervals, the readings of four telephone message registers, each operated by a cup-contact anemometer with a battery of wet Leclanché cells (to avoid having to remove accumulators for charging at frequent intervals), and also the reading of a remote-indicating wind direction instrument (see p. 121). For the purpose the pointer of the latter was replaced by a thin aluminium disc marked with a 360° scale at 5° intervals.

A light-tight box contains the direction disc and five message registers—one for timing—as well as two lamps, supplied from an accumulator, and switched on by time switch for six seconds every thirty minutes. The time switch also moves forward the timing counter by one digit at each exposure. The light reflected from the counters and direction indicator dial (illuminated through a slot to pick out the angle of wind direction) passes through a prism and lens to photographic recording paper which is driven at constant speed (by clockwork) inside a detachable camera.

The half-hourly wind runs at the four anemometers are obtained by subtraction of each row of figures from those in the next row on the record. The paper runs for one week before it has to be changed. The recorded wind directions are, of course, merely those at the time of exposure, i.e. at the end of each half hour, but the general direction of the wind does not usually change greatly during such a period of time so that such a record may be sufficiently accurate.

Measurement of wind direction

An essential part of wind direction measuring equipment is a pivoted wind vane which quickly follows direction changes without overshooting. The direction may be merely observed from the vane's position in relation to fixed arms pointing to the main points of the compass; it may be indicated on a remotely situated, dial-pattern instrument or, again, it may be continuously recorded.

Direction indicators. These may be directly operated mechanically from the vane, under which they are located, but more usually, and conveniently, they are operated electrically. The latter type require a transmitter driven by the vane—which must be capable of producing the necessary torque—and a receiver to act as indicator.

In the Meteorological Office pattern (Fig. 45) the vane spindle carries moving electrical contacts which, moving over fixed contacts in the transmitter, cause a corresponding rotation of the indicator through a Desynn system of transmission (Ref. 11). A drawing of the complete instrument is shown in Fig. 46.

A Swedish indicator (Ref. 25) uses a Selsyn transmission system as shown in Fig. 47. Both the transmitter and receiver have two-pole stator windings with three Y-connected windings, displaced 120°, on their rotors the windings of which are connected externally, through slip rings, as shown. The rotors move in synchronism with each other and provide a simple means of transmitting motion but the system demands an alternating current supply and is probably more expensive than that previously described. A similar system is used in the American Bendix Frieze wind direction indicators.

Direction recorders. Direction recording by the Dines Anemometer (see p. 104) and by the E.R.A. photographic recorder have already been described.

Chart recorders. Hartley (Ref. 11) describes several alternative methods of recording through mechanical transmission from the wind vane spindle which carries a cam driving a pen over a recorder chart.

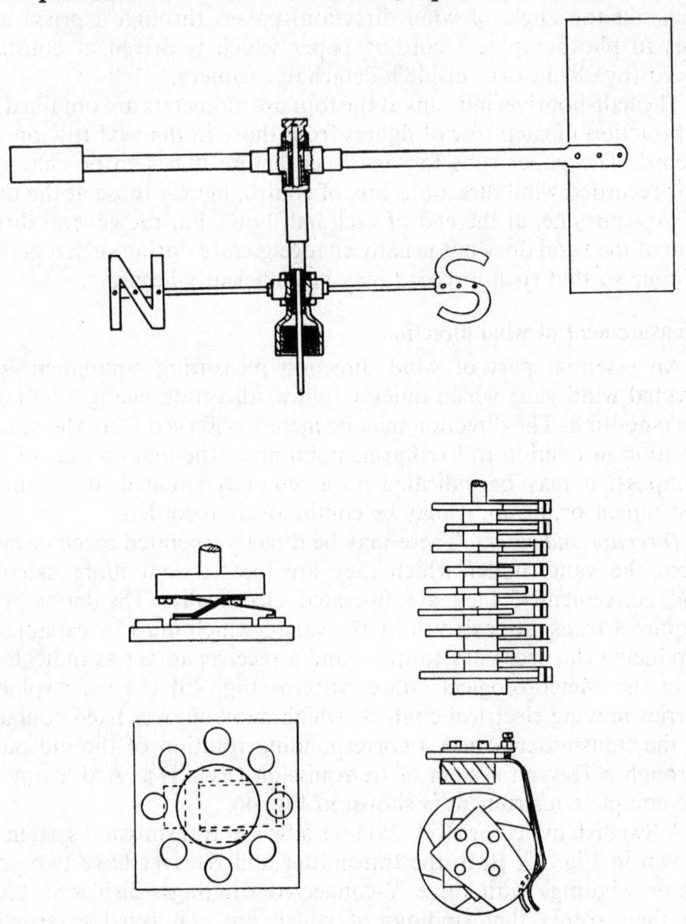

FIG. 45. *Details of wind direction indicator*

(Crown copyright)

He also describes a bi-directional wind vane used in wind recording gear which, in addition to measuring wind speed in the horizontal plane, records wind direction in both horizontal and vertical planes.

In this recorder Magslip transmitters and receivers are used.

Direction analysers. At the Building Research Station a direction recorder, or analyser, has been developed (Ref. 23) for use with a cup-contact anemometer and a wind vane. The vane moves a contact arm over an eight-way switch, each of the eight contacts corresponding to the main points of the compass at 45° intervals. Connected to the contacts are eight circuits containing telephone message registers (electromagnetically operated counters) so that the run of wind, in miles, from each main direction in any given period of

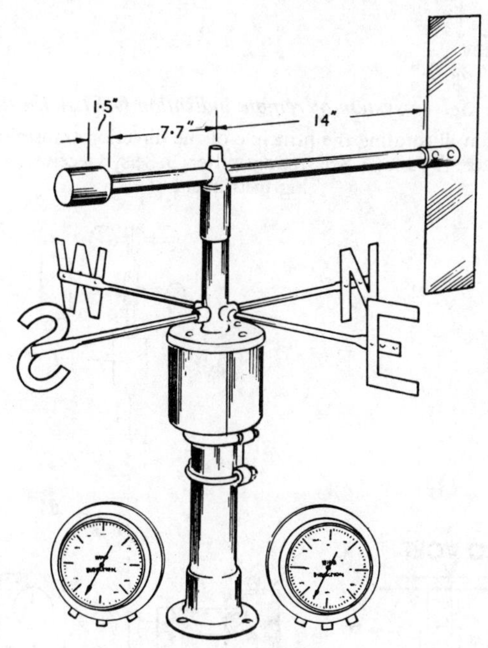

FIG. 46. *Construction of wind direction indicator*
(*R. W. Munro Ltd.*)

time is measured. A further counter gives the total wind run in the same period.

An adaptation of the German Summenschreiber instrument already described (see p. 119) goes rather further and measures the wind runs in each of eight main directions for each hour or each minute. An eight-segment commutator is driven by a wind vane, used with a cup-contact anemometer as in the English instrument just described, and nine complete recording Summenschreiber elements and charts are fitted (see Fig. 48). The ninth gives total wind run records.

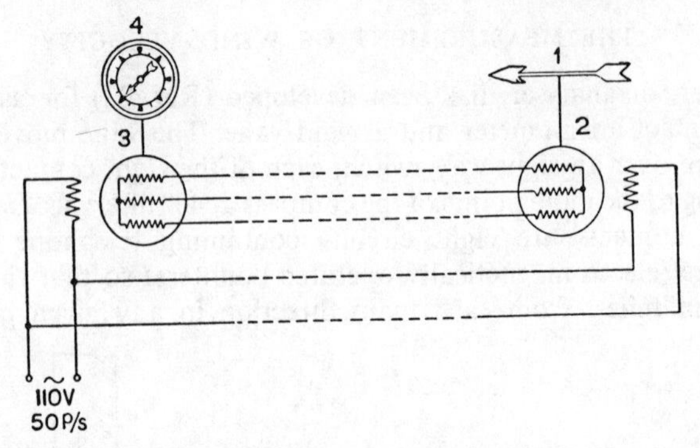

FIG. 47. *Selsyn system of remote indication* (*ASEA Electric Ltd.*)
Diagram illustrating the principle of the direction transmitter and indicator. 1. Vane; 2. Selsyn transmitter; 3. Selsyn receiver; 4. Reading instrument

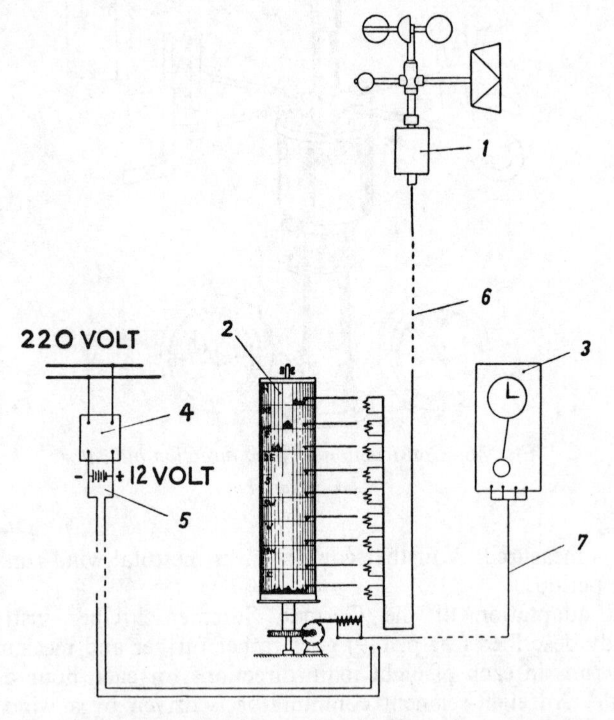

FIG. 48. *Diagram illustrating the operation of an anemometer with a Summenschreiber instrument* (*R. Fuess*)
1. Transmitter; 2. Rewinder; 3. Time switch; 4. Rectifier; 5. Accumulator; 6. Connecting lead; 7. Connecting lead

Other specially-designed, wind-measuring instruments

Some instruments used for the measuring and recording of wind direction and velocity not already included in the descriptions of types are worthy of mention. Especially so, as bearing directly on the wind power problem, are the instruments used in connection with the Grandpa's Knob, 1250 kW aerogenerator project.

Smith-Putnam (Grandpa's Knob) Project. P. C. Putnam (Ref. 26) describes fully the assumptions, regarding wind behaviour, which he and his associates made at the outset of this project in 1939 and also the methods used to test their validity.

At the start of their survey in the Green Mountains of Vermont they used Robinson cup anemometers with four $2\frac{5}{8}$ in. cups but troubles from icing led them to design a cup anemometer with a heated rotor. This was based partly on similar designs in Norway and at Mount Washington observatory.

(Mr. Odd Dahl of the Chr. Michelsens Institutt, of Bergen, designed a heated-cup anemometer for use at meteorological stations in Norway. Somewhat surprisingly, icing troubles have not made it necessary to use heated anemometers in Norway except at a station in the central mountain range at an altitude of about 6000 ft.)

Some of the American instruments were gas heated and some electric. In addition to wind tunnel calibration tests on both Robinson and heated-rotor instruments, the latter were calibrated also, as a check, by mounting them on a truck which was then driven at steady speeds in conditions of dead calm weather. The heavy, heated rotors were found to overrun the Robinson cup instruments and to give some 4 per cent greater indication over a long period of measurement (forty-four months).

Measurements of wind run were integrated, in half mile steps, and recorded on the chart of a clockwork-driven drum-type chronograph recorder. Two other recorders used at the Grandpa's Knob site were a U.S. Weather Bureau, category-type integrating recorder and one of the demand-meter type. The former records the total minutes of wind in each of ten different speed categories and the latter prints, on a slowly moving tape, the mileage of wind passing every fifteen minutes.

Other instruments. A range of wind-measuring instruments by a German manufacturer, R. Fuess, includes a Universal wind recorder which gives instantaneous wind velocity, run of wind and wind direction, all on a single chart. It has a rotating cup rotor and wind vane as well as a pressure tube and is made with either mechanical transmission—when the recorder is situated beneath the transmitter—or semi-electric transmission for longer distances.

Another interesting German instrument is the A.E.G. Ferrari recorder. The main feature is a solenoid-operated punch which makes holes in a paper ribbon driven at constant speed by an electrically-wound clockwork mechanism. The instrument is a multi-purpose one depending upon the source of the impulses energizing the solenoid. H. Christaller, of Energie-Versorgung Schwaben AG, has suggested that if the impulses originate from an anemometer a hole can be made for each 5000 m (or any other convenient flow) of wind.

The punched paper is passed through an analysing machine, with twenty-four counters, which records the numbers of spaces, between holes, of each of twenty-four different lengths. In this way—the lengths being proportional to the different times for the passage of the selected flow of wind—data for the construction of a wind velocity duration curve, in twenty-four steps, can be collected. The use of such a recorder could save much time in analysing wind records and introduces also the possibility of obtaining records of wind velocities occurring at the same hour on a number of days. The main difficulty may prove to be the cost of such equipment.

R. Vezzani, Chief Civil Engineer to the Department of Public Works in Italy, has devised a wind measuring instrument intended especially for wind power purposes. It records the angle of inclination of the wind stream to the horizontal as well as measuring wind speeds from any point of the compass. Details of the instrument have not, however, been published.

Another instrument which has been developed especially to investigate wind power potentialities at a particular site has been described (Ref. 27) by J. Ackeret and J. Egli of the Institut für Aerodynamik of the Technical College of Zurich. It is an ingenious and very elaborate instrument working on the pressure-tube principle but incorporating an electrical counting system which divides the integrations of wind pressure (proportional to the square of the wind speed) into ten groups of ascending magnitude so that a kind of 'modified power-duration curve' can be plotted from their readings. The instrument has many refinements, including heating to avoid icing troubles, and gave satisfactory results over a trial period of one year. Its complex, and obviously expensive, construction does not, however, render it suitable for wind survey work in general.

BIBLIOGRAPHY

(1) ABBE, CLEVELAND. *Treatise on meteorological apparatus and methods.* Government Printing Office (Washington, 1888).

(2) WOOD, LOUVAN E. Meteorological instruments. *Handbook of Meteorology,* Section VIII, pp. 531–572. McGraw-Hill Book Co. Inc. (New York, 1945).

(3) *Dictionary of Applied Physics,* Vol. III. Meterological instruments (1923).

(4) OWER, E. *The measurement of air flow*. Chapman & Hall (1949).
(5) FERREL, W. Relation of the pressure to the velocity of the wind. *American Meterological Journal*, Vol. 4, p. 173 (1887).
(6) ROBINSON, T. R. *Proceedings of the Royal Irish Academy*, Ser. II, Vol. II, p. 427 (1875). *Transactions of the Royal Society*, Vol. 169 (1878).
(7) FERGUSSON, S. P. Experimental studies of anemometers. *Harvard Meteorological Study*, No. 4, Oxford University Press (1939).
(7*a*) FERGUSSON, S. P. The sensitiveness of anemometers. *Bulletin of the American Meteorological Society*, p. 95 (April 1934).
(8) PATTERSON, J. The cup anemometer. *Transactions of the Royal Society of Canada*, Section III, Vol. XX (January 1926).
(9) DINES, J. S. *London Meterological Magazine*, 72, p. 3 (1937).
(9*a*) DINES, J. S. *London Meteorological Magazine*, 71, p. 133 (1936).
(10) SCRASE, F. J. and SHEPPARD, P. A. The errors of cup anemometers in fluctuating winds. *Journ. Sci. Insts.*, Vol. 21, p. 160 (1944).
(11) HARTLEY, G. E. W. The development of electrical anemometers. *Journal of the Institution of Electrical Engineers*, Vol. 98, Part II, p. 430, No. 64 (August 1951).
(12) BRAZIER, C. E. Recherche expérimentale sur les moulinets anémométriques. *Annales du Bureau Central Météorologique de France*, *Mémoires*, p. 157 Année 1914).
(13) SCHRENK, O. *Zeit. tech. Phys.*, Vol. 10, p. 57 (1937).
(14) SHEPPARD, P. A. *Journ. Sci. Insts.*, Vol. 17, p. 218 (1940).
(15) DEACON, E. L. The over-estimation error of cup anemometers in fluctuating winds. *Journ. Sci. Insts.*, Vol. 28, p. 231 (1951).
(16) GOLDING, E. W. and STODHART, A. H. The selection and characteristics of wind-power sites. Electrical Research Association, *Technical Report*, Ref. C/T108 (1952).
(17) CRAWFORD, S. G. A simple form of sensitive electric contact cup anemometer. *Journ. Sci. Insts.*, Vol. 28, No. 2, p. 36 (February 1951).
(18) LANGLO, K. Anemometers for the Norwegian-British-Swedish Antarctic expedition 1949–52. *Journ. Sci. Insts.*, Vol. 29, p. 60 (February 1952).
(19) CONOVER, J. H. Tests of the Friez aerovane in the natural wind at Blue Hill observatory. *Bulletin of the American Meteorological Society*, p. 523 (November 1946).
(20) AILLERET, P. L'énergie éolienne: sa valeur et la prospection des sites. *Revue générale d'électricité* (March 1946).
(21) BIBBY, J. R. *Quarterly Journal*, Royal Meteorological Society, 306, p. 277 (1944).
(22) LONG, I. *Journal of the Institution of Electrical Engineers*, Vol. 98, Part II, p. 548 (1951).
(23) PRATT, A. W., COLLINS, B. G. LACY, R. E. and SPINK, E. W. Some meteorological instruments used in building research. *Journal of the Institution of Electrical Engineers*, Vol. 98, Part II, No. 64, pp. 474–479 (August 1951).
(24) DEACON, E. L. *Porton Report* No. 43 (16th February 1943).
(25) OHLUND, R. E. Wind metering equipment. *ASEA Journal*, p. 39 (March–April, 1948).
(26) PUTNAM, P. C. *Power from the wind.* Van Nostrand (1948).
(27) ACKERET, J. and EGLI, J. Ein Windzähler zur Bestimmung der Bruttoleistung des Windes. *Elektrizitätsverwertung* (August 1946).
(28) DINES, W. H. Anemometer comparison. *Quarterly Journal*, Royal Meteorological Society, Vol. 18, p. 165 (1892).

(29) FERGUSSON, S. P. *Experimental studies of anemometers.* Oxford University Press (1939).

(30) SHEPPARD, P. A. Anemometry: a critical and historical survey. *Proceedings of the Physical Society*, Vol. 53, p. 361 (1941).

(31) ROSENBROCK, H. H. and TAGG, J. R. Wind- and gust-measuring instruments developed for a wind power survey. *Journal of the Institution of Electrical Engineers*, Vol. 98, Part II, p. 438 (August 1951). (Electrical Research Association, *Technical Report*, Ref. C/T104.)

(32) HALSTEAD, M. H. The relationship between wind structure and turbulence near the ground. *Supplement to Interim Report* No. 14. The Johns Hopkins University, Laboratory of Climatology, Seabrook, New Jersey (30 June 1951).

(33) Johns Hopkins University. Micrometeorology of the surface layer of the atmosphere. *Interim Report* No. 18. Laboratory of Climatology, Seabrook, New Jersey (1 April 1952 to 30 June 1952).

(34) BERRY, F. A., BOLLAY, E. and BEERS, NORMAN R. *Handbook of meteorology.* McGraw-Hill Book Co. Inc. (New York, 1945).

(35) BREVOORT, M. J. and JOYNER, U. T. Aerodynamic characteristics of anemometer cups. *N.A.C.A. Tech. Note* No. 489 (November 1933).

(36) BRADFIELD, F. B. Aerodynamic properties of a hemispherical cup. With application to the hemispherical cup windmill and anemometer. *R. & M.* No. 712, *British A.R.C.* (1919).

(37) PICK, W. H. *A short course in elementary meteorology.* Meteorological Office (1938).

(38) HAWKINS, A. E. A simple vane anemometer giving continuous and direct indication of wind velocity. *Journ. Sci. Insts.*, Vol. 31, No. 11, p. 404 (November 1954).

(39) COX, W. J. G. A cup-type transmitting anemometer with improved response to decreasing gusts. Royal Aircraft Establishment, *Technical Note* No. INSTN. 140 (July 1954).

Chapter 9

WIND STRUCTURE AND ITS DETERMINATION

In 1883 Osborne Reynolds published the results of his experiments on fluid flow and showed that it may be 'laminar' or 'turbulent'. When applied to wind, laminar flow means that the air flows along streamlines without any whirling motion and so without any mixing of the air in different layers of the stream. In turbulent flow there is a whirling motion and the layers of air mix. Reynolds showed that, when a fluid flows in a pipe, the transition from laminar to turbulent flow depends upon the dimensionless number—since called the Reynolds number—which is given by $\frac{\mu r}{\nu}$ where μ = average velocity of the fluid, r = radius of the pipe and ν = the kinematic viscosity of the fluid (Ref. 1).

Winds in the open are almost always turbulent although when they flow over hills of good aerofoil shape there may be compression of the streamlines which damps out the turbulence and permits their being considered, for some wind power calculations, as having laminar flow. Following the conception of Reynolds, the variations in the steady flow of the wind consist of "gusts and lulls accompanied by small changes in direction. These variations are commonly ascribed to eddies embedded in the general steady flow, and in so far as these eddies are visualized they are supposed to consist of more or less circular disturbances travelling with the wind. When the flow in an eddy coincides with the wind direction the velocity is increased and a gust is experienced; while when the flow in the eddy is against the wind direction a lull occurs" (Ref. 2).

These eddies have been shown by G. I. Taylor, F. J. Scrase, A. C. Best and other investigators (Refs. 3, 4, 5, 6, 7) to have axes which are orientated in all directions so that the variations in wind flow from a steady value also occur in all directions.

The departures of the wind from steady flow are referred to under the general term 'structure' of the wind.

The interest in wind structure from the viewpoint of wind power arises from the influence which it may have;

(i) on the energy output of a wind-driven machine;
(ii) on the stresses set up in the blades of a wind rotor.

The first of these influences has already been considered in Ch. 3.

Here we are concerned rather with the nature of the wind speed variations which might occur to bring about fluctuating, and possibly dangerous, stresses in the blades and with measurements which can be made to evaluate these speed variations.

Gust forces acting on the blade system

Consider the forces acting on a blade element, length *dr*, at radius *r*. Referring to Fig. 49 these are the lift and drag forces, ΔL and ΔD respectively (see Ch. 13) with, a resultant ΔF. The blade angle is β

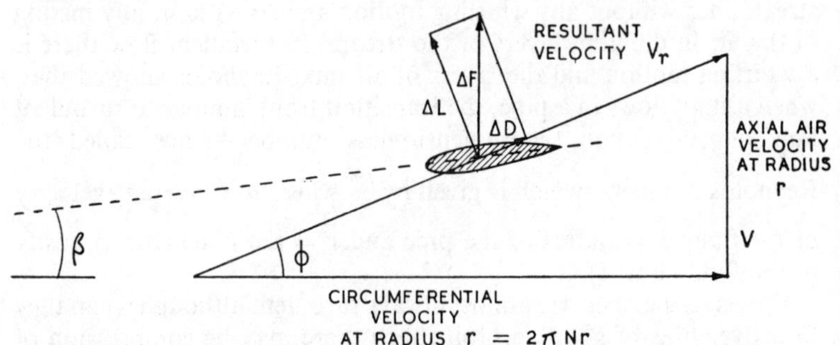

FIG. 49. *Forces acting on a blade element*

and ϕ is the angle which the resultant wind makes with the blade. The resultant wind V_r is the resultant of the circumferential velocity $2\pi Nr$ of the element ($N =$ speed of revolution of the rotor) and the axial velocity V of the air meeting the blade system. This axial velocity is assumed to be uniform over the whole area swept by the rotor.

It can be shown (see Ref. 8) that $\Delta F \propto V_r^2(\phi - \beta)$.

Now consider the effect of a rapid increase in the axial air velocity from V to V_1 supposing that this occurs so suddenly that neither the circumferential velocity nor the blade angle β can change.

The resultant force on the blade element will become,

$$\Delta F_1 \propto V_{r1}^2(\phi - \beta)$$

where
$$V_{r1} = \sqrt{(2\pi Nr)^2 + V_1^2}$$

as compared with the original value $V_r = \sqrt{(2\pi Nr)^2 + V^2}$

The new angle ϕ_1 is given by

$$\tan \phi_1 = \frac{V_1}{2\pi Nr}$$

whereas
$$\phi = \tan^{-1} \frac{V}{2\pi Nr}$$

In the War Production Board Report (Ref. 8), where the question of gust forces is discussed, an example is taken in which the blades are 100 ft long and rotate at 30 r.p.m. while the initial angle of attack $\phi - \beta$ is constant at 2° for the whole of the blade length.

When V changes suddenly from 30 m.p.h. to 60 m.p.h.;

(*a*) at the blade tip: V_r^2 increases from 130500 to 136300, and ϕ increases from 8° to 15·6°, so that, β remaining the same at 6°, the value of $\phi - \beta$ increases from 2° to 9·6°.

Thus ΔF increases in the ratio

$$\frac{136300\ (9{\cdot}6)}{130500\ (2)} = \frac{5{\cdot}01}{1}$$

(*b*) at one quarter of the blade length from the hub (i.e. $r = 25$ ft): V_r^2 increases from 9973 to 15781 and the angle of attack $\phi - \beta$ from 2° to 21°. At this radius, therefore, the force on a blade element is increased, by the sudden gust, in the ratio

$$\frac{15781\ (21)}{9973\ (2)} = \frac{16{\cdot}8}{1}$$

Although some approximations and assumptions are involved in these calculations they are sufficiently exact to indicate the order of magnitude of the gust forces which act on a blade and to show how very high stresses are set up, particularly near the blade root, when a steep-fronted gust meets it. The method of calculation used above takes account of the fact that a gust may take the form of a change in wind speed, or direction, or of both.

The influence of wind structure on the design of wind rotors

Except for blade elements at only a very short radius from the hub (where, in fact, the above calculations are inapplicable because of the effect of the hub fairing, or the nacelle, which usually blanks off the inner radii) the circumferential velocity is so high in relation to the wind speed that it is the change in the angle of attack ($\phi - \beta$), rather than that in V_r^2, which causes the great increase in the forces acting on the elements.

Clearly, if the blade angle β could be altered sufficiently quickly to keep pace with the change in wind speed, the gust forces could be much reduced; hence the arrangements for rapid pitch-changing which frequently form a feature of the design of large wind rotors. Relief from the bending moment due to a gust can also be obtained by allowing the blades to 'cone'. Coning hinges are fitted and, when the gust occurs, the whole blade bends back like the rib of an umbrella.

The bending moments due to the aerodynamic and centrifugal forces are in opposition and, with free coning of the blade, the coning angle taken up is that which makes these two moments exactly balance one another.

An alternative method of construction which has been used, for example, in the Danish machine mentioned in Ch. 13, p. 214, is to brace the blades to enable them to withstand the gust stresses without pitch-changing though it is not yet known how far this construction would be practicable for very large rotors.

Whatever the construction adopted, information on the actual behaviour of the wind under gusty conditions is essential for design calculations to be made. Important questions are:

(*a*) What rates of change of velocity are likely to occur in gusts?

(*b*) What are the probable distributions of gust velocities in relation to the area swept by the blades?

(*c*) What climatic or topographical conditions are likely to bring about gusts creating dangerously high stresses?

A considerable amount of information on gusts is, of course, available, and may be studied in the papers referred to in the bibliography at the end of this chapter, but this is not always in a form suited to the requirements of the designer of wind power plant. Until much more experience in the design and operation of large wind rotors has been obtained gust measurements on the actual site of the plant must be regarded as a necessary part of a large-scale project.

Some relevant information can be given from past work, and methods of making gust measurements will be described later in the chapter. In a work of this kind, however, it is impossible to do more than to indicate some of the problems posed by gusts. Their precise effects cannot be separated from those of the other forces which act upon a system of blades rotating under the conditions which apply to a windmill rotor; nor can these effects be dissociated from the design of the machine itself. There is no doubt (see Refs. 8 and 9) that the calculation of the stresses in the blades, taking into account the vibrations which may arise when pitch-changing and coning are used, presents a complex problem which has delayed the development of a satisfactory large aerogenerator. Vibrations of the blades might be caused also by the natural increase of wind speed with altitude. If this were very marked the wind pressure during the upper half of the rotation would be greater than for the lower half causing alternating stresses of frequency equal to that of the rotation. Fortunately, as shown in Ch. 7, the vertical wind gradient on a hill summit which may be chosen as a good wind power site is probably less than that over level ground though this effect must be considered when choosing the height of the tower. In gusty winds the direction, as well as the speed, may change rapidly and, if the rotor were to be designed to follow such rapid changes of direction by re-orientation, high gyroscopic forces would be brought into play.

The magnitude of gusts

P. A. Sheppard (Ref. 10) has discussed the changes of speed which occur in natural winds with reference to different time scales and has shown that, whether the units on the time scale be seconds, minutes, hours, days, or even years, the graph of wind speed can always be represented as a curve fluctuating relatively slowly with a more rapid fluctuation superimposed on it. From the point of view of wind power, the slower variations are of interest only in so far as they affect the energy output; rapid changes, occurring in times of one second or less, are particularly important for calculations of blade stresses. The maximum rate of change in Sheppard's highest speed record (obtained by a hot-wire anemometer in conjunction with a cathode-ray oscillograph) appears to be 5 m.p.h. in 0·1 sec. The mean wind speed was about 6 m/sec (13·5 m.p.h.) measured at a height of 2 m over downland on a summer afternoon.

L. F. G. Simmons and J. A. Beavan (Ref. 11) used a hot-wire anemometer and hot-wire direction meter to obtain records of the speed and direction, in a vertical plane, of gusts occurring at a height of 64 ft above ground. The highest mean wind speed during their tests was about 70 ft/sec (47·5 m.p.h.).

In one test when the mean speed was 55 ft/sec, a gust lasting for about 0·8 sec showed a change in the horizontal wind speed from 55 ft/sec to 125 ft/sec and back again to 50 ft/sec. while the vertical component of the wind speed changed from + 50 ft/sec to − 50 ft/sec in 0·4 sec. The authors suggest that these changes were due to an eddy of about 40 ft diameter, probably caused by local obstructions, passing within 10 ft of the anemometers. Another, high-speed, record shows velocity changes of about 15 ft/sec in $\frac{1}{25}$ sec.

Fig. 50 shows the traces of two records obtained from a gust anemometer (see p. 148) developed by the Electrical Research Association. The records refer to measurements of the horizontal component of wind velocity made at the top of a 30 ft tower at the summit of Costa Hill, Orkney, during a period when the wind was very gusty, following the passage of a fast-moving cold front. An adjacent cup anemometer mounted at the same height indicated the mean speed during the test period as just over 40 m.p.h. with maximum and minimum speeds of 84 m.p.h. and 20 m.p.h. It will be seen from record (*a*) that the highest rate of change was from 52 m.p.h. to 85 m.p.h. in $\frac{1}{4}$ sec. Record (*b*) shows the rise and decay of a gust, with a maximum horizontal speed of 70 m.p.h., during some 4 sec following a period of very steady wind speed.

F. J. Scrase (Ref. 5) used a Dines anemometer, a bi-directional

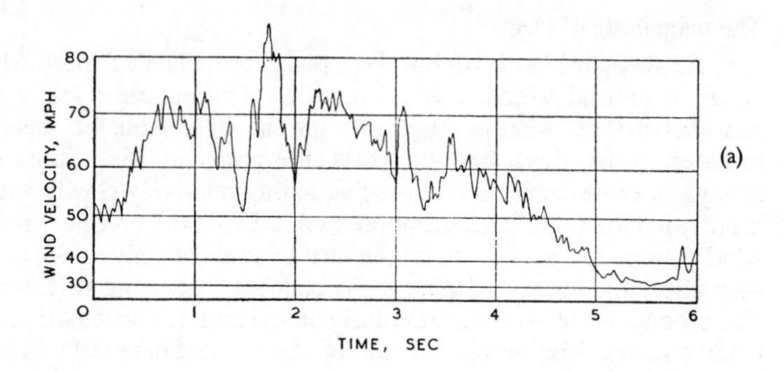

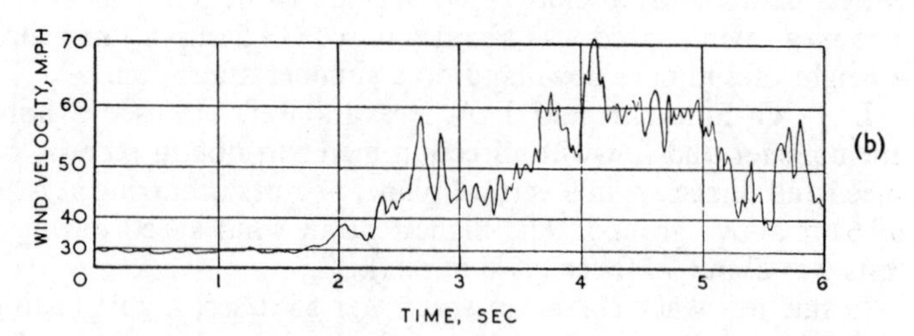

FIG. 50. *Gust records from Costa Hill, Orkney*

vane and a swinging-plate anemometer together with a kinematograph camera to record wind eddies at heights above ground from $1\frac{1}{2}$–19 m. His results are expressed in terms of the three components of velocity along the x axis in the mean direction of the wind, the y axis in the horizontal direction perpendicular to the mean wind, and the z axis in the vertical direction. The components of eddy velocity along these axes are u, v and w respectively. He introduces a 'gustiness factor' which, for the horizontal wind along the x axis is

$$\frac{u}{U} = \frac{U_{max} - U_{min}}{U_{max} + U_{min}}$$

where U is the mean wind speed, U_{max} and U_{min} the maximum and minimum values and μ is the eddy component which is equal to the semi-width of the trace of the speed, as obtained from a Dines pressure tube anemometer. Gustiness factors can be calculated similarly for the other components. (In view of the importance of gustiness in the wind, when considering either the output of a wind-driven generator or the stresses set up in the blades, a 'gustiness factor' g might be introduced. There is, however, some doubt about the best definition to adopt for g. Alternatives are

(i) The mean deviation

$$\frac{1}{n\bar{V}}\sum^{n}(V_i - \bar{V})$$

(ii) The standard deviation σ which is

$$\sqrt{\frac{1}{n\bar{V}^2}\sum^{n}(V_i - \bar{V})^2}$$

(iii) $$\frac{1}{(n-1)\bar{V}}\sum^{n}(V_{i+1} - V_i)$$

where V_i = instantaneous velocity

$\bar{V}$ = mean velocity.

V_{i+1} is the velocity at one unit of time subsequent to that corresponding to V_i so that $V_{i+1} - V_i$ represents the change in velocity from (say) 1 sec to the next.

In all three expressions all the differences are reckoned as positive. Only the third expression takes into account the frequency of the velocity fluctuations.)

Scrase, in summarizing his results, divides them into those for large-scale turbulence over an interval of 1 hr, intermediate-scale turbulence over an interval of the order of a few minutes, and small-scale turbulence (which is of most interest when considering the production of gust stresses in windmill blades) for a period from a few seconds to 1 min. For the last he states that:

(*a*) most of the small-scale eddies probably have periods of the order of 1 sec;

(*b*) the x, y and z components of eddy velocity are in the ratio 1·0:1·16:0·75 at a height of 1·5 metres above ground;

(*c*) the x component is nearly constant over the height interval 1·5 m to 19 m;

(*d*) at 19 m the *y* and *z* components decrease to about $\frac{2}{3}$ of their value at 1·5 m.

Some examples of rapid changes of wind speed and direction taken from Dines anemometer records are given in Table XVIII.

TABLE XVIII

Change in wind speed (m.p.h.)	Change in wind direction (degrees)	Time for the change (minutes)	Meteorological station	Reference
+ 65	160	6	Croydon	*Met. Mag*. Vol. 76, p. 184 (1947)
+ 33	65	7	Bell Rock	Ref. 12
+ 43	190	3	Lerwick	Ref. 12
+ 34	190	1½–2	Lerwick	Ref. 12
+ 35	90	8	Bell Rock	Ref. 12
+ 38	80	7	Cranwell	Ref. 12
− 16	130	1	Cardington	Ref. 2

Although these changes can be classed as rapid they fall in Scrase's classification of intermediate-scale turbulence and are of interest for wind power mainly in relation to yawing (or re-orientation) requirements rather than for their bearing on blade stresses.

Maximum gust speeds

Records from Dines anemometers at Meteorological Office stations in Great Britain have frequently shown gust speeds over 100 m.p.h. including one of 110 m.p.h. at St. Ann's Head (Pembrokeshire). Abroad, even in much less windy areas than Great Britain, gusts of over 90 m.p.h. have been measured. Examples are 98 m.p.h. at Godavari Bridge, near Rajahmundry, India, in May, 1924, and 94 m.p.h. at Juhu (Bombay). In Australia gusts up to 95 m.p.h. (in the south) and up to 100 m.p.h. (in the north) have been recorded. Gust speeds up to 95 m.p.h. occurred during the period 1939–47 in South Africa.

Sir Douglas Mawson, in his account (Ref. 13) of the Australian Antarctic Expedition 1911–14, to Adelie Land, gives some interesting descriptions of their experiences in very high winds. Measurements of hourly wind speeds were made continuously with a Robinson cup anemometer and from among the records given may be selected the following instances:

11th May, 1912. Average speed 80 m.p.h. for 24 hr.

15th May, 1912. Average speed 90 m.p.h. for 24 hr, with gusts of the order of 200 m.p.h.

1st–7th February, 1913. Continuous gale for a week with a hurricane in which a wind speed of approximately 80 m.p.h. was recorded.

6th July, 1913. Hourly average speed of 116 m.p.h. with an average of 107 m.p.h. for 8 hr.

Monthly averages: March, 49 m.p.h.; April, 51·5 m.p.h.; May, 60.7 m.p.h.

Yearly average: 50 m.p.h.

Clearly there is justification for Mawson's description of the climate as "little more than one continuous blizzard the year round; a hurricane of wind roaring for weeks together, pausing for breath only at odd hours". Even so there were occasional short periods of complete calm.

The highest gust speeds recorded in Great Britain (so far as is known) were measured by the Electrical Research Association at Costa Hill, Orkney, on the morning of 31st January, 1953 (Ref. 14). The brief history of these measurements is as follows:

On 15th January, 1952, a heavy south-westerly gale occurred in Orkney but the maximum gust speed at the E.R.A. Costa Hill station was not measured because the pen of the Dines anemometer recorder (which was the only instrument then in continuous operation capable of measuring gusts) was driven over the top of the chart drum by a gust estimated at $>$ 126 m.p.h.

To ensure that such a loss of record should not be repeated in future storms a standard cup-generator anemometer, specially strengthened and provided also with an alternative set of cups for measurements up to 200 m.p.h. (if necessary) was calibrated, with its recorder, in a wind tunnel at air speeds up to 150 m.p.h. It was mounted on a very stiff 30 ft mast on Costa Hill summit. This anemometer, together with a Dines anemometer with its head at 80 ft above ground, and four cup-contact anemometers at heights of 50, 80 and 110 ft above ground, was in operation on 31st January, 1953. Table XIX shows the wind speed readings which were obtained.

A further period of 24 hours elapsed before the peak gust speeds dropped below 60 m.p.h. During the period indicated in the table an instantaneous wind speed of 107 m.p.h. was observed at the meteorological station at Grimsetter airport some twenty miles from Costa Hill.

The calibration of the cup-generator anemometer used at Costa was subsequently checked and found to be sensibly the same as before its installation on the hill. Plate XXI shows a photograph of its record for the worst part of the gale

TABLE XIX

Wind records, Costa Hill, Orkney, 31st January 1953

Time	Peak gusts		Mean wind speed
	Dines anemometer at 80 ft (m.p.h.)	*Cup-generator anemometer at 30 ft (m.p.h.)*	*Average of four cup-contact anemometers at 50, 80 and 110 ft (m.p.h.)*
0300 to 0400	27	27	14
0400 to 0500	28	30	20
0500 to 0600	38	42	22
0600 to 0700	57	52	32
0700 to 0800	92	88	46
0800 to 0840	$>$ 116	110	83
0840 to 1000	approx. 126†	123	no reading*
1000 to 1100	no reading	125	"
1100 to 1200	"	114	"
1200 to 1300	"	107	"

* Owing to failure of electricity supply to the instruments.
† Recorder pen carried over the chart drum at this point.

Distribution of wind speed over the area swept by a wind rotor

The instantaneous distribution of wind speed over the rotor circle is important from two points of view—that of the designer considering possible blade stresses and that of persons concerned with relating the electrical output of an aerogenerator to the corresponding wind speed. The first wants to know how variable the distribution can be and the second how uniform it can be—and for how long. For reasons which will appear in the following discussion, neither question can be answered with exactitude but, nevertheless, sufficient information is available to give useful indications on the order of the answers.

R. H. Sherlock and M. B. Stout (Refs. 15, 16 and 17), using a number of pressure-plate anemometers having a quick response, have given the results of measurements during winter storms on a gently sloping ridge at Ann Arbor, Michigan. Their instruments were mounted on 50 ft poles arranged 60 ft apart in a line across the prevailing direction of storm winds and also at 50 ft and 25 ft intervals of height on a 250 ft tower. Diagrammatic records are given (Ref. 15) in the form of 'iso-velocity contours' to show the wind structure during intervals of a few seconds when the conditions were of special interest. In Tables XX, XXI, XXIII and XXIV a few

TABLE XX

Wind speeds (in m.p.h.), at $\frac{1}{2}$ sec intervals, for positions shown in Fig. 51

A	*B*	*C*	*E*	*D*
35	35	35	33	38
34	36	36	37	37
33	34	35	44	36
31	33	36	49	37
29	39	37	48	37
31	39	37	50	42
32	30	38	51	43
32	33	39	48	43
32	32	40	44	41
34	33	42	44	38
38	46	38	46	37
45	42	36	44	38
45	46	42	44	37
48	44	41	44	37
49	48	40	43	40
50	46	44	44	43
51	46	39	46	46
52	45	39	46	44
51	46	39	46	43
51	46	43	46	45
50	43	45	45	46
50	45	42	47	44
50	48	42	46	44
47	45	44	47	45

TABLE XXI

Wind speeds (in m.p.h.), at $\frac{1}{2}$ sec intervals, for positions shown in Fig. 51

A	*B*	*C*	*E*	*D*
45	44	42	39	44
45	44	42	39	43
45	45	41	39	42
44	44	41	42	43
45	45	43	44	42
43	43	42	44	41
45	42	42	42	41
44	41	42	43	43
42	41	41	43	44
41	42	41	43	43
42	41	41	42	42
42	41	40	43	41
43	41	40	42	41
43	39	42	41	40

of the wind speed values given on these diagrams are presented. These have been selected so that they may bear as closely as possible upon the wind rotor problem. The measurements were not made at the summit of a steep hill nor were the storms during which they were made of the violence experienced at especially windy sites, but they show what wind conditions may be expected to occur—probably fairly frequently—at such sites.

The key to the positions to which the measured values in Tables XX and XXI apply, is given in Fig. 51.

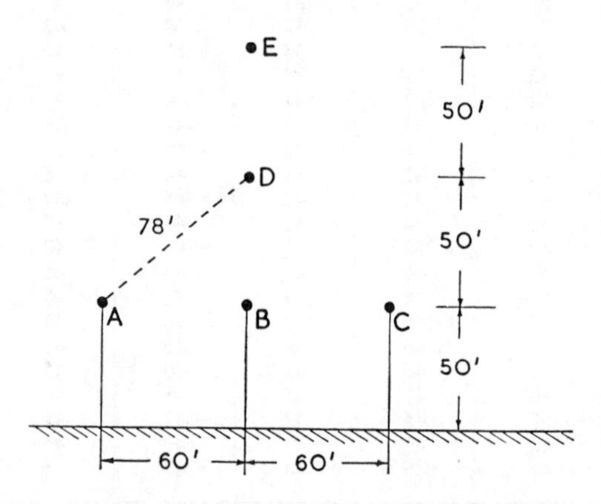

FIG. 51. *Arrangement of gust measuring instruments*

In Table XX the figures show the passage of a gust which is first evidenced by a wind speed of 44 m.p.h. at position *E*—the highest point. This rises and falls away again gradually and it can be seen that it reaches position *A* 4 or 5 seconds after its arrival at *E*. Wind speeds level out at around 45 m.p.h. after about 11 seconds.

Table XXI is included to show a set of wind speeds which remain steady and approximately the same (42 m.p.h. $\pm$ 3 m.p.h.) for all five positions for a period of 7 seconds, a period long enough to allow a steady output from a wind-driven generator to be obtained and measured.

The wind speeds given in these tables are, of course, those in the wind, uninfluenced by the presence of a windmill rotor. But, to form an idea of the varying wind speeds which a blade may encounter during its rotation when a gust passes through its swept circle, consider a blade 78 ft long rotating at 30 r.p.m. with its hub at point *D*, i.e. with a supporting tower 100 ft high. (The length 78 ft is chosen because it happens to be equal to the distance DA (= DC). The rotational speed of 30 r.p.m. is a reasonable one for a windmill having a rotor 156 ft diameter.)

Fig. 52 shows a series of positions of the blade, with corresponding wind speeds, during one revolution taking 2 seconds to complete, these 2 seconds being the second and third seconds referred to in Table XX.

In this one revolution the blade would thus be affected by the following variations of wind speed:

TABLE XXII

Position	Wind speed at the root (m.p.h.)	Wind speed at the tip (m.p.h.)
(i)	36	44
(ii)	37	30
(iii)	37	39
(iv)	40	37
(v)	43	55

TABLE XXIII

Wind speeds, at ¼ sec intervals, measured at three heights in a vertical line

Height above ground		
75 ft (m.p.h.)	*50 ft (m.p.h.)*	*25 ft (m.p.h.)*
20	21	22
21	19	22
21	21	22
20	18	22
20	20	23
20	19	22
20	18	22
21	18	21
21	18	20
20	18	20
21	18	21
21	20	20
20	19	20
19	19	20
18	19	20

There is here some confirmation from actual measurement that the example, quoted on p. 131, of a sudden increase of wind speed from 30 m.p.h. to 60 m.p.h. at a certain blade section was not unreasonable. The change from 37 m.p.h. to 55 m.p.h. at the blade tip, shown in Fig. 52 takes place in a little over $\frac{1}{2}$ second so that

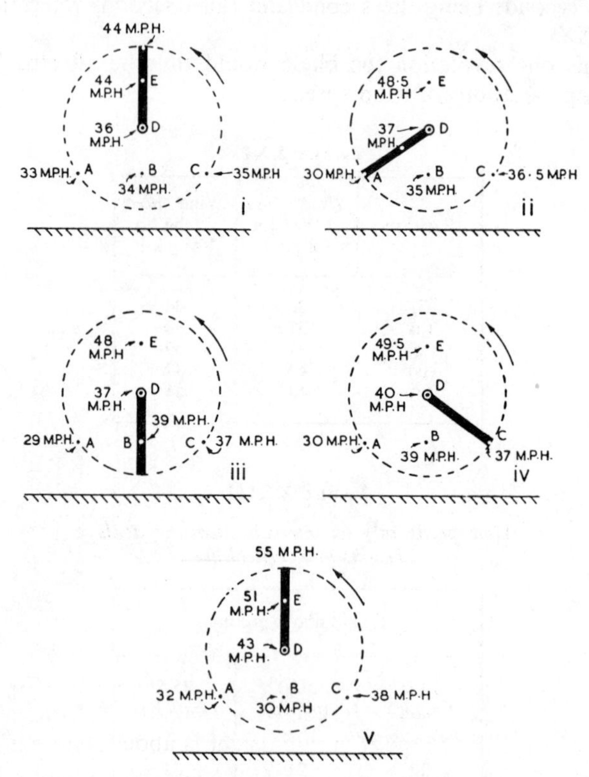

FIG. 52. *Gust speeds during one revolution of a blade*

pitch changing must be rapid if gust stresses are to be relieved.

Table XXIII, the figures in which are obtained from one of Sherlock and Stout's diagrams, show the variations of wind speed at 75, 50 and 25 ft for $\frac{1}{4}$ second intervals. This series has been selected, again, to show the steadiness of wind speed (during 4 seconds in this instance) which can occur even during gusty conditions.

On the other hand, the wind speed figures shown in Table XXIV were measured some 15 seconds later in the same record.

To form an idea of the effect of these variations on a rotating blade, imagine a blade 25 ft long rotating at 120 r.p.m. with its hub at a height of 50 ft. Since it will make half a revolution in $\frac{1}{4}$ second the wind speeds marked with brackets in columns (i) and (iii) will be those

TABLE XXIV

Wind speeds, at $\frac{1}{4}$ sec intervals, measured at three heights in a vertical line

Height above ground		
(i) 75 ft (m.p.h.)	*(ii) 50 ft (m.p.h.)*	*(iii) 25 ft (m.p.h.)*
(34)	27	22
33	29	(25)
(36)	32	21
38	28	(21)
(38)	28	23
35	27	(21)
(39)	25	23
44	23	(28)
(42)	28	24
38	28	(27)
(37)	38	34
32	41	(34)
(36)	38	33

at the blade tip for successive top centre and bottom centre positions as it rotates; the speeds in column (ii) will be those at the hub.

The highest rate of change of wind speed approaching the blade tip is from 21 m.p.h. to 39 m.p.h. in $\frac{1}{4}$ second (half a revolution) during which time the speed at hub height is about 26 m.p.h.

Fig. 53 gives, in smoothed form, two small parts of the records from four gust anemometers mounted on the measuring mast on Costa Hill, Orkney. They were on the perimeter of a 60 ft diameter circle with centre 80 ft above ground: *A* at 110 ft, *B* at 80 ft right extreme, *C* at 80 ft left extreme, and *D* at 50 ft. The wind speed during the recording time varied from 35 m.p.h. to 55 m.p.h.

Fig. 53(*a*) shows one of many instances of the same wind speed, within 2 m.p.h., at all four points while Fig. 53(*b*) shows the highest variation over the circle appearing in the record, which covered a total period of 24 min.

A section of the recorder chart for this test is shown in Fig. 54.

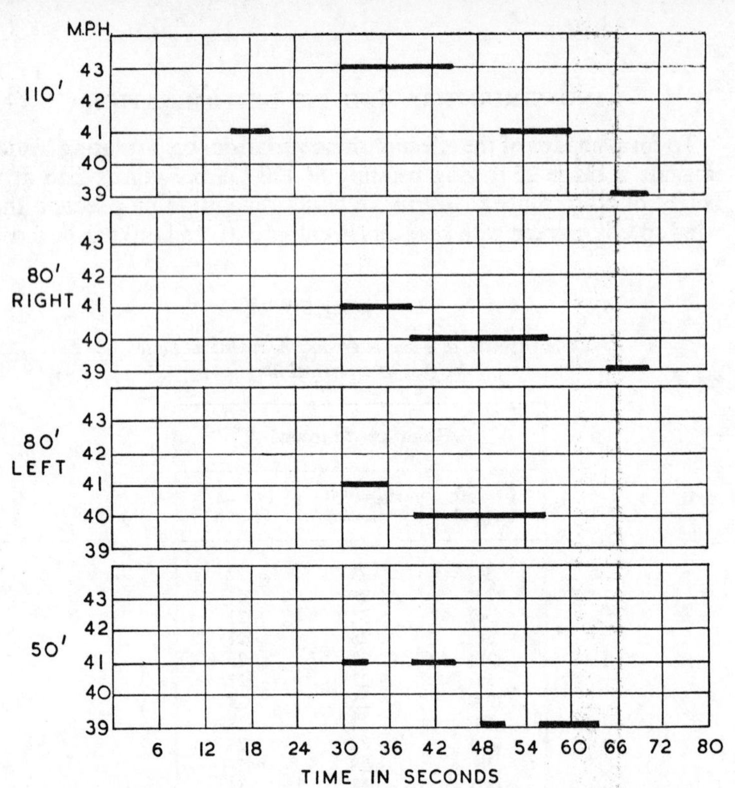

FIG. 53 (a). *Occurrence of gust speeds at different heights above ground*

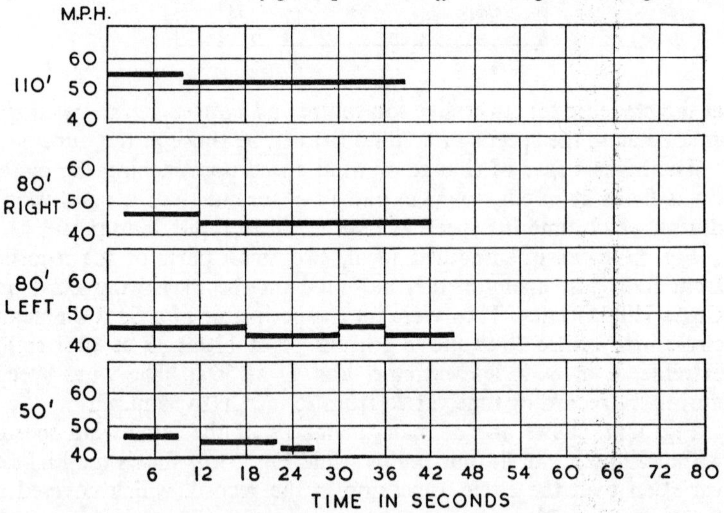

FIG. 53 (b). *Occurrence of gust speeds at different heights above ground*

Notes. Lines represent period when the maximum variation from the mean does not exceed $\pm$ 1 m.p.h. for a period of 2 sec. or more. The gaps are for periods when the variation was greater than $\pm$ 1 m.p.h.

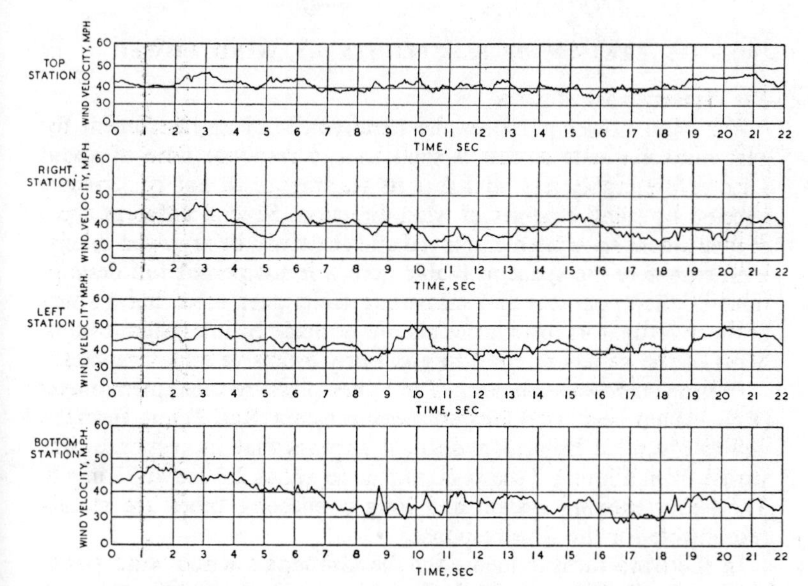

FIG. 54 (a). *Gust records for different heights above ground*
Extracts from gust anemometer records at four positions on 60 ft. diameter swept circle

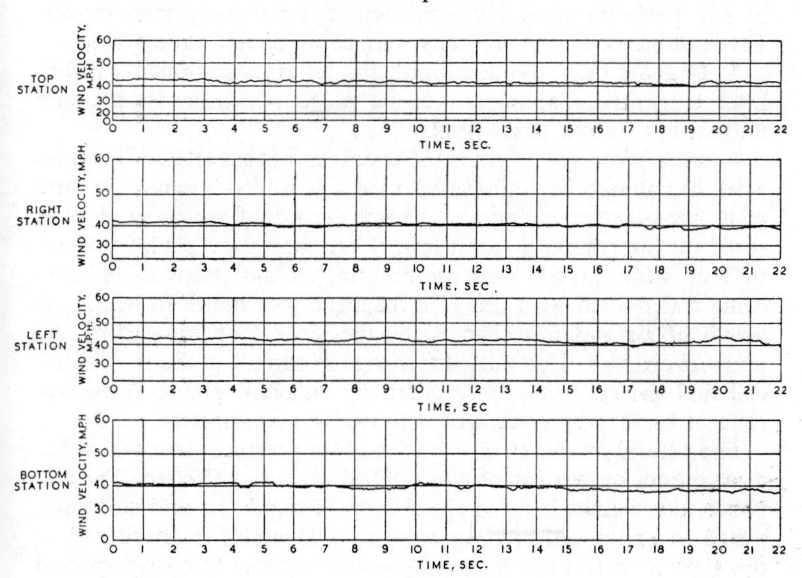

FIG. 54 (b). *Gust records for different heights above ground*
Extracts from gust anemometer records at four positions on 60 ft. diameter swept circle

The measurement of gusts

For wind power purposes the requirement of an instrument for gust measurements is that it shall have a response time of about $\frac{1}{4}$ sec or preferably less, and that its readings shall not be seriously affected by small changes of wind direction. Several different types of instrument have been employed and these will be reviewed briefly. Unfortunately the space available here will not permit full descriptions of the apparatus and measuring techniques used, but a study of the investigators' own accounts can be made by interested readers. Most of the papers referred to give fairly extensive bibliographies.

Pressure-tube anemometers. The Dines pressure-tube anemometer (Ref. 18) has been used for gust measurements (Ref. 2) but, from the figures given on its performance, it appears that its response time varied from about 1·7 sec at 60 m.p.h. to about 2·3 sec at 21 m.p.h. The heavy moving system and long connecting tubes are mainly responsible for the slow response.

In the wind measurements for the Grandpa's Knob wind power project P. C. Putnam (Ref. 9) used a pressure-type instrument, designed by K. O. Lange, having a pitot head turned into wind by twin tail vanes. Pressures were transmitted down the measuring mast by $\frac{3}{8}$ in. diameter tubes, 310 ft in length, to 4 mm water manometers. The readings of the manometers were continuously photographed.

A. Graham (Ref. 19) developed, for his investigations on turbulence, a sensitive diaphragm pressure gauge to measure the pressure fluctuations from a Dines head.

Pressure plate anemometers. Sherlock and Stout (Ref. 17), whose work has already been discussed, used a specially-designed pressure plate anemometer giving an accurate record of the average wind speed for intervals of $\frac{1}{4}$ sec or less. It had a pressure plate 9 in. high by 8 in. wide, hinged near the bottom, the variations in pressure being transmitted by a rod to a magnetic transmitter in which the length of air gap was changed by the movement initiated by the pressure changes. This caused a change in current in the transmitter coil and the current was recorded by an oscillograph. A twelve-element oscillograph was used with twelve anemometers.

The natural frequency of vibration of the pressure plate, which has a maximum movement of about 0·005 in., was 115 c/sec. The response to a change in wind speed from 80 m.p.h. to zero was completed in about 0·12 sec. An error of 1·25 m.p.h. is introduced in this type of instrument due to lag in the response to changing wind direction and other errors may be caused by vertical components of wind velocity and by vibrations of the plate, in a gusty wind, at a frequency of 115 c/sec. The moment of inertia about a vertical axis must be small to allow the tail to function quickly in turning the instrument into wind.

Hot wire anemometers. The principle of using the wind to cool a thin wire, heated by current passing through it, and included in one arm of a Wheatstone bridge network so that the effect of the cooling upon the wire's resistance can be measured, has been used by many investigators.

J. T. Morris (Ref. 20 and 21) used the method to study the changes in air speed close to a circular rod placed in an air stream.

L. F. G. Simmons and J. A. Beavan (Ref. 11) whose work has already been mentioned (p. 133) used the hot-wire anemometer for the measurement of wind speed and direction. A. C. Best (Ref. 6) used it with a Cambridge Unipivot micro-ammeter to measure the out-of-balance current in the Wheatstone bridge network as the sensitive hot-wire element was cooled by the wind. To improve the scale of the instrument, which is normally very open at low wind velocities, Best used a vacuo thermo-junction in the galvanometer branch of the bridge when measuring high wind speeds.

(In addition to the hot-wire instrument, to indicate the gustiness, Best also used a very light bi-directional vane, with a universal pivot which could take up the wind direction rapidly in both vertical and horizontal planes. Its movements were recorded by a pen on a chart, the record taking the form of traces consisting of masses of continuous lines covering areas roughly oval in shape.)

Although the hot-wire anemometer is a valuable instrument, especially for laboratory measurements of low wind speeds, (a Simmons hot-wire anemometer is manufactured by Messrs H. Tinsley & Co., Ltd.) it has some disadvantages for gust measurements in the open air under the severe conditions encountered in wind power investigations. These are that its calibration does not remain constant for a long period and that the hot wire is thin and fragile—it may even be broken by a rain drop falling on it.

Windmill anemometers for gust measurements. An anemometer having a very light windmill rotor driving a small electric generator can be used for high speed measurements (Refs. 22, 23 and 24). The blades of the rotor may be of balsa wood or of a plastic material.

The wind mill anemometer is a convenient type for wind power investigations and could perhaps be made to measure gusts with a time response of 0·1 sec. There may be doubt, however, about the ability of the tail to keep the instrument pointing into wind sufficiently accurately with rapid changes of wind direction.

A windmill anemometer having four laminated balsa wood blades driving the rotor of a small alternator, has been used by the Electrical Research Association (see also p. 112). The calibration is linear above about 15 m.p.h.; below this speed the voltage of the output must be amplified to operate the electronic counter which is used in conjunction with the anemometer.

The intention is to use the apparatus, in conjunction with a recording wattmeter measuring the machine's output, for performance tests on a wind-driven generator during short periods of steady wind.

G. Dady (Ref. 25) has described the use of two windmill-type anemometers, with fixed orientations at right angles, to trace rapid changes in the direction of the horizontal component of the wind. Their output voltages were fed to the two pairs of plates of a cathode-ray oscillograph and the movements of the spot during periods of up to 5 min duration were recorded photographically.

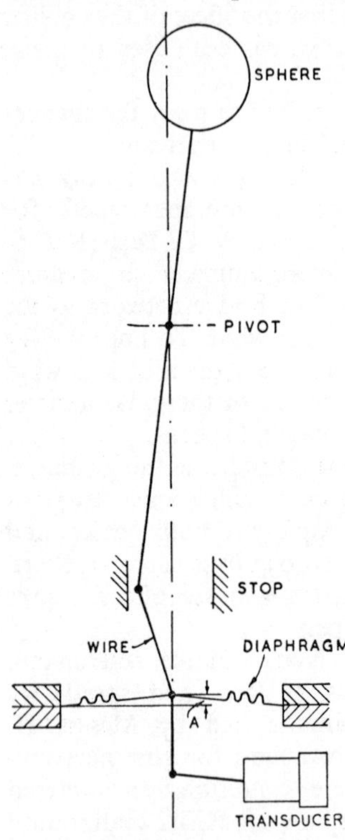

FIG. 55. *Principle of E.R.A. gust anemometer*

E.R.A. gust anemometers. When investigations on gust velocities were started by the Electrical Research Association as part of their wind power research programme there appeared to be no measuring instrument available which would be well suited to the particular conditions applying in such work. It was therefore decided to develop a suitable instrument.

The time response required was chosen as 0·1 sec on the basis that, in a 70 m.p.h. wind—the maximum speed in which an aerogenerator is likely to continue operating—10 ft of air would flow through the rotor in approximately that time. And 10 ft was supposed to be the smallest dimension for a body of air capable of having a significant effect on the rotor.

The principle of the instrument is shown in Fig. 55 (Refs. 26 and 27). Details of the design and performance of a range of instruments of this type for the measurement of (*a*) horizontal components of wind velocity without regard to direction (*b*) velocity components in two horizontal directions at right angles and (*c*) vertical components have been given by H. H. Rosenbrock (Ref. 27).

Referring to Fig. 55, a hollow brass or duralumin sphere (about $1\frac{1}{2}$ in. diameter and perforated with holes $\frac{1}{16}$ in. diameter to ensure that the force on the sphere is proportional to the square of the wind speed) is mounted on a vertical arm pivoted so that the sphere can move in any horizontal direction. The lower end of the arm is attached to a thin steel wire the lower end of which is fixed to the centre of a spring diaphragm. When the sphere is deflected by the pressure of the wind the diaphragm is thus deflected and its movement is communicated to the anode pin of a mechano-electronic transducer valve. The output voltage of this valve is almost directly proportional to the angular movement of the anode pin and the sensitivity, at 40 V for 1° movement, is high.

The output is recorded by a quick-response recording voltmeter (Evershed & Vignoles Ltd.) the two servo-operated pens of which are driven by a valve amplifier. The recorder, which can thus be used with two gust anemometers (one for each pen) has four chart speeds, namely $\frac{1}{2}$, 1, 3 and 6 in./sec.

To damp the motion of the sphere, and to prevent oscillation of the wire, the wire and the lowest part of the operating arm are immersed in a damping fluid.

Photographs of two forms of the anemometer, for horizontal and for vertical components, are shown in Plates XXII and XXIII

BIBLIOGRAPHY

(1) Reynolds, Osborne. On the dynamical theory of incompressible viscous fluids and the determination of the criterion. *Phil. Trans. Royal Soc. (London)*, Ser. A, Vol. 186, p. 123 (1895). (Also, *Scientific Papers*, Vol. 2, p. 535 (1891–1900).

(2) Giblett, M. A. The structure of wind over level country. *Meteorological Office Geophysical Memoir* No. 54, Vol. VI (1932).

(3) Taylor, G. I. Eddy motion in the atmosphere. *Phil. Trans. Royal Soc. (London)*, A., 215, p. 1 (1914).

(4) Taylor, G. I. *London Proc. Roy. Soc.*, A., 94, p. 141 (1917).

(5) Scrase, F. J. Some characteristics of eddy motion in the atmosphere. *Meteorological Office Geophysical Memoir* No. 52 (1930).

(6) Best, A. C. Transfer heat and momentum in the lowest layers of the atmosphere. *Meteorological Office Geophysical Memoir* No. 65 (1935).

(7) Sutton, O. G. *Atmospheric turbulence*. Methuen & Co. Ltd. (1949).

(8) Final report on the wind turbine. *Research Report*, PB25370, Office of Production, Research and Development, War Production Board, Washington, D.C. (1946).

(9) Putnam, P. C. *Power from the wind*. Van Nostrand (1948).

(10) Sheppard, P. A. Atmospheric turbulence. *Weather*, Vol. 6, No. 2, p. 42 (February 1951).

(11) SIMMONS, L. F. G. and BEAVAN, J. A. Hot-wire type of instrument for recording gusts. *A.R.C.R. & M.* No. 1615 (February 1934).
(12) GOLD, E. Wind in Britain. *Quarterly Journal*, Royal Meteorological Society (April 1936).
(13) MAWSON, SIR DOUGLAS. *The home of the blizzard.* Hodder & Stoughton, 6th Edition (1938).
(14) DOUGLAS, C. K. M. Gale of January 31st, 1953. *Meteorological Magazine*, Vol. 82, pp. 97–100 (1953).
(15) SHERLOCK, R. H. and STOUT, M. B. Wind structure in winter storms. *Journal of the Aeronautical Sciences*, 5, p. 53 (1937).
(16) SHERLOCK, R. H. and STOUT, M. B. Characteristics of wind gusts. *N.E.L.A. Bulletin* (January 1932).
(17) SHERLOCK, R. H. and STOUT, M. B. An anemometer for a study of wind gusts. *Engineering Research Bulletin* No. 20, University of Michigan (May 1931).
(18) DINES, W. H. Anemometer comparisons. *Journal of the Royal Meteorological Society* (July 1892).
(19) GRAHAM, A. Structure of turbulence in a natural wind, with a description of a sensitive pressure gauge. *R. & M.* No. 1704, *A.R.C. Tech. Report* (1936).
(20) MORRIS, J. T. The distribution of wind velocity in the space surrounding a circular rod in a uniform current of air. *Engineering* (8th August 1913).
(21) MORRIS, J. T. The electrical measurement of wind velocity. *Paper to the British Association for the Advancement of Science, Engineering* (27th December 1912).
(22) FERGUSSON, S. P. Gustiness under various weather conditions. *Bulletin of American Meteorological Society*, 24, p. 22 (1943).
(23) FERGUSSON, S. P. Sensitiveness of anemometers. *Bulletin of the American Meteorological Society*, 15, p. 95 (1934).
(24) CONOVER, J. H. Tests of the Friez aerovane in the natural wind at Blue Hill observatory. *Bulletin of the American Meteorological Society*, 27, p. 523 (1946).
(25) DADY, G. Etude photo-oscillographique de la composante horizontale du vent. *Comptes Rendus des Séances de l'Académie des Sciences*, Vol. 238, No. 3 (18 January 1954).
(26) ROSENBROCK, H. H. and TAGG, J. R. Wind- and gust-measuring instruments developed for a wind-power survey. Electrical Research Association, *Technical Report* C/T104 (1951).
(27) ROSENBROCK, H. H. The design and development of three new types of gust anemometer. Electrical Research Association, *Technical Report* C/T106 (1951).
(28) CARRUTHERS, N. Variation in wind velocity near the ground. *Quarterly Journal*, Royal Meteorological Society (October 1943).
(29) MORGANS, W. R. Relation between ground contours, atmospheric turbulence, wind speed and direction. *A.R.C.R. & M.* No. 1456 (December 1931).
(30) OWER, E. *Measurement of air flow.* Chapman & Hall (1949).
(31) LAMB, SIR HORACE. *Hydrodynamics.* Cambridge University Press, 5th Edition (1924).
(32) BRUNT, SIR DAVID. *Physical and Dynamical Meteorology.* Cambridge University Press, 2nd Edition (1952).
(33) PRANDTL, U. and TIENTJENS, O. G. *Applied hydro- and aeromechanics.* Translated by J. P. Den Hartog (1934).

(34) *Aerodynamic theory*. Ed. W. F. Durand, Vol. III, Sect. G (1934).
(35) Burgers, J. M. Experiments on the fluctuations of the velocity in a current of air. *Versl. d. Kon. Akad. v. Wetensch* (Amsterdam), Vol. XXIX, No. 4, pp. 547–558 (1926).
(36) Simmons, L. F. G. and Bailey, A. Note on a hot-wire speed and direction meter. *A.R.C.R. & M.* No. 1019 (1925).
(37) Simmons, L. F. G. and Salter, C. Velocity variations in turbulent flow. *Proceedings of the Royal Society* (*London*), Ser. A, Vol. 145, p. 212 (1934).
(38) Simmons, L. F. G. and Salter, C. An experimental determination of the spectrum of turbulence. *Proceedings of the Royal Society* (*London*), Ser. A, Vol. 165, pp. 73–87 (18 March 1938).
(39) Von Karman, Th. The fundamentals of the statistical theory of turbulence. *Jour. Aero. Sci.*, Vol. 4, No. 4, pp. 131–138 (Feb. 1937).
(40) Von Karman, Th. and Howarth, L. On the statistical theory of ísotropic turbulence. *Proc. Roy. Soc.* (*London*), Ser. A, Vol. 164, No. 917, pp. 192–215 (January 1938).
(41) Taylor, G. I. The spectrum of turbulence. *Proc. Roy. Soc.* (*London*), Ser. A, Vol. 164, pp. 476–490 (18 February 1939).
(42) Taylor, G. I. Statistical theory of turbulence. *Part I, Proc. Roy. Soc.* (*London*), Ser. A, Vol. 151, No. 873, p. 421 (1935).
(43) King, L. V. On the convection of heat from small cylinders in a stream of fluid: determination of the convection constants of small platinum wires with applications to hot-wire anemometry. *Phil. Trans. Roy. Soc.* (*London*), Ser. A, Vol. 214, p. 373 (1914).
(44) Dryden, H. L. and Kuethe, A. M. The measurement of fluctuations of air speed by the hot-wire anemometer. *N.A.C.A. Rep.* No. 320 (1929).
(45) Press, H. and Thompson, J. K. An analysis of the relation between horizontal temperature variations and maximum effective gust velocities in thunderstorms. *N.A.C.A. Technical Note* 1917 (July 1949).
(46) Donely, P. Summary of information relating to gust loads on airplanes. *N.A.C.A. Technical Note* 1976 (November 1949).
(47) Corrsin, S. Extended applications of the hot-wire anemometer. *N.A.C.A. Technical Note* 1864 (April 1949).
(48) Fiat. *Review of German science* 1939–1946. Meteorology and Physics of the Atmosphere. Ratje Mugge and others (1948).
(49) Cox, W. J. G. A cup-type transmitting anemometer with improved response to decreasing gusts. Royal Aircraft Establishment, *Technical Note* No. INSTN. 140 (July 1954).

CHAPTER 10

WIND DATA AND ENERGY ESTIMATION

The method of estimation

In previous chapters the selection of wind power sites and the methods followed to determine their wind régimes have been discussed. Unless they are to be used simply to place the sites in order of windiness, the wind speed measurements must be interpreted to form the basis of estimation of the annual energy which might be extracted by a wind-driven generator, of given size and type, installed on those sites.

Although some experience of the commercial operation of such plant of significant size has been gained, e.g. in Denmark, U.S.A. and Russia, little information is available on the relationship between data on the wind régime and the actual annual output of energy supplied to a network capable of receiving this energy as and when it is generated.

This relationship will depend upon both the actual wind behaviour, in detail, at the site and upon the design of the machine. But these factors cannot be taken into account in energy estimations to be made from wind survey data giving only hourly wind speeds. The procedure which can be followed, and which is not likely to lead to any misconceptions provided that its limitations are well recognized, is as follows:

(*a*) Relate the results of the wind speed measurements made during a relatively short period—perhaps one or two years—to the long-term wind régime at the site so that a velocity duration curve can be drawn. This can be accepted as a basis for estimations referring to the period of life of the wind power plant.

(*b*) Cube the ordinates of this curve to give the power duration curve for the site (since Power $\propto$ (Wind speed)^3).

(*c*) Estimate the specific output, T_s, (expressed in kilowatt hours per annum per kilowatt of installed capacity) corresponding to a given rated wind speed, V_p. The 'rated' wind speed is that at which the wind-driven generator gives its full rated power. The method of making the estimate is explained below.

Before further discussion it is well to note that the area under the power-duration curve is proportional to the annual energy, *available in the winds at the site*, for a given area to be swept by the rotor of the machine. For reasons which will be given later, only a small fraction of this energy can be extracted. Theoretically (see p. 192) a maximum of 59·3 per cent of the power in the wind at any moment can be extracted by a wind rotor but, in practice, this percentage will be reduced by losses in the rotor, in the transmission and in the generator, to something between 25 and 45 per cent. It is not necessary, however, to be concerned with this figure in energy estimations—indeed it would be impossible to take it into account without an exact knowledge of the characteristics of the actual machine to be installed. It can be accepted that the inefficiency is counterbalanced, in the construction of the machine, by making its rotor correspondingly larger than it would be for 100 per cent efficiency. On the assumption that the efficiency is constant for all wind speeds between the cut-in point and the rated wind speed, the estimation of the annual energy to be obtained can be based entirely on the power-duration curve, disregarding the design of the machine. The assumption introduces some error but this is not likely to be of sufficient magnitude to affect the estimation materially.

Other influences, dealt with later in this chapter, may have a greater —and less predictable—effect.

Estimation of specific output

Suppose that the power-duration curve for a given site, obtained by cubing the ordinates of the velocity-duration curve, is as shown in Fig. 56. It would be uneconomic to attempt to build a machine which could extract full power at very high wind speeds, so that a rated wind speed V_p (somewhere between 25 and 35 m.p.h. for sites in Great Britain) is chosen. The output is controlled at the full load value for all wind speeds above this until some wind speed, V_f, (the furling speed) is reached when it becomes advisable to shut down the plant to avoid damage. Line *cb* in Fig. 56 represents the period of operation at full output. Below rated wind speed the output falls until, at the cut-in speed, V_c, it is no longer worth while running the machine, the effective output being zero. The shaded area *bcfgh* represents the annual output of energy to the same scale as the rectangle *adeo* represents the annual output if the plant were running at full rated power throughout the entire year. The ratio $\frac{\text{area } bcfgh}{\text{area } adeo}$ is the annual plant load factor and multiplication of this by 8760 (i.e. the number of hours in one year) gives the specific output, T_s, in kilowatt-hours per annum per kilowatt.

The specific output is thus a characteristic of the power-duration curve for the site. Its value varies, however, with the rated wind speed and other operating points chosen and this choice affects the economy of the machine (see Ch. 15).

During the year there will be a certain number of hours of full output and a further period when the output varies from zero to the full value. It is sometimes convenient to think of the specific output as the equivalent number of hours of full-load operation.

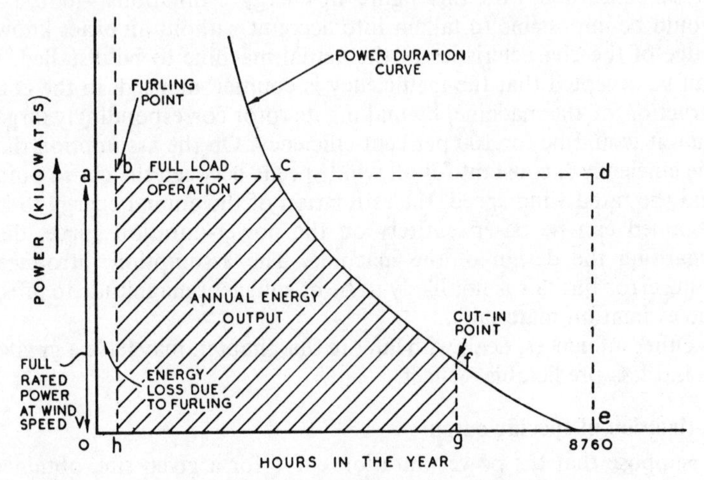

FIG. 56. *Estimation of annual energy output*

Estimations of specific outputs for selected sites in Great Britain

Table XXV, abstracted from E.R.A. Technical Report Ref. C/T108, gives values of specific output which have been estimated for a number of sites, of varying degrees of windiness, in Great Britain. The estimates have been made from the measured velocity-duration curves for these sites following the method just described. Three rated wind speeds were considered, namely, 30 m.p.h., 25 m.p.h. and 20 m.p.h. with corresponding cut-in speeds of 17 m.p.h., 13 m.p.h. and 10 m.p.h. respectively. The furling speed assumed was 60 m.p.h. in all three cases.

The first six lines of the table refer to sites selected, from wind surveys, as potential wind power sites; hence the high specific outputs. The remainder are Meteorological Office stations which are located to give wind records typical of the surrounding district so that the estimated outputs from them are much lower. In fact, the outputs shown near the bottom of the table would be quite uneconomic for the purpose of large-scale generation in Great Britain.

TABLE XXV

Site	Annual mean wind speed, V_{AM} (m.p.h.)	Specific output (kWh/year/kW) for rated wind speed shown		
		V_P = 30 m.p.h.	V_P = 25 m.p.h.	V_P = 20 m.p.h.
Mawr (1950)	25·3	4420	5520	6070
Costa	24·6	4350	5400	6200
Rhossili	23·7	4000	4960	5930
Mawr (1949)	23·5	4000	5100	6000
St. Agnes (1950)	21·7	3600	4850	6050
St. Agnes (1949)	21·0	3300	4300	5600
Bell Rock	18·0	2550	3700	4850
Lerwick	17·6	2425	3530	4790
Butt of Lewis	17·6	2300	3500	4600
St Ann's Head	16·2	2100	3000	4100
Stornoway	15·6	1900	2900	4040
Scilly	15·4	1850	2800	4000
Tiree	14·9	1750	2700	3800
Kirkwall	14·7	1560	2550	3730
Southport	13·7	1350	2100	3100
Fleetwood	12·9	1150	1700	2800
Plymouth	11·2	800	1400	2200
Eskdalemuir	11·0	800	1450	2350
Cranwell	9·9	650	1200	2050
Aberdeen	9·7	340	750	1600
South Shields	9·6	530	1050	1870
Yarmouth	9·4	450	900	1600
Birmingham	8·8	195	620	1350
Catterick	7·1	250	500	1000
Leicester	6·2	60	200	500

In Fig. 57 the values from the table are plotted and it can be seen that the individual points lie sufficiently close to the mean curves for the latter to be used in obtaining the relationship between mean annual wind speed and specific output for sites having similar wind régimes.

Results of wind velocity surveys

Measurements to determine wind power possibilities on selected sites in various parts of Great Britain have been made by the Electrical Research Association since the summer of 1948. These surveys have followed the methods already described in Ch. 6. The general principle adopted has been to install a sufficient number of recorders, giving mean wind speeds hour-by-hour, to provide data on the wind régimes at good sites in each of the main districts concerned. At the remainder of the sites cheaper, and more easily installed, measuring equipment has been used to measure simply the run of wind (in miles) from which monthly or annual mean wind speeds can be obtained, these in turn being used for estimations of specific output from the curves of Fig. 57.

Table XXVI, gives a summary of the results obtained up to January, 1954.

Among the sites are a few meteorological stations which acted as a link with the long-term wind records for the various districts included in the surveys.

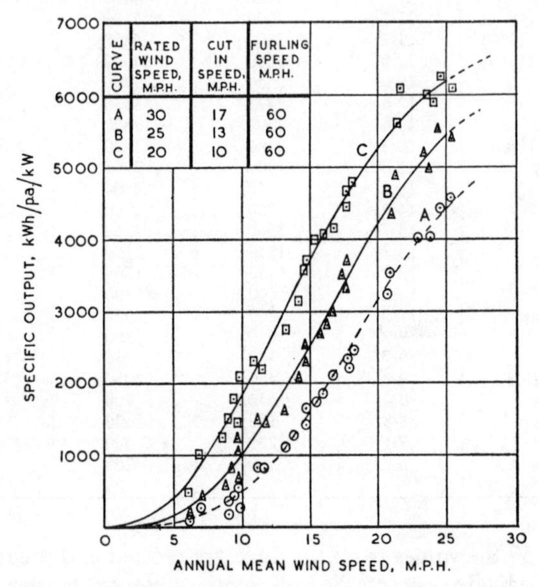

FIG. 57. *Relationships between specific output and annual mean wind speeds*

Out of sixty-five hill sites thirty-nine have been shown to have annual average wind speeds exceeding 20 m.p.h. and should give specific outputs between 3000 and 4750 kWh per annum per kilowatt for wind driven plant with a rated wind speed of 30 m.p.h.

Total potential installed capacity

While it may appear premature, in the present stage of development, to estimate the possible total wind power capacity which could only be realized at a much later stage, nevertheless it is useful to consider the question in order to place this source of energy in its proper perspective. Political and economic conditions in different parts of the world differ so widely that a degree of wind power utilization which may not be feasible or desirable in one country may be quite practicable in another. A misleading impression of the potentialities could be gathered from figures referring only to a few exceptionally windy sites and, for that reason, a wider assessment is necessary to enable a judgment to be made upon whether fuller consideration of the question is justifiable.

From the curves given in Fig. 57 it follows that a knowledge of the annual mean wind speeds of all the practicable hill sites in any district will be sufficient to enable an estimate to be made of the total capacity which could be installed on them if certain assumptions are made concerning the eventual size and performance of economic aerogenerators. The method of estimation suggested is outlined below.

Suppose that the diameter of the disc swept by the rotor of the future aerogenerators is 200 ft (this is probably a reasonable value for the maximum diameter which could be used in the present state of knowledge of blade construction). Take the over-all power coefficient of the machines to be 0·4.

Then, from the formula for power in the wind, kilowatts $= 0{\cdot}0000053\ AV^3$ (see Ch. 3.) and applying the coefficient, we have for the rated capacity of the aerogenerators

$$\text{kilowatts} = 0{\cdot}4\left[0{\cdot}0000053\,\frac{\pi}{4}\,.\,200^2\,.\,V_P{}^3\right] = \frac{V_P{}^3}{15}$$

where V_P is the rated wind speed in miles per hour. Suppose also that, at each site, the rated wind speed for the aerogenerator installed is chosen so that the specific output is 4000 kWh per annum per kilowatt. We can then compile Table XXVII for corresponding capacities, rated wind speeds and annual mean wind speeds for the sites, the last being obtained from the curves of Fig. 57 extrapolating when necessary.

Then, in the district concerned, select the hills on which it would be practicable to install 200 ft diameter wind turbines, taking into account also the number of machines which could be erected on each hill—a flat topped hill or ridge might accommodate several with a spacing of about $\frac{1}{4}$ mile. For each hill estimate, in the light of measurements on adjacent sites its annual mean wind velocity.

If n_{28}, n_{26} etc. are the numbers of machines which could be installed on sites having the annual mean wind speeds given in column three of Table XXVII then the total possible installed capacity is obviously

$$P = 2850n_{28} + 2300n_{26} + 1800n_{23\cdot5} + 1380n_{21\cdot5} + 1040n_{19\cdot5} + 760n_{17\cdot5} + 533n_{15\cdot5}$$

Table XXVI

Wind survey sites and results

Site			Altitude above sea level (feet)	Height of instrument above ground (feet)	Duration of observations (months)	Estimated long-term average wind speed, V_{AM}	
No.	*Name*	*District*				*Miles per hour*	*Metres per second*
Orkney and Shetland							
1	Costa Hill	Orkney Mainland	500	10	37	25·0	11·2
2	Vestra Fiold	Orkney Mainland	430	10	7	23·0	10·3
3	Bignold Park	Kirkwall, Orkney	130	35	37	14·8	6·6
4	Shurton Hill	Shetland	576	10	18	19·5	8·7
5	Ward of Scousburgh	Shetland	863	30	18	24·5	11·0
6	Sandness Hill	Shetland	817	10	18	22·8	10·2
Mainland of Scotland							
7	Dunan Mor	Sutherland	523	10	30	19·1	8·6
8	Meall an Fheadain	Ross-shire	663	10	30	22·0	9·9
9	Macrihanish	Kintyre	100	30	44	13·7	6·2
10	Cnoc Moy	Kintyre	1462	15	17	24·8	11·1
11	Torr Mor	Kintyre	1358	10	44	25·0	11·2
12	Pinbain Hill	Ayrshire	734	15	43	23·0	10·3
13	Downan Hill	Ayrshire	348	10	17	22·0	9·9
14	Newlaw Hill	Kirkcudbright	600	10	24	14·4	6·4
Western Isles							
15	Chaipaval	Isle of Harris	1203	10	30	25·7	11·5
16	Ben Tangaval	Isle of Barra	1092	10	30	23·2	10·4
17	Ben Geary	Isle of Skye	929	10	30	21·0	9·4
18	Crauchan Treshnish	Isle of Mull	704	10	30	19·5	8·7
19	Maol Buidhe	Islay	542	10	14	17·4	7·8
20	Sgarbh Breac	Islay	1192	10	14	27·2	12·2
Wales and Isle of Man							
21	Dolgarrog	Caernarvonshire	1390	30	13	17·0	7·6
22	Mynydd Rhiw	Caernarvonshire	990	15	11	20·5	9·2
23	Mynydd Mawr	Caernarvonshire	524	30	53	25·0	11·2
24	Mynydd Anelog	Caernarvonshire	628	10	50	25·5	11·4
25	Mynytho Common	Caernarvonshire	600	10	21	18·0	8·1
26	Holyhead Mountain	Anglesey	431	10	24	19·0	8·5
27	Foel Eryr	Pembrokeshire	1535	15	48	25·0	11·2
28	Mynydd Castlebythe	Pembrokeshire	1137	10	23	19·0	8·5
29	Rhossili Down	Glamorgan	633	15	40	24·0	10·7
30	Garth Mountain	Glamorgan	1000	30	13	13·5	6·1
31	South Barrule	Isle of Man	1585	10	17	22·5	10·1
32	Snaefell	Isle of Man	2034	30	17	27·0	12·1

Cornwall							
33	The Lizard	Cornwall	100	10	16	14·5	6·5
34	Carn Bean	Cornwall	665	10	11	17·0	7·6
35	Watch Croft	Cornwall	827	10	6	20·0	9·0
36	St. Agnes Beacon	Cornwall	629	15	17	22·0	9·9
37	Carn Brea	Cornwall	657	30	17	22·0	9·9
Rest of England							
38	Portland Bill	Dorset	141	10	10	13·6	6·1
39	Firle Beacon	Sussex	600	10	8	13·2	5·9
40	Telegraph Hill	Norfolk	250	10	13	13·2	5·9
Channel Islands							
41	Les Landes	Jersey	250	10	10	15·9	7·1
42	Brecqhou	Channel Islands	150	10	17	19·5	8·7
Northern Ireland							
43	Divis Mountain	Co. Antrim	1574	30	15	22·0	9·9
44	Evish Hill	Co. Antrim	1100	10	8	17.0	7·6
45	Big Collin	Co. Antrim	1163	10	10	17·0	7·6
46	Carnanmore	Co. Antrim	1253	10	25	25·0	11·2
47	Knocklayd	Co. Antrim	1695	30	11	22·0	9·8
48	Chimney Rock	Co. Down	2153	10	16	26·0	11·6
49	Slieve Gullion	Co. Armagh	1894	10	26	25·0	11·2
50	Binevenagh	Co. Londonderry	1260	30	9	18·0	8·1
51	Dooish	Co. Fermanagh	1100	30	11	19·0	8·5
Irish Republic							
52	Leahan	Co. Donegal	1418	30	10	22·0	9·8
53	Slieve Tooey	Co. Donegal	1515	30	38	25·5	11·4
54	Bloody Foreland	Co. Donegal	1038	10	37	27·0	12·1
55	Crocknafarragh	Co. Donegal	1707	10	15	24·0	10·7
56	Ardmalin	Co. Donegal	362	10	11	17·0	7·6
57	Blue Stack Mountains	Co. Donegal	2219	10	10	22·0	9·8
58	Glencastle Hill	Co. Mayo	760	10	10	18·4	8·3
59	Gortbrack South	Co. Mayo	791	10	12	23·7	10·6
60	Knockmore	Co. Mayo	1119	10	23	22·6	10·1
61	Tully Mountain	Co. Galway	1172	10	11	24·0	10·7
62	Errisbeg	Co. Galway	987	10	14	22·4	10·0
63	Knockanore	Co. Kerry	880	10	25	16·4	7·3
64	Mount Eagle	Co. Kerry	1696	10	18	21·5	9·6
65	Knockgour	Co. Cork	1589	10	18	21·5	9·6
66	Mount Gabriel	Co. Cork	1339	10	12	19·2	8·6
67	Verschoyle's Hill	Co. Dublin	1093	10	18	14·3	6·4
68	Three Rock Mountain	Co. Dublin	1497	10	8	20·3	9·1

Notes on table

Column 2. The Lizard is a Meteorological Office Station. Bignold Park and Macrihanish are one-time Meteorological Office stations at which measurements have been discontinued.

Column 7. The long-term averages have been estimated by comparison of the observations with long-term averages for the nearest Meteorological Office stations.

It is important to note that the 'target' figure of 4000 kWh per annum per kilowatt used above was arbitrarily chosen to illustrate the method of estimation rather than because it will necessarily apply to the most economic generation. Considerations of economic design given in Ch. 15, may show that acceptance of a lower value for the specific output will lead to a lower cost per kilowatt-hour generated. This is accompanied by choice of a higher rated wind speed than that given in Table XXVII for a given annual mean wind speed, with consequent increase in the capacity of the aerogenerator for a given diameter of propeller.

TABLE XXVII

Relationship between capacity, rated wind speed and annual mean wind speed

Wind turbine 200 ft diameter, overall power coefficient 0·4, specific output 4000 kWh p.a./kW

Capacity of aerogenerator (kW), P_G	Rated wind speed (m.p.h.), V_P	Annual mean wind speed (m.p.h.), V_{AM}
2850	35	28
2300	32·5	26
1800	30	23·5
1380	27·5	21·5
1040	25	19·5
760	22·5	17·5
533	20	15·5

It is, of course, equally easy to follow the outlined method of estimation taking 3000 kWh per annum per kilowatt or 2000 kWh per annum per kilowatt for the specific output instead of 4000.

Probable relationship between actual and estimated outputs

The method of output estimation described on p. 153 is sufficiently accurate for the purpose of assessing the relative potentialities of different sites and of forming a realistic conception of the total annual energy which might be obtained from a group of aerogenerators. When a generating unit is installed and operated on a particular site it becomes possible to check the actual output with that estimated. It is clearly too much to expect that the two will agree exactly and, regardless of whether the actual output obtained is greater or less than was expected, it is important to be able to judge why the discrepancy has arisen. Only by doing so can the test results be interpreted to afford useful knowledge for application in future design and operation as well as in estimations based on wind data.

Possible causes of discrepancy

The discrepancies between estimated and actual values of output will be greatly influenced by the adequacy of the methods used to determine the wind régime, by the application of the results of wind measurements and by the characteristics of the wind-driven generator itself. Some of the more obvious causes of discrepancy are:

(i) *Wind distribution.* Estimations of the energy in the wind must be based on measurements of wind speed by some form of anemometer. Those of annual energy available are made by using hourly mean wind speeds following the method described earlier in this chapter. In testing the performance of an actual generating unit, shorter period measurements using, perhaps, several quick response anemometers would be used. Even so, the assumption must be made that the measured values of wind speed apply either to the whole, or to the major portion of, the circular area swept by the wind turbine rotor, whereas in fact they only apply strictly to a very small area where the anemometer is located.

If the estimate is to be made from anemometer records for a low height above ground a correction must be made for the increase of wind speed with height (see Ch. 7). But even if the anemometer is located at hub height, or several anemometers at different heights are used and a mean value of wind speed is taken, it is unlikely that the calculated power output will agree precisely with that from the non-uniformly distributed wind which will occur in actual operation.

Unless the hill-top site of the wind turbine is very peaked, the wind speed at a given height above ground will probably not vary significantly from point to point on the hill top but this does not apply to the vertical distribution of wind speed.

(ii) *Wind direction.* Cup anemometer readings are little affected by the direction of the wind. The instrument reading is entirely independent of wind direction in the horizontal plane while the vertical component of wind velocity only becomes important when it is sufficiently large for the total wind direction to depart from the horizontal by a fairly large angle. On the other hand, the effect of wind direction, and of changes in it, on the output of a wind turbine will depend upon the coning angle of the blades and their rate of response in changing this angle as well as upon the rate of reorientation of the machine as the horizontal wind direction changes.

The inclination to the horizontal of the wind over a hilltop depends upon the shape of the hill and upon the height above ground (see also Ch. 7). It may well be a few degrees, especially at heights up to 30 ft and on a very peaked summit, but at greater heights and with flatter summits the wind direction is not likely to depart significantly from the horizontal. The effect on output, depending, as it does, so much on the design of the turbine, is not predictable precisely but it is not likely to be serious.

Slowness in response of the turbine when the horizontal wind direction changes would cause some reduction of output especially in a gusty wind having rapid changes of direction.

(iii) *Hourly mean wind speed and the overall power coefficient of the machine*. In estimating energy output by a machine from the power duration curve for the site on which it is to be installed, and in default of information on its performance, it must be assumed that a constant fraction of the power in the wind will be extracted over the operating range from cut-in to rated wind speed after which the output will be maintained at full rated value. This fraction is C_{op}, which may be called the 'overall power coefficient' so that

$$C_{op} = \frac{\text{electrical power output}}{\text{power in the wind for the swept area}}$$

In practice the wind turbine may be designed so that the blade pitch remains fixed until full output is reached, and this will cause a reduction in C_{op} for lower power outputs as will a fall in generator efficiency at fractions of full load. If this were not so the curve of output power would be of the same shape as the curve *bcfg* in Fig. 56 as indicated in the dotted curve of Fig. 58 but its ordinates would be reduced in the fixed ratio C_{op}:1. Actually C_{op} will probably vary with wind velocity as shown in Fig. 59 which gives its ratio to that at rated wind speed. Using such variable values of C_{op} applied as factors to arrive at the curve of actual output from the wind-driven generator we obtain the full line curve of Fig. 58. On this account, therefore, the annual energy output will be slightly less than that estimated from wind survey results.

The fact that C_{op} for full output is much less than unity—of the order of 0·4—is allowed for in the design of the plant by increasing the length of the blades to compensate for it by sweeping a greater area. The fall in the value of C_{op} at the lower wind speeds must, however, be accepted as causing some loss of output. Fortunately, at a good site more than half of the total annual energy is obtained from wind speeds equal to, or above, the rated value so that this loss may not be serious.

(*iv*) *Changes in wind speed*. The results of wind surveys, using cup anemometers, are expressed in terms of hourly mean wind speeds: even if the recorder indicates shorter period mean speeds the labour which would be involved in analysis makes it impracticable to take

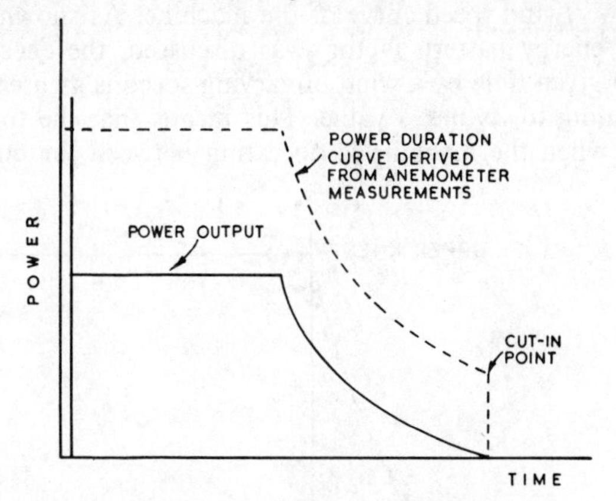

FIG. 58. *Estimated and actual power output curves*

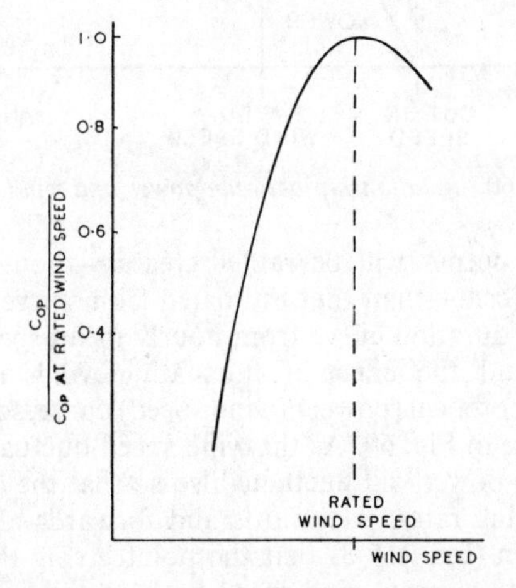

FIG. 59. *Variation of power coefficient with wind speed*

cognisance of them. But, in fact, the wind speed during the hour is continuously changing. The change is relatively slow and of only a small percentage in a 'steady' wind, with faster and greater variations in a 'gusty' wind. The effects produced depend upon the rates of change of wind speed in relation to the response of the control gear of the wind turbine and must, therefore, be dealt with separately.

(*a*) *Slow wind speed changes*. These are supposed to take several seconds—slow enough to be followed accurately by the control gear so that the power output at any moment corresponds to that given by the power/wind speed curve of the machine. As shown in Ch. 3, in which 'energy pattern factor' was discussed, the energy contributed in a given time by a wind of varying speed is greater than that corresponding to its mean value. This means that the total energy produced when the machine is operating between the cut-in point

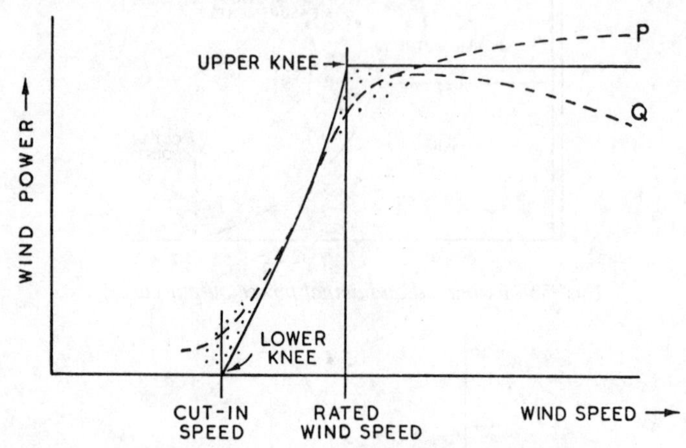

FIG. 60. *Relationship between power and wind speed*

and full rated output will be rather greater—though probably by only a few per cent—than that estimated from survey results which give the power duration curve from hourly mean speeds.

To understand the effect of fluctuating winds more precisely, consider the theoretical (power)/(wind speed) curve, for steady winds, given in full line in Fig. 60. As the wind speed fluctuates about some mean value the power will fluctuate also so that the operating point moves, at varying rates, backwards and forwards along the curve. It can be shown (see Ref. 3) that the point giving the mean power corresponding to a mean wind speed for a period of time will lie at the centre of gravity of that part of the curve (having varying mass, to take into account the varying rate of movement of the operating point along the curve) moved over during the time concerned. Over the flat portions of the curve the point connecting mean power and mean wind speed will lie very close to the curve but, when the mean wind speed is near either the cut-in speed or the rated speed ('lower knee' or 'upper knee' in Fig. 60) the true operating point will lie off the curve. Near cut-in it will lie above the curve and near the rated wind speed it will lie below.

This effect is shown by the clouds of points in the figure, the individual points corresponding to different degrees of fluctuation about the mean value. Thus, if the design is such that the power output is rigidly controlled to follow the (power)/(wind speed) curve the 'knee effect' is to reduce the mean power when the hourly mean wind speed is close to the rated wind speed and to increase it when the hourly mean wind speed is close to the cut-in value.

Otherwise expressed, fluctuations of wind speed above the rated wind speed do not lead to an increase in the power to counter-balance the effect of those below the rated value so that there is a net reduction of power. The reverse is true at the cut-in point.

The power curve in actual operation will thus be a distortion of the theoretical curve. The amount of the distortion depends upon the amplitude of the wind speed fluctuations and upon the mean value of the wind speed during the time considered.

(*b*) *More rapid changes of wind speed.* These are supposed to be slow enough to overcome the inertia of the machine but too rapid for the control gear to follow them completely unless its operation is practically instantaneous.

Their effect is most marked when the wind speed is above the rated value. First, in a gusty wind, the r.m.s. value of the output current—and therefore the heating of the machine—will be greater than that of the mean output current. This may lead to overheating unless forced cooling is employed. Otherwise the output power may have to be reduced below the full load value in spite of the mean wind speed being above the rated value. Secondly, the controls may operate to reduce output, and so avoid excessive stresses, more quickly when the wind speed suddenly rises than when it falls. This will also lead to a reduction of output.

The actual power curve for wind speeds above the rated value may thus lie between the limits indicated by the dotted curves P and Q in Fig. 60. Gustiness of the wind will certainly influence the energy output for a given hourly mean wind speed but precise estimates of this influence are impossible without knowledge of the rate of response of the control mechanism.

(*c*) *Very rapid changes of wind speed.* High speed gusts will change the aerodynamic conditions at each blade section but, except under unusually gusty conditions, they are usually not of great amplitude nor are they likely to occur uniformly over the whole of the swept area. The inertia of a large rotor will probably be great enough for such rapid changes to have only a negligible effect on the energy output but they are very important in setting up stresses and vibrations in the wind turbine.

(*d*) *Errors in wind speed measurements.* Estimations of energy output based on long term records from cup anemometer measurements could be erroneous if the anemometers themselves were inaccurate or if they differed from those used in making performance tests on an installed aerogenerator. An analysis (see Ref. 3) of the probable discrepancies due to this cause has shown, however, that they are likely to be negligible.

The French wind survey

Starting with the first installation in October 1946, the Research Department of Electricité de France has since continued a survey (see also Ch. 6) of potential wind power sites in different parts of France but more particularly round the coasts, around Paris and at a few distributed sites in the centre of the country (Ref. 5).

The instrument used has been that described in Ch. 8 (p. 114) which gives an integration of kilowatt-hour per square metre although it should be noted that the Betz coefficient 0·593 is included in the calibration of the instrument which therefore measures, not the energy *in the wind*, but that theoretically obtainable from the wind by an ideal windmill.

About 100 stations have been in operation, the instruments being mounted in a few cases on 40 m masts but otherwise on lighthouses and other convenient structures so that their heights above ground vary from 8 m to 305 m (at the Eiffel Tower).

The annual energy per square metre measured at the top of the Eiffel Tower is from 3000 to 3500 kWh. This site, for which long-term wind records exist, has been used as a basis of comparison for the other measuring sites the results from which range from 0·158 to 2·392 times the Eiffel Tower value. The highest measured value (7460 kWh/m^2) is for Fort Réal, at the eastern end of the Franco-Spanish frontier, where there are other good sites. The Brittany coast and the north coast of France also have windy sites.

Unfortunately it is not possible to make a direct comparison between the French results, giving an integration of the wind energy, and those made in Great Britain based on the hourly wind speeds with a specified operating range of 17–30 m.p.h. Thus, for example, using the formula

$$kW = 0{\cdot}37\left(\frac{V}{10}\right)^3 . A$$

(V = metres per second, A = square metres)

for the power extracted by an ideal windmill, that for 30 m.p.h. and a swept area of 1 m^2, would be 0·894 kW which would imply the incredibly high value of

$$\frac{7460}{0{\cdot}894} = 8344 \text{ kWh per annum per kilowatt}$$

for the site at Fort Réal. The apparent explanation is that the French instrument measures much more energy than could be extracted by a windmill with a reasonable operating range of wind speeds. By dividing the kilowatt-hours per square metre, measured in a given period of time, by 0·37, the integration of $\left(\frac{V}{10}\right)^3$ is obtained, from which it is possible to calculate the cube root of the mean cube of the wind speed. But this will be a higher figure than the mean wind speed for the same period of time (see 'Energy pattern factor' p. 28). To obtain an approximate value for the mean wind speed at a measuring site from readings of the French instrument requires, therefore, previous calibration against an anemometer measuring mean wind speed (or total run of wind) and a subsequent assumption of an 'energy pattern factor'.

To illustrate the difficulty in comparing the results of measurements of kilowatt-hours per square metre with those of mean wind speed, some comparable figures are given in Table XXVIII. These were obtained in some recent tests with a cup-counter anemometer and a French anemometer mounted close together at the same site (Bull Point Lighthouse, North Devon). Readings were taken at the same time each day so that the intervals between readings are of 24 hr duration.

The values of kilowatt-hours per square metre in column, (4) calculated from the 24 hr mean wind speeds in column (3) are, as might be expected, somewhat lower than those given by the French instrument which takes count of the energy content of fluctuating winds. By its cubic calibration the latter emphasizes the effects of high wind speeds. In the same way the cube-root-mean-cube figures in column (6) are generally higher than the mean speeds of column (3). The agreement between the two instruments is fair for the first period of time (16th August, to 24th August), in the table. When the wind speed was high (47·1 m.p.h.) in the second of the periods considered, the reading of the French instrument was lower than the calculated value in column (4); this may be explained by the fact that the French anemometer reads low at high wind speeds.

The apparently large discrepancy between the two instruments

TABLE XXVIII

Run of wind readings (miles)	Cup counter anemometer: 24 hr mean wind speed $\bar{V}$ (*m.p.h.*)	Cup counter anemometer: 24 hr mean wind speed $\bar{V}_1$ (*m/sec*)	Cup counter anemometer: Calculated* kWh/m² for 24 hr	French anemometer: kWh/m² readings	French anemometer: Calculated 24 hr† mean wind speed $\bar{V}_3$ (m/sec)
16–24 Aug., 1953					
9,033					
	14·5	6·5	2·45	2·62	6·65
9,382					
	17·7	7·9	4·40	6·88	9·00
9,807					
	12·2	5·45	1·45	1·01	4·85
10,100					
	19·2	8·6	5·65	7·97	9·65
10,561					
	19·2	8·6	5·65	8·71	9·90
11,023					
	29·2	13·1	20·0	27·03	14·5
11,729					
	18·7	8·4	5·25	8·06	9·68
12,179					
	23·6	10·5	10·2	14·49	11·8
12,747					
20–23 Sept. 1953					
22,896					
	25·5	11·4	13·1	20·48	13·2
23,508					
	47·1	21·0	82·0	76·67	20·5
24,639					
	14·5	6·5	2·45	4·77	8·13
24,987					
11–13 Nov., 1953					
41,150					
	28·8	12·9	19·0	48·5	17·6
41,841					
	32·9	14·7	28·5	34·15	15·7
42,631					

* Using the formula $\text{kWh/m}^2 = 24\left[0{\cdot}37\left(\frac{\bar{V}_1}{10}\right)^3\right]$

† Using the formula $\bar{V}_3 = \sqrt[3]{\frac{1000(\text{kWh/m}^2)}{(24)(0{\cdot}37)}}$

shown in the penultimate line of the second part of the table may have been caused by very variable wind speeds during the 24 hr period concerned.

Smith-Putnam wind survey

As a preliminary to the construction of the 1250 kW, Smith-Putnam aerogenerator in Central Vermont during World War II, studies were made of the wind behaviour in the mountains of New England (see also Ch. 6). Certain assumptions concerning this behaviour, and particularly relating to the accelerating effects of different shapes of hills, were made as described by Putnam (Ref. 4, Ch. 2).

Wind tunnel tests on hill models proved unsuccessful in checking these assumptions as did also measurements of vertical wind gradient using floating balloons at one selected site. It thus became necessary to institute a wind research programme using anemometric and ecological methods.

Beyond its main purpose of investigating potential wind power sites in addition to that at Grandpa's Knob, which was chosen as the site for the 1250 kW unit before anemometer measurements were started, the survey was aimed at the exploration of the relationship between the geometry of hill profiles and their effects upon wind flow.

Measurements were made at twenty sites most of which were on hilltops above 1400 ft in altitude. Cup anemometers, at first of standard pattern but later of a new design with heated cups to eliminate icing troubles, were used. It was found that, through over-running, these new anemometers, having heavy rotors, recorded more miles of wind over a given period than the lighter cup anemometers.

The anemometers were mounted at heights above ground varying from 40–185 ft and at several sites three or more instruments at different heights were installed to determine the vertical wind gradient. A significant feature of the survey is the short time over which measurements were made: only at the Grandpa's Knob test site did they continue for more than a year.

In addition to measurements of run-of-wind by integrating recorders, horizontal wind directions were recorded at Grandpa's Knob and occasional measurements were made of the vertical direction of the wind.

The ecological research programme was concerned with observations of the different kinds of deformation of trees by the wind and correlation of these with wind behaviour. Thus, a rough but useful yardstick for estimation of wind strength was developed. It is perhaps worth mentioning here that the use of such ecological methods is not generally applicable: so often at potential wind power sites there are no trees on the hills.

The results of the survey measurements enabled conclusions to be drawn (see Putnam, Ref. 4, Ch. 4) on the frequency distribution of wind direction and wind velocity, upon vertical wind gradient and other features of wind behaviour. These are discussed in other chapters but here we are most interested in the average annual wind velocities measured and in the estimations of energy output therefrom.

The average annual velocities referred to a height of 140 ft above ground, varied from 12·3 m.p.h. to 44 m.p.h. (the latter being at 4100 ft above sea level). The estimated annual energy outputs from a wind turbine of the Smith-Putnam design, having a rated wind speed of 30 m.p.h. were plotted against annual average wind speed for three different heights above sea level. The curve for sea level is in substantial agreement with the curve for 30 m.p.h. rated wind speed, which has been drawn for sites in Great Britain (see Fig. 57).

The need for such surveys and for careful anemometer measurements over a sufficiently long period is emphasized by the fact that the Grandpa's Knob site which was originally estimated to have an average wind speed of 24 m.p.h. (before measurements were made) proved to have an average of only 16·7 m.p.h. Indeed, had this been known earlier, the wind turbine would not have been built on the site.

BIBLIOGRAPHY

(1) Golding, E. W. The potentialities of wind power for electricity generation (with special reference to small-scale operation). Electrical Research Association, *Technical Report* W/T16 (1949).

(2) Golding, E. W. Large-scale generation of electricity by wind power—preliminary report. Electrical Research Association, *Technical Report* C/T101 (1949).

(3) Golding, E. W. and Stodhart, A. H. The selection and characteristics of wind-power sties. Electrical Research Association, *Technical Report* C/T108 (1952).

(4) Putnam, P. C. *Power from the wind.* Van Nostrand (1948).

(5) Ailleret, P. L'énergie éolienne: sa valeur et la prospection des sites. *Revue générale d'électricité* (Mars 1946).

(6) Juul, J. Investigation of the possibilities of utilization of wind-power. *Elektroteknikeren*, Vol. 45, No. 20, pp. 607–635 (22nd October 1949).

(7) Organization for European Economic Co-operation. Committee for Productivity and Applied Research, Working Party No. 2 (Wind Power) (1954).

(8) Thomas, P. H. *Electric power from the wind.* Federal Power Commission (1945).

(9) Karmishin, A. V. Small-power wind-electric generating units. *Priroda* (*Nature*), No. 11, pp. 24–31 (November 1949).

(10) Problemas del aprovechamiento de la energia eolica en la Republica Argentina. *Revista Electrotecnica*, Vol. 38, pp. 1–20 (January 1952).

(11) Vezzani, R. An aero-electric pilot plant of medium power. (From *L'Elettrotecnica*, Vol. 37, No. 9, September 1950, pp. 398–419, 30 illustrations.) *The Engineers' Digest*, Vol. 12, No. 5 (May 1951).

(12) Brun, E. A. and Oniga, T. Utilizacao da energia dos ventos. *Dados gerais*. A situacao no Brasil. Instituto Nacional de Tecnologia (Rio de Janeiro, 1952).

(13) Sektorov, V. R. Operating conditions and types of wind power installations for rural districts. *Elektrichestvo*, No. 10, pp. 33–37 (October 1949).

(14) Sil, J. M. Windmill power. *The Indian Journal of Meteorology and Geophysics*, Vol. 3, No. 2 (April 1952).

(15) Cambilargiu, E. *La Energía del viento en el Uruguay*. Instituto de Máquinas, Publication No. 12, Montevideo (Uruguay, 1953).

(16) Studiengesellschaft Windkraft e.V., Stuttgart. *Windkraft Mitteilungen*, No. 3 (1st April 1954).

(17) Serra, L. Le vent en France et ses possibilités d'utilisation. *La Météorologie* (October–December 1953), pp. 273–292.

CHAPTER 11

THE TESTING OF WIND-DRIVEN A.C. GENERATORS

IN the preceding chapter we were concerned with estimations, based on wind speed measurements, of the energy which might be obtained at a particular site from a wind-driven generator installed there and assumed to be operating over a given range of wind speeds. Here we shall suppose that the machine has been built and consider the tests which might be made to check its performance in service. This may appear to be a simple matter but, with so little being known about the effects of the great number of variable factors which may enter into the performance, it is in fact very complex.

With small wind-driven generators (see Ch. 14) for individual premises there has been sufficient experience to indicate the testing which should suffice and perhaps, also, the same applies to the direct-current machines, of rather larger size, which have been used for many years in Denmark. But we are at the outset of the development of large aerogenerators for connection to alternating-current supply networks and, apart from the information which has been given on the testing of the 1250 kW machine built during World War II at Grandpa's Knob in Central Vermont, U.S.A., we have little to guide us. It would thus be quite unrealistic to suggest a testing procedure which could be regarded in any way as a matter of routine; the tests must be an essential part of the research and development. Nevertheless it may, at least, clarify the matter if we attempt to formulate the questions which have to be answered and to suggest what measurements might throw some light on them.

A convenient classification is (*a*) Energy output tests, (*b*) Power generation tests and (*c*) Tests concerned with research and development.

(*a*) Tests of energy output

These tests are the simplest in their object which is to determine whether the methods of estimation described in Ch. 10 are justified by the results of practical operation. The question to be asked is, "Does a wind-driven generator, with a given rated wind speed, give an annual specific output which is substantially in accordance with the curves of Fig. 57 when it is located at a site with a known annual average wind speed?" Or, since these curves are rather for general application than for a particular case, "Does the annual output correspond closely to that estimated from the power duration curve for the site?"

The obvious method to adopt in making such a check is to continue the wind speed measurements under exactly the same measuring conditions, for height above ground and instrumentation, as used in providing data for the estimation, and to measure, with a kilowatt-hour meter, the energy output for a full year's operation. In practice, however, the check would not be quite so simple because (*a*) inevitably there would be some stoppages during the year, either for maintenance of the wind-driven generator itself or through power failures on the line to which it is connected; (*b*) there may be temporary failures in the measuring instruments; (*c*) the assumption, made in the estimation of output, that the efficiency of the machine is the same at lower loads as it is at full load is not quite justified, and (*d*) the estimations are based on the assumption of perfectly steady wind speed, of the hourly average value, during each hour.

There is, also, the assumption that the machine is so designed that it follows the operating schedule of wind speeds for cut-in, full power output and furling, exactly as used in the estimations of energy output.

Some discussion of the several complicating factors is perhaps worth while.

Machine efficiency. The first point concerns the efficiency of the machine at varying loads and the influence, upon total output, of reduced efficiency at the lower wind speeds in the operating range.

Consider the figures given in Table XXIX which are taken (with some rearrangement and additions) from the War Production Board report (Ref. 1).

Column 2 gives the calculated values of power, available in a circle of diameter 60 ft, for the wind speeds in column 1. In column 3 are given power coefficients. These are obtained from the blade characteristics for tip speed ratios (see p. 208) corresponding to varying wind speeds and a fixed rotational speed of 91 r.p.m. The figures in column 4 are the products of those in columns 2 and 3. The net outputs in column 7 are the difference between columns 4 and 5. Overall efficiency (column 6) is obtained by dividing net output (column 7) by the power in the wind (column 2). The figures in column 8 are taken from a velocity frequency curve for the site on which the machine is supposed to be erected; incidentally the mean annual wind speed for this site is found, by calculation, to be almost exactly 17 m.p.h.

The figures in the final column are the products of those in columns 7 and 8. (It is worth noting that some 42 per cent of the total

TABLE XXIX

Wind-driven generator giving 100 kW at 29 m.p.h.: Propeller diameter 60 ft rotating at 91 r.p.m.

1	2	3	4	5	6	7	8	9
Wind speed (m.p.h.)	Power in the wind (kW)	Rotor power coefficient C_P	Power output from the wind rotor (kW)	Power losses in generator and power transmission (kW)	Overall efficiency (per cent)	Net power output (kW)	Annual wind speed duration (hours)	Annual energy output (kWh)
15	49	0·220	11	9	4·1	2	375	750
16	59	0·289	17	9	13·6	8	385	3,080
17	71	0·340	24	10	19·7	14	400	5,600
18	84	0·375	32	10	26·2	22	380	8,360
19	99	0·398	39	11	28·3	28	365	10,220
20	115	0·410	47	12	30·5	35	350	12,250
21	133	0·413	55	12	32·4	43	335	14,405
22	153	0·411	63	13	32·6	50	310	15,500
23	175	0·405	71	14	32·6	57	295	16,815
24	199	0·399	79	15	32·2	64	270	17,280
25	225	0·390	88	16	32·0	72	255	18,360
26	253	0·380	96	18	30·8	78	225	17,550
27	283	0·372	105	19	30·4	86	200	17,200
28	316	0·360	114	20	29·8	94	190	17,860
					Full power output	100	1,260	126,000
							Total annual output	301,230

energy is produced by the machine running at full output; had the mean wind speed at the site been higher this percentage would have been greater).

It is seen from column 6 that, in this instance, the maximum overall efficiency does not coincide with the full output but occurs at about half load.

Instead of taking varying efficiencies as in the table, assume a constant efficiency of 29 per cent corresponding to that at the full output of 100 kW which is produced at 29 m.p.h. (This is the method used in energy estimations from wind survey results.) Recalculating the values for column 9 by using this efficiency to obtain net outputs from the power input values in column 2, we find a total annual energy output of 297,450 kWh which is in quite close agreement with the total in the table and errs slightly on the low side.

Effect of unsteady winds. J. Juul (Ref. 2) has described the testing methods used with a SEAS* experimental wind-driven asynchronous generator of 13 kW capacity, the wind rotor of which had two blades and a diameter of 8 m, and has given some of the results obtained. His report is especially interesting because he compares actual energy outputs with estimated outputs and offers an explanation for the discrepancies between the two sets of values. The estimations were made in a way very similar to that discussed above. Juul used a wattmeter to measure the power output of his generator for various values of wind speed indicated by an anemometer of the windmill type mounted, at hub height, close to the machine. He chose periods of steady wind for his measurements so that there should be good agreement between them and so obtained a (power)/(wind speed) curve. Using this curve, he estimated the monthly output of energy from the durations of the measured hourly wind speeds in the month. The estimated outputs, and the outputs actually measured when the machine was run connected to the a.c. network, are compared in Table XXX.

The actual energy production for the eight months, being 10 per cent higher than the estimated total, may be regarded as confirmation that the machine has operated satisfactorily but, clearly, monthly divergences of up to 50 per cent between actual and estimated output need to be explained. Juul gives the explanation—and expresses it with some confidence—that outputs appreciably higher than those estimated are produced when the wind blows from landward directions, and is gusty; winds from the sea (the mill is located near the shore), which are steadier, produce outputs in close agreement with estimates. The winds were blowing mainly from over the land in

* SEAS = Sydøstsjaellands Elektricitets Aktieselskab.

TABLE XXX

Estimated and actual monthly outputs of energy from a 13 kW wind-driven generator: constant rotor tip speed 38 m/sec

Month	Estimated monthly output (kWh)	Measured monthly output (kWh)	Percentage error of estimation (+ means actual greater than estimated)
December 1950	1,623	1,810	+ 10·3
January 1951	1,081	1,827	+ 40·8
February 1951	1,373	1,937	+ 29·2
March 1951	3,048	2,915	− 4·5
May 1951	1,472	1,815	+ 18·9
June 1951	1,213	821	− 47·8
July 1951	1,890	1,666	− 13·5
August 1951	939	1,256	+ 25·2
Totals	12,639	14,047	+ 10

January, February and May: hence the high production in those months. He suggests that the negative divergences for the months of June and July are due to the fact that light winds in those months introduced errors in wind speed measurements.

Juul's explanation is interesting if not, perhaps, entirely convincing. There is, of course, a sound reason for a higher output in a gusty wind because the 'energy pattern factor' (see p. 28) is greater with such a wind than it is for a steady wind. The design of the machine, especially the rate of response of the speed control or pitch changing mechanism, and the shape of the (power coefficient)/(tip speed ratio) curve, will exert an important influence, however, and it is to be noted that Juul's machine has fixed-pitch blades and a rather flat-topped power coefficient curve. Tests of output over a shorter period of a few hours in gusty, and in steady, winds could perhaps prove this point more conclusively.

(*b*) Tests of the generated power

A wind-driven generator having been designed and built to produce a specified power output at a certain rated wind speed, and to give a falling output at wind speeds diminishing from the rated value to the cut-in point, an obvious step in the performance tests is to obtain its (power)/(wind speed) curve.

If the machine could be subjected to a completely steady and uniformly distributed wind, as for example in a very large wind tunnel, the determination of this curve would be straightforward.

it is for this reason that I have decided to undertake an independent study on either the generation of electrical power from a windmill, or another field involving nuclear energy.

The first project would involve...

RMVILLE, VIRGINIA

ter

______________ **Date** ______________

nt records for the student's inspection. The law also permits the ndation. The applicant's signature below constitutes a waiver; *no* er of rights to inspect this form also waives the right to inspect all

'e ______________

	Excellent	Above Average	Average	Below Average	No Basis to Judge

But even medium-sized machines, having rotor diameters between 20 and 100 ft, cannot be tested conveniently or cheaply in this way, much less the large aerogenerators with rotor diameters up to perhaps 200 ft. If they could be so tested the value of the test result might be somewhat doubtful; we have here to consider the purpose of power measurements and it seems clear that the results obtained, if not so precise, are more realistic if they refer to actual operating conditions in open air. The difficulty then lies, of course, in expressing the wind speed to which the measured power output corresponds.

The nature and extent of gusts or eddies in wind flow have been discussed in Ch. 9 and it can easily be understood that a completely steady wind of uniform speed over a circular area of more than 100 ft diameter is a rare occurrence. Nevertheless, evidence has been given (see pp. 144 and 145) that an approximation to this can be obtained though perhaps for only a few seconds at a time and at widely spaced intervals. With smaller machines, having rotor diameters of (say) 20 or 30 ft diameter, there appears to be a better chance of uniform wind distribution over the swept circle.

The procedure which has thus been followed in power tests on machines of this size has been to measure, with a cup-generator or windmill type of anemometer (see Ch. 8 and 9), the indicated wind speed at hub height at a point close to the machine and to measure simultaneously, with a wattmeter, the indicated output power. It is then implied that the wind speed at the anemometer is the same as that approaching the wind rotor and that this wind speed is uniformly distributed over the swept circle. By choosing periods of steady winds of different speeds—an elaboration would be to install a gust anemometer and quick response recorder to check this steadiness—and by taking a sufficiently large number of observations, a (power)/(wind speed) curve can be obtained with fair accuracy. Even within the short periods of time needed to read the instruments the wind will vary in speed, however steady it may seem to be, but the effects of its variations may be masked by the inertia of the generator and the slow response of the measuring instruments. These effects will appear, however, in the scatter of the observed values when these are plotted, with several slightly different power outputs apparently corresponding to the same wind speed. In general, the smaller the machine the greater the probable scatter of the plotted points.

J. Juul (Ref. 3) has used this method in testing the 13 kW asynchronous generator already mentioned on p. 175 connected to an a.c. network and driven by a propeller of 8 m diameter running at a constant tip speed of 38 m/sec. His results are summarized in Fig. 61 which gives curves for the theoretical power output per square metre (59·3 per cent of the power in the wind), for the actual power output of the generator and for its efficiency, all plotted against wind speed and tip-speed ratio. The shape of the power curve, showing the levelling off at high wind speeds, is interesting. The blades are of fixed pitch and the output control is through the reduction of the power coefficient at high wind speeds when the

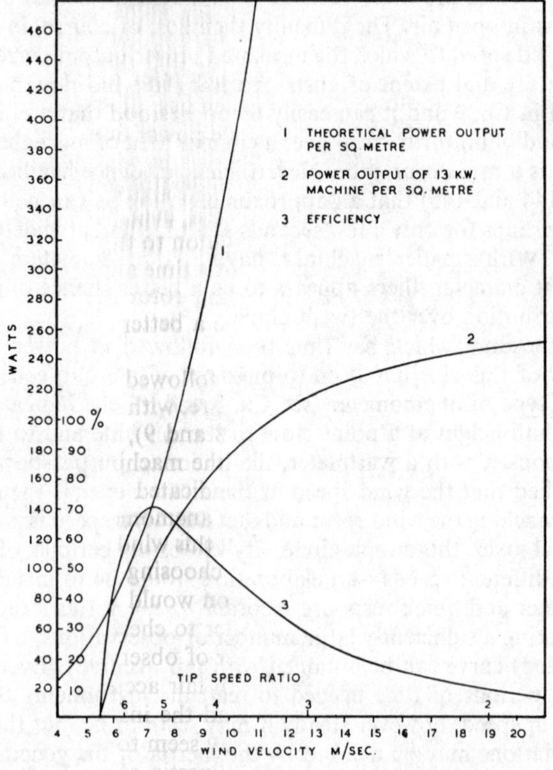

FIG. 61. *Power output and efficiency curves for* 13 *kW Danish aerogenerator*

tip speed ratio is falling more and more below its optimum value of about 5·2.

Juul also used recorders for his anemometer and wattmeter measurements and two examples of records obtained are shown in Fig. 62. In the upper records the wind is gusty and it is difficult to determine with precision the value of the power corresponding to a given wind speed. In the lower records, on the other hand, the wind being much steadier, it is easier to obtain corresponding figures. Thus,

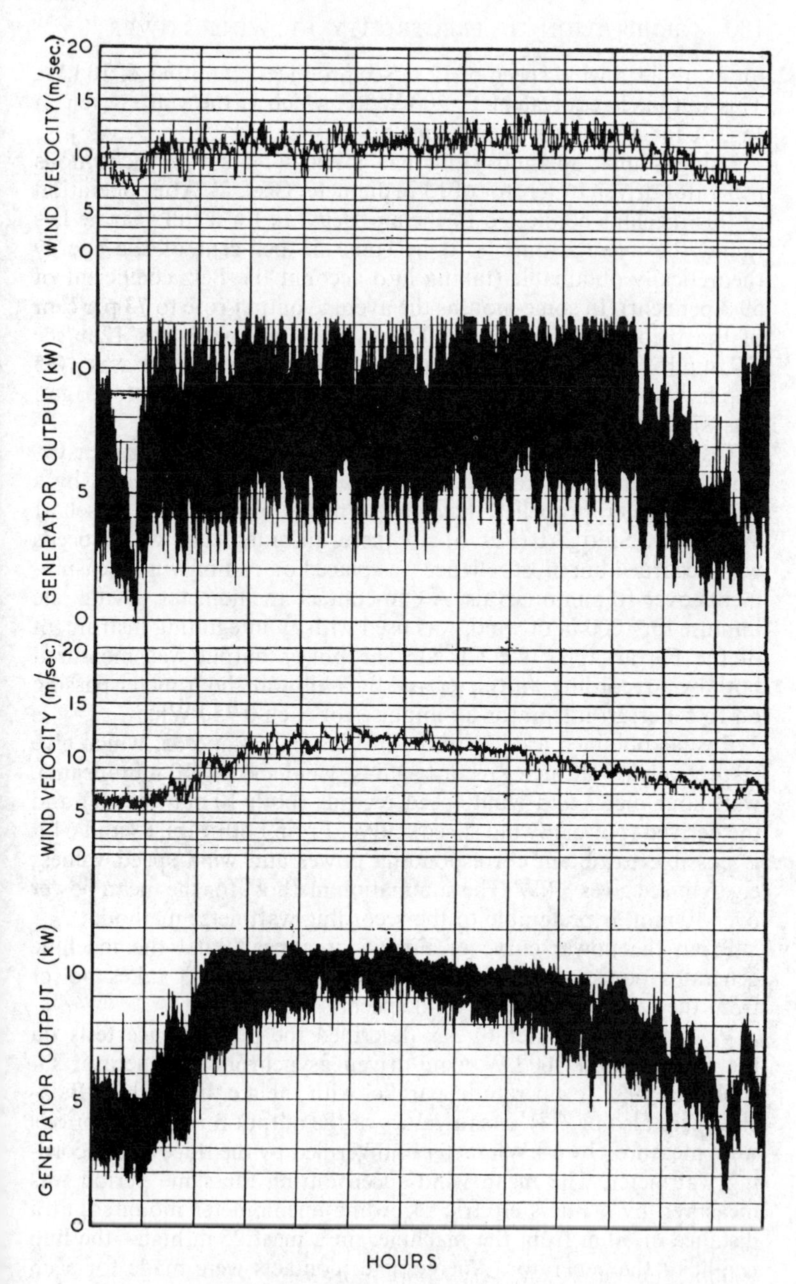

FIG. 62. *Power output and wind speed records*

for example, a wind speed of 12 m/sec produces an output of 10 kW. This output is equivalent to 200 W/m^2 which is the value given in Fig. 61 for a wind speed of 12 m/sec.

Using similar methods Juul has tested a 45 kW asynchronous generator driven by a rotor of 13 m diameter (Ref. 2). After operation of this machine connected to the a.c. network for a full year he has found the average output to be some 53 per cent of the energy theoretically obtainable (taking into account the Betz coefficient of 59·3 per cent). In some months the average output rose to 73 per cent of the theoretical maximum. The rated wind speed was 12 m/sec (27 m.p.h.) and the mean annual wind speed at the site was 10·5 m.p.h. while the total output, allowing for losses due to stoppages, was 87,170 kWh in the year.

Tests on an Allgaier wind power plant having an 8·8 kW asynchronous generator, connected to an a.c. network and driven by a 10 m diameter propeller, have been made by Studiengesellschaft Windkraft, Stuttgart (Ref. 5). In these, instantaneous wind speeds were recorded but most reliance was placed on run-of-wind measurements over 10 min intervals. A cup-contact anemometer, giving one impulse for 1000 m of wind, was used with an integrating instrument of the Ferrari type (see Ch. 8). The power output was measured both by a recording wattmeter and by a kilowatt-hour meter passing on to a Ferrari integrator an impulse for each 0·75 kWh.

Results obtained are illustrated in Figs. 63, 64 and 65, which give respectively a wattmeter record against wind speeds (10 min means), watt-hour meter and wind speed records (both 10 min means) and the derived (power)/(wind speed) curve. From both Figs. 63 and 64 it is possible to obtain corresponding power and wind speed values; e.g. 6 m/sec gives 3 kW. The summation method, for the mean power over 10 min, is preferable to the recording wattmeter method.

From the power curve of Fig. 65, it appears that the machine generates power at wind speeds above 3·5 m/sec; it takes power from the network at lower wind speeds.

V. R. Sektorov (Ref. 6) has described the performance tests on the experimental 100 kW wind-driven asynchronous generator installed in 1931 to operate in parallel with the a.c. network at Balaclava (see also p. 218). Mean values of the output for 20 min periods were measured by a kWh meter and verified by the trace of a recording wattmeter. The mean wind speed during the same period was measured by a Fuess electric recording anemometer mounted, at a distance of 50 m from the machine, on a mast 25 m high—the hub height of the generator. Anemometer contacts were made for each 500 m run of wind. Tests were not made for wind directions such that there was danger of the wind rotor screening the anemometer.

The tests proved the power output of the generator to be 100 kW in a mean wind speed of 11 m/sec.

(*c*) Tests for research and development purposes

The energy output and power generation tests discussed in the preceding paragraphs are sufficient to check the validity of the

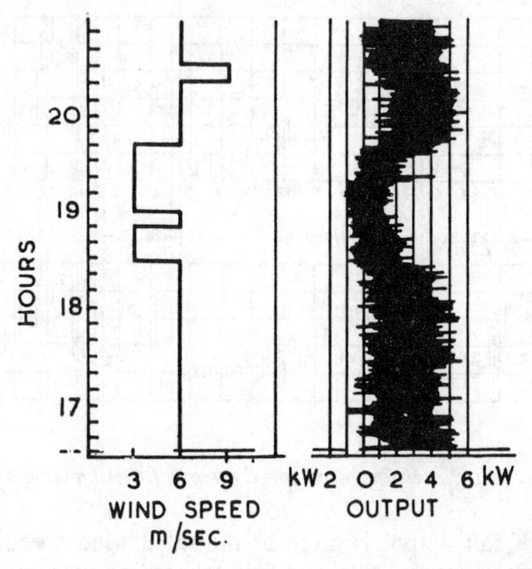

FIG. 63. *Wind speed and output curves for Allgaier machine*

methods used to estimate output from wind measurements and to determine the adequacy of the performance of small or medium-size machines.

In the development of large wind-driven generators, for use in connection with networks, there must be several stages, each employing a pilot plant, or prototype, the size of which will increase as progress is made towards the ultimate optimum size of perhaps two or three thousand kilowatts. Clearly, as much information as possible must be obtained from tests on such pilot plants so that test data can be used by the designers for the next upward stage.

One group of tests will be concerned with the details of operation of the particular machine concerned; they will be rather similar to acceptance tests on more conventional machines. Some will be, indeed, merely visual observations of the behaviour of the plant under different operating conditions. The efficacy of the control systems is important and it will be necessary to determine at the outset:

(i) If the machine 'cuts-in' at the correct wind speed, continuing to produce an increasing amount of power as the wind speed rises above that value and stopping when the wind falls below the cut-in speed;

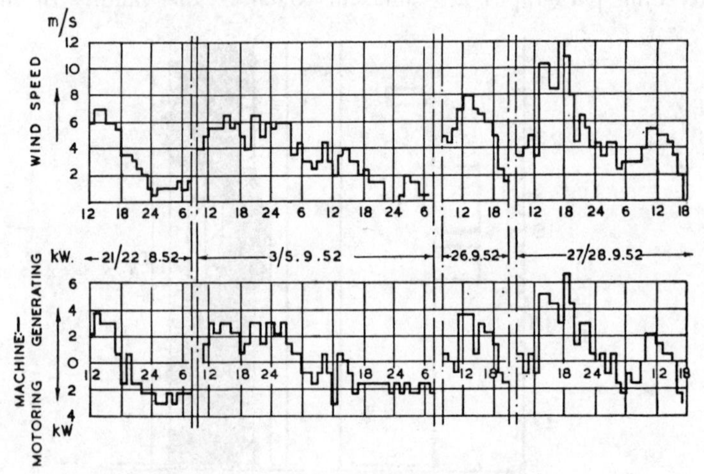

FIG. 64. *Power and wind speed records for Allgaier machine*

(ii) if full output is given at the rated wind speed and if the output is properly controlled for higher wind speeds;

(iii) if the furling mechanism operates correctly, shutting down the plant at the designed wind speed and bringing it back into operation as the wind speed falls again;

(iv) if the running of the machine is 'smooth' i.e. without appreciable vibration of the blades, as they rotate, or of the tower through a resonance effect. (There may be pulsating forces on the rotor the period of which matches the natural period of oscillation of the tower structure);

(v) if the yawing device operates satisfactorily, changing the orientation of the head of the machine as the wind direction changes;

(vi) if the protective devices function properly to safeguard the machine in the event of overspeeding, or the blade system becoming out-of-balance (due to breakage of a blade or, perhaps, to icing), or against failure of the electricity supply from the network.

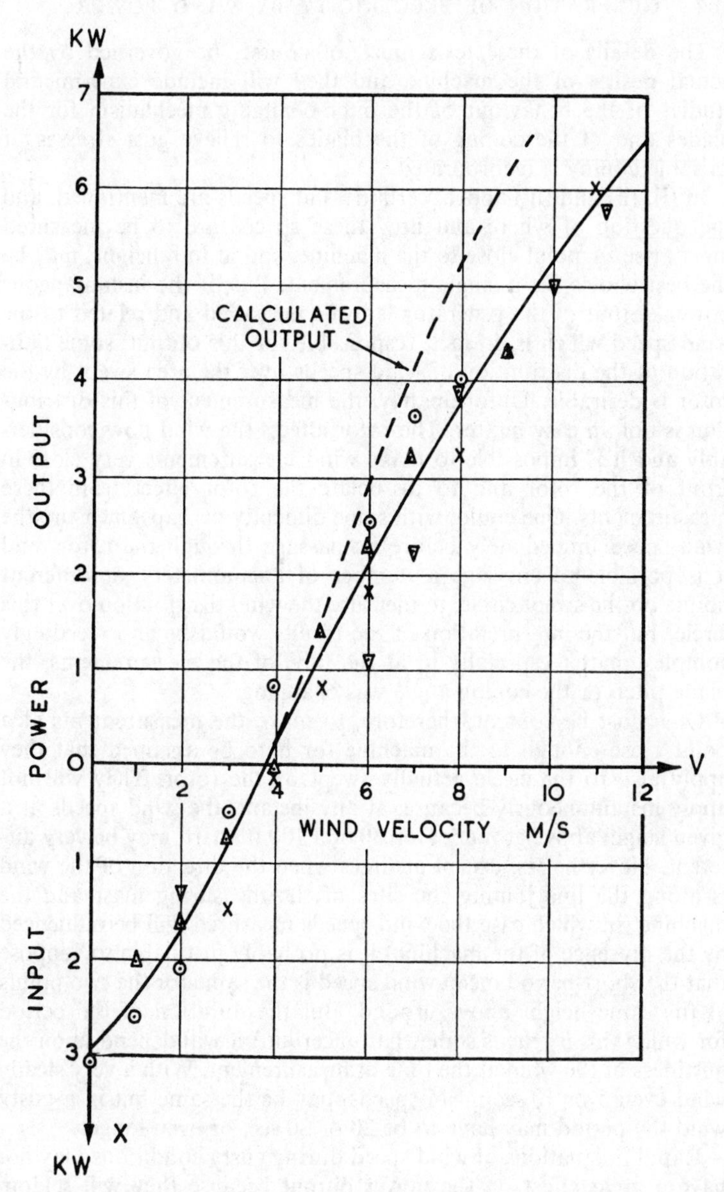

FIG. 65. *Relationship between power and wind speed for Allgaier machine*

The details of these tests must, of course, be governed by the actual design of the machine and they will include experimental studies of the behaviour of the pitch-changing mechanism for the blades and of the coning of the blades to relieve gust stresses, if variable coning is incorporated.

In (i), (ii) and (iii) above, certain wind speeds are mentioned, and the question of where and how these speeds are to be measured must arise. A point close to the machine, and at hub height, may be the best choice for a single measurement. But, if the instantaneous power output of the generator is to be measured and related to the wind speed which is, in fact, responsible for this output, some indication of the distribution of wind speeds over the area swept by the rotor is desirable. Unfortunately, the measurement of this distribution is not an easy matter. The rotor affects the wind flow considerably and it is impossible to make wind measurements very close in front of the rotor and to dissociate the rotor effect from these measurements. One could, with some difficulty perhaps, measure the wind speed immediately before its passage through the rotor, and it is possible to envisage a number of anemometers, at different points on the swept circle, to measure the wind distribution over this circle, but the interpretation of the results would be an exceedingly complex matter especially if, at the time of the measurements, the blade pitch or the coning angle was changing.

One must be content, therefore, to make the measurements at a point close enough to the machine for it to be assumed that they apply also to the circle actually swept by the rotor. They will not apply instantaneously because, at any instant, the wind speeds at a given height above ground, even 50 or a 100 ft apart, may be very different. Nevertheless, except at times when the direction of the wind is along the line joining the sites of the measuring mast and the machine (in which case the wind speeds measured will be influenced by the presence of the machine) it is probably justifiable to suppose that the short-period mean wind speed is the same for the two points at the same height above ground. But the duration of the period for which this is true is somewhat uncertain; it will depend upon the gustiness of the wind at the time of measurement. With a very steady wind even 5 or 10 sec mean speeds may be the same but in a gusty wind the period may have to be 20 or 30 sec, or even longer.

Rapid fluctuations of wind speed during gusty conditions may not have a great effect on the power output because they will seldom occur at all parts of the swept area at the same time and also because the inertia of a large rotor will not allow it to follow them. Their importance will lie mainly in the blade stresses which they cause and, if strain gauge measurements on the blades are to be made, a knowledge of the magnitudes of these fluctuations will be essential.

Clearly, quick-response instruments are needed for both wind and power output measurements and, as described in Ch. 9, gust anemometers and recorders with a response time of 0·1 sec have been developed in Great Britain.

Short period measurements of energy output (of the order of 10–60 sec) can be made together with similar measurements, at the same time, of the mean wind speed at the adjacent measuring site. Even for the same mean wind speeds, these outputs will vary with the gustiness and the result will be a cloud of points on the (wind speed)/(power) curve. Putnam (Ref. 7, Ch. 8) discusses this point in his description of the testing and operation of the 1250 kW Grandpa's Knob aerogenerator. In his Fig. 72 he gives power output records for wind speeds both above and below the rated speed and for gusty conditions.

In addition to measurements to determine the distribution of wind speeds over a swept circle corresponding to that of the windmill rotor, it may be important to measure the vertical and horizontal components of the wind speed. If the vertical component is appreciable it may have a considerable influence both on the power output of the machine and on the stresses in the structure. The rate of change of wind direction should be measured also because of its possible influence when the rate is too high to be followed by the orientation of the machine.

The measurements mentioned above have been catered for in the testing programme for the pilot plants, of 100 kW capacity and upwards, which are being installed in Great Britain, and the arrangements made can be seen in Fig. 66.

There is a 120 ft mast, which is rotatable so that the plane containing the mast and cross arm can be parallel to that swept by the windmill rotor, for wind measurements which can be related to the machine's performance, at a distance of 80 yd from it. Some 200 yd further away are 70 ft and 10 ft masts carrying anemometers so that the records from anemometers on the 120 ft mast can be linked up with those previously obtained at the site over a period of several years.

The instruments, which are indicated by numbers in Fig. 66 are listed in Table XXXI. Those on the 120 ft mast are placed mainly at heights of 110 ft, 80 ft and 50 ft, the middle group being on a 60 ft long cross arm so that instruments are located at four points on the circumference of a circle, of 60 ft diameter, with its centre at 80 ft. (This circle corresponds to that swept by the windmill rotor.)

TABLE XXXI

No. in Fig.	Instrument	Range	Form of output
1	Cup-contact anemometer	2 m.p.h. upwards	Contacts, up to 1/sec in impulse recorder circuit
2	Cup-generator anemometer	5 m.p.h. upwards	0 to 20 volts
3	Dines (pressure tube) anemometer	0 to 110 m.p.h.	Direct chart record
4	Wind direction indicator	—	Recorded photographically
5	Gust anemometer for horizontal component	30 to 75 m.p.h.	0 to 20 volts
6	Gust anemometer for horizontal component	60 to 150 m.p.h.	0 to 20 volts
7	Gust anemometer for 2 horizontal components at right angles	20 to 70 m.p.h.	0 to 20 volts
8	Gust anemometer for vertical components	—	0 to 20 volts
9	Temperature recorder	—	—
10	Balsa wood, windmill type, anemometer	2 to 150 m.p.h.	Alternating voltage of frequency up to 1000 cycles per second

For use in conjunction with the instruments listed in Table XXXI, there are recorders of several different types. The purposes for which they are used are given briefly below:

(*a*) *Photographic recorder*. This apparatus (see Ch. 8) photographs, at half-hourly intervals, the (run-of-wind) readings of four counters operated by cup-contact anemometers and also the indication (at the instant of photographing) of the wind-direction indicator located at the top of the 120 ft mast. It has been used for several years with three of the anemometers on the 70 ft mast and that on the 10 ft pole.

(*b*) *Impulse recorder*. This recorder (see Ch. 8) is located in the cabin on the 120 ft mast and is used to record the run-of-wind measured by three cup-contact anemometers mounted on this mast. The recorder has four channels so that it can also record impulses from a kilowatt-hour meter, fitted with a contact device, to register short-period outputs of energy by the generator.

The mast is rotated, through a motor drive controlled by the yawing vane *YV* (in Fig. 66), to face the wind in synchronism with the orientation of the windmill itself. The cabin, mounted on the mast, houses the impulse recorder and the recording mechanism of the Dines anemometer thus avoiding the need for slip rings and other

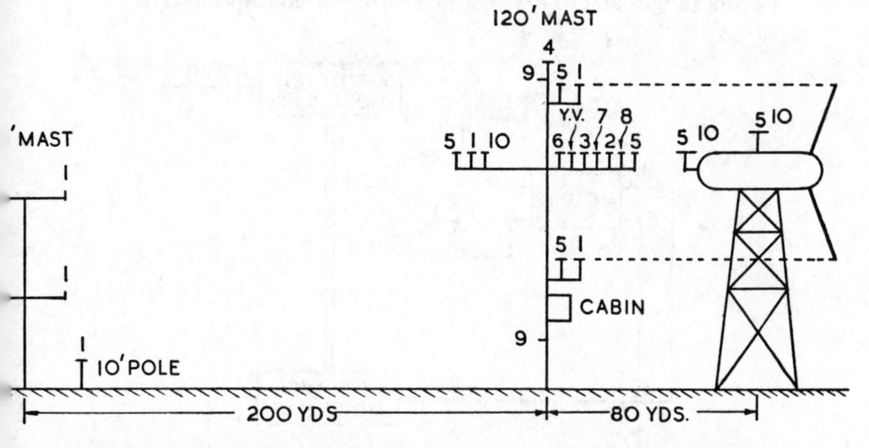

FIG. 66. *Distribution of measuring instruments at Costa Hill, Orkney*

flexible connections to stationary recorders used with moving instruments.

(*c*) *Quick-response recorder*. Two of these instruments, each with two channels, are used to record the outputs from four of: the cup-generator anemometer (No. 2 in Fig. 66) which has special light aluminium cups; the gust anemometers (Nos. 5, 6, 7 and 8); and an auto-repeater which measures power output from the generator. (This is described later.)

The recorders have a response time of the order of 0·1 sec and the four synchronized records which they give afford a simultaneous comparison of the wind flow at various points on the 120 ft mast or on the windmill nacelle with the power output of the generator. Four chart speeds, $\frac{1}{2}$ in., 1 in., 3 in. or 6 in./sec, can be used.

(*d*) *Electronic counter*. This receives the pulses (15 per revolution of the anemometer) from the balsa wood windmill-type anemometer so that it is possible to count the revolutions of the anemometer—and, therefore, to measure the run-of-wind—for short intervals between 1 and 30 sec.

(*e*) *Milliammeter recorder*. This has a chart speed of 1 in./hr and receives the output from a cup-generator anemometer with specially strengthened cups.

(*f*) *Other recorders*. In addition to the instruments for specialized purposes there are self-recording, voltmeter and frequency recorders having ranges 360 to 480 V and 47 to 53 c/sec respectively.

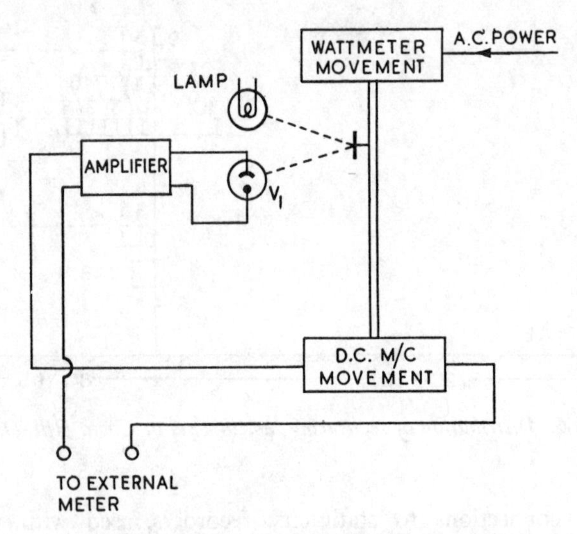

FIG. 67. *Principle of operation of auto-repeater*

The auto-repeater mentioned under (*c*) above operates on the principle illustrated in Fig. 67. The power to be measured is passed into a wattmeter movement mounted on a shaft which also carries a small mirror and a d.c. moving-coil movement. A small deflection of the system causes reflected light from the mirror to fall on a photo-electric cell the output from which is amplified and passed on to the moving-coil movement which acts as a control for the moving system. The direct current needed to furnish this control is a measure of the a.c. power and this direct current is measured on an external meter. The net result is that a.c. power up to 150 kW is measured in terms of a small direct current which can be recorded by the quick-response recorder.

BIBLIOGRAPHY

(1) War Production Board. Final report on the wind turbine. *Research Report*, PB25370, Office of Production, Research and Development, War Production, Washington, D.C. (1946).

(2) JUUL, J. Report of results obtained with the SEAS experimental wind power generator. *Elektroteknikeren*, Vol. 47, pp. 5–12 (7th January 1951). (In Danish.)

(3) JUUL, J. Supplement to report on results obtained with the experimental windmill of Sydøstsjællands Elektricitets Aktieselskab—SEAS. *Elektroteknikeren*, Vol. 48, No. 4, pp. 65079 (22nd February 1952).

(4) JUUL, J. Conversion of wind power plant from direct current to alternating current output. Organization for European Economic Co-operation, Report of the Committee for Productivity and Applied Research (P.R.A.) Working Party No. 2 (Wind Power). *Technical Paper* No. 39 (1954).

(5) Parallelbetrieb einer 8,8 kW-Windkraftanlage mit einem Stromversorgungsnetz. Studiengesellschaft Windkraft e.V., Stuttgart. *Windkraft Mitteilungen*, Nr. 1 (1 April 1953).

(6) SEKTOROV, V. R. The present state of planning and erection of large experimental wind power stations. *Elektrichestvo*, No. 2, pp. 9–13 (1933). (In Russian.)

(7) PUTNAM, P. C. *Power from the wind.* Van Nostrand (1948).

(8) ANDRIANOV, V. N. and BYSTRITSKII, D. N. Parallel operation of a wind power station with a powerful grid. *Elektrichestvo*, No. 5, pp. 8–12 (1951).

CHAPTER 12

WIND-DRIVEN MACHINES

The extraction of power from the wind

THE power in the wind at any moment is the result of a mass of air moving at speed in some particular direction. To capture this power—or a fraction of it, since it is not practicable to extract the whole—it is necessary to place in the path of the wind some machine which retards it and so brings about a transfer of power from the wind to the machine. Wind power being proportional to the cube of the wind speed, reduction of this speed diminishes the power in the wind; the power output of the machine is that lost by the wind except for the inevitable losses which must occur in the process of transference.

The 'machine' must either move along with the wind against a back pressure as in a sailing ship or, remaining stationary, must be caused to rotate on an axle carrying a load which exerts a braking effect. We are more concerned with the second kind of machine and, in this chapter, with the different forms it may take.

If we consider details of design these forms are legion but, in general principle, they can all be placed in one of two main classes, (see p. 195) each class having clearly recognizable advantages and disadvantages which are inseparable from it in spite of the hundreds, perhaps even thousands, of designs which have been used, or suggested, since the first windmill was invented several thousand years ago. This is not to say that there are not specific advantages of one design as compared with others in the same class, but much time and effort could be saved by would-be inventors if they recognized the limitations of the class of windmill within which their particular design falls.

We are indebted to G. Lacroix (Ref. 1) for an excellent review of the possible types of windmill giving a very clear exposition of their relative merits. This review, and other comprehensive studies which have been made in Great Britain and other countries, indicate clearly that the propeller type of windmill—or, rather more generally, the horizontal axis type with radial blades or sails—is likely to have the highest efficiency and to prove the most economical for power production, especially in large units. But both climatic and economic conditions vary so greatly in places where wind power could be utilized that it would be wrong to dismiss all other types as inferior.

To take one example by way of illustration of this point, the crude Chinese and Persian windmills used for water pumping were of the vertical-axis type with sails made of any suitable materials which happened to be available. They were very simply constructed, were robust enough to withstand the winds to which they were likely to be subjected and could be maintained and repaired with very little trouble. Like the old pair of shoes which had been given five new soles and two new uppers, their 'life' was indefinitely long: so what matter if their efficiency, expressed in engineering terms, was probably extremely low? They served their purpose.

Other times and other purposes demand different, and more efficient, constructions but nevertheless it is well to be prepared to adapt ideas to changing circumstances in various parts of the world.

The transference of power

A. Betz, of the Institute of Göttingen, in his studies of the windmill published in 1927 (Ref. 3) applied the simple momentum theory, established by W. J. M. Rankin and W. E. Froude (Ref. 5) for the ship's propeller, to the windmill. The following brief outline covers the more essential points.

The retardation of the wind passing through a windmill occurs in two stages, one before and one after its passage through the windmill rotor.

Thus V_1 = wind speed at a considerable distance upwind,
V = wind speed actually passing through the rotor,
V_2 = wind speed at a considerable distance down wind from the rotor.

If M is the mass of air passing through the rotor in unit time, the rate of change of momentum is $M(V_1 - V_2)$ which is equal to the resulting thrust. The power absorbed is, therefore, $M(V_1 - V_2)V$.

Again, the rate of change of kinetic energy in the wind is $\frac{1}{2}M(V_1^2 - V_2^2)$.

These two expressions are obviously equal so that

$$M(V_1 - V_2)V = \tfrac{1}{2}M(V_1^2 - V_2^2)$$

from which

$$V = \frac{V_1 + V_2}{2}$$

Thus the retardation of the wind $V_1 - V$ before the rotor is equal to the retardation $V - V_2$ behind it. (Here the assumptions are made that the direction of wind velocity through the rotor is axial and that the velocity is uniform over the area A. The argument holds good however without these assumptions.)

The power extracted by the rotor is

$$P = \rho AV(V_1 - V_2)V$$

where ρ = air density and A = the area swept by the rotor.

$$P = \rho AV^2(V_1 - V_2) = \rho A \left(\frac{V_1 + V_2}{2}\right)^2 (V_1 - V_2)$$

$$= \rho \frac{AV_1^3}{4} [(1 + \alpha)(1 - \alpha^2)]$$

where $$\alpha = \frac{V_2}{V_1}$$

It can easily be shown, by differentiation, that the power P is maximum when $\alpha = \frac{1}{3}$, i.e. when the final wind velocity V_2 is one third of the upwind velocity V_1. Thus, the maximum power which can be extracted is $\rho AV_1^3 \times \frac{8}{27}$, as compared with $\frac{\rho AV_1^3}{2}$ in the wind originally, i.e., an ideal windmill could extract $\frac{16}{27}$ (or 59·3 per cent) of the power in the wind.

Power coefficients

This simple theory takes no account of the aerodynamic losses in the rotor and it also assumes that the air velocity is uniform over the swept area and that the slipstream does not mingle with, but remains separate from, the surrounding air. The supposition is that the column of air, of cross section A, approaches the rotor, is retarded, and passes on as before but with a lower speed. It has been suggested that the column of air tapped by the windmill rotor is greater in cross section than the swept area A so that the power extracted will be greater than $\frac{16}{27} \rho AV_1^3$. On the other hand, whatever the type of rotor, its effect upon the air flow will not be so simple as the above theory assumes and power losses will occur which will certainly reduce the fraction of power extracted.

Referring to a propeller-type windmill, the uniform air flow mentioned above could be obtained only by using an infinite number of frictionless blades of negligible thickness. With a small number of blades, with friction, the power extracted must be less than the maximum. There is, in this type of rotor, a component of air flow along the radius adjacent to the blade paths. This results in an exchange of air flow round the blade tips with a consequent 'tip-loss' which, with frictional losses, may reduce the power extraction to only 70 to 80 per cent of the maximum. This tip loss has been fully investigated by S. Goldstein (Ref. 6) in his important work on the vortex theory of screw propellers and by C. N. H. Lock, R. C. Pankhurst and J. F. C. Conn (Ref. 7). These analyses followed earlier work by L. Prandtl and A. Betz (Ref. 8) and by H. Glauert (Ref. 9).

Corrections can be made for tip losses in calculating windmill performance.

The fraction of power extracted is known as the 'power coefficient' C_p and its value, in practice, will vary with the design of the rotor. From the expression for the power, developed from the simple momentum theory,

$$C_p = \frac{\rho \frac{A}{4} V_1^3[(1+\alpha)(1-\alpha^2)]}{\rho \frac{A}{2} V_1^3} = \frac{(1+\alpha)(1-\alpha^2)}{2}$$

An alternative expression which may sometimes be more convenient is $C_p = 4a(1-a)^2$, where a, called the 'interference factor', is the fractional retardation of the wind speed V_1 when it approaches the rotor, i.e. $V = V_1(1-a)$, or the retardation before the rotor (equal to that behind it) is aV_1. It is easy to show that $a = \frac{1-\alpha}{2}$.

Using the same form, the axial thrust, tending to overturn the windmill, is $\rho \frac{A}{2} V_1^2 - 4a(1-a)$ which is maximum when $a = \frac{1}{2}$. When the rotor drives an electric generator, probably through gearing, there will also be mechanical and electrical power losses which will reduce the power output of the machine. The ratio of this output to the power $\rho \frac{A}{2} V_1^3$ in the wind upstream from the windmill may be called C_{op} (the 'overall power coefficient'). This is a more exact expression for the ratio than 'efficiency' which should, strictly, refer only to the ratio of power extracted to that actually passing through the rotor but which is difficult to determine.

An interesting theory concerning the power which may be extracted by a windmill is that of the Russian Professor G. Kh. Sabinin (Ref. 10). His theory of the ideal wind wheel, which is well expounded by E. M. Fateev (Ref. 11) is based upon the vortex produced when the wind passes through the wheel and shows that the flow distortion results in an induced air velocity behind the wheel. There is a depression behind the wheel and a mass of air is sucked into the vortex from the surrounding air flow with a consequent increment of momentum (see Fig. 68). Sabinin states the effective swept area as the arithmetic mean of the cross sections of the air column before and behind the wheel. The final result of this theory is to increase the maximum power coefficient from 0·593—in the simple momentum theory—to 0·687.

Again, the War Production Board report (Ref. 12), which is concerned with the design of a propeller-type windmill, while accepting the momentum theory as a basis for rotor design provided

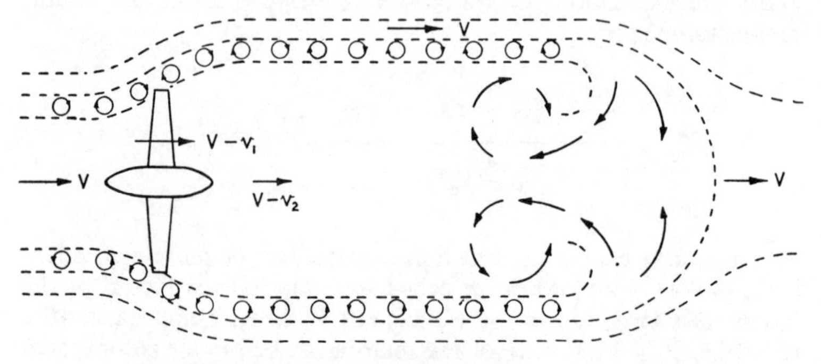

FIG. 68. *Sabinin's theory of wind flow past a propeller*

proper allowance is made for frictional and tip losses in the propeller, points out apparent discrepancies which might arise from applying it indiscriminately. The report suggests that the expression $4a(1 - a)^2$ for C_p might be modified to $\frac{4}{3}(1 - a)[1 - (1 - a)^3]$ since this leads to an expression for the thrust on the rotor which more closely approaches practical experience.

The maximum value of C_p then becomes 0·63 instead of 0·593.

Another suggestion that the power coefficient in practice is greater than 0·593 comes from P. H. Thomas (Ref. 13) who states that, owing to power extracted by the windmill from air which does not pass through the area swept by its rotor, there is a "possibility of a magnitude of output far beyond that possible from the energy content of the air passing through the wheel circle". This increased power is thought by Thomas to result from the non-uniform velocity of the air over the propeller disc.

H. H. Rosenbrock (Ref. 15) has studied the implications of the momentum theory and considered some of the reasons given for accepting a higher maximum power coefficient than 0·593. He concludes, from his analysis, that the 0·593 figure is justified and that the reasons usually given for its modification point to a decrease, rather than an increase, in its value.

Clearly experimental research is needed to clarify this question of the maximum power coefficient; there is not yet confirmation that higher outputs than those indicated by the simple momentum theory can be obtained. Indeed, direct determination of the power coefficient under practical operating conditions with a large machine would be exceedingly difficult, not only because the wind speed, and thus the output power, continually change but because it is almost impossible to measure the true up-wind velocity V_1 to be used in calculating the input power.

Types of wind-driven machines

Two main classes of machines which may be distinguished are;

(*a*) machines having the effective moving surfaces of their rotors moving in the direction of the wind;

(*b*) machines whose rotors move in a plane, or planes, perpendicular to the direction of the wind.

Class (*a*) *machines.* These usually have rotors which run about a vertical axis and which respond equally to all wind directions though there have been exceptions to these general rules. They are characterized by simple construction and by a lower power coefficient than class (*b*) machines. The principle of the machines of this class, sometimes called 'panemones' is outlined below.

When a stationary surface of area A is struck perpendicularly by a wind of speed V and air density ρ the force on the surface is $C_F\,(\frac{1}{2}\rho AV^2)$ where C_F is a coefficient of pressure whose value varies with the form, nature and size of the surface. If the surface itself moves with a speed v, in the same direction as the wind, the relative speed is $(V - v)$ and the force is then $C_F\,.\,\frac{1}{2}\rho A(V - v)^2$. The power given to the moving surface is thus $C_F\,.\,\frac{1}{2}\rho A(V - v)^2 v$ which can easily be shown to be maximum when $v = \frac{V}{3}$, i.e. when the speed of the moving surface is one third that of the wind. The power is then $\frac{2}{27}\,C_F\,.\,\rho AV^3$ which is only one third of the power $(= C_F\frac{1}{2}\rho A\,(V - v)^2 V)$ exerted by the wind, the remaining two thirds being lost in the form of eddies behind the moving surface or being carried away in the wind which passes round the surface. This means that the maximum possible power coefficient for a simple panemone is $\frac{1}{3}$.

Other disadvantages of a wind-driven machine constructed on this principle are (i) that the moving surface must always move at less than the speed of the wind; the speed of revolution of the rotor is thus low and this implies the need for expensive gearing when a high speed electrical generator is to be driven: (ii) the surfaces, which move with the wind during one half of their revolution about their vertical axis, must move against the wind during the other half revolution and this involves a further considerable loss of power.

Various devices have been employed to overcome this second disadvantage but, even so, for the reasons given and from the facts that the surfaces, moving in a circular path, are not all subjected, at the same time, to pressure from the wind acting perpendicular

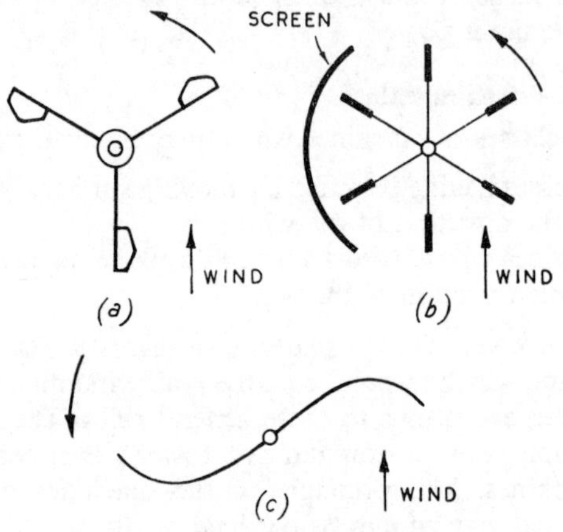

FIG. 69. *Three forms of panemone*

to them, and also to the screening effect between them, it is certain that the power coefficient, in practice, will be much less than the theoretical maximum value of $\frac{1}{3}$.

Fig. 69 shows, in simple outline, three different forms of panemone with fixed paddles. The first (*a*) is the now familiar construction used in the cup anemometer (see Ch. 8). A resultant rotating torque is produced because the wind pressure on the concave surfaces of the paddles, or cups, is greater than that on the convex surfaces presented to the wind as they move against it. In (*c*) a bent metal sheet is used instead of individual paddles but the principle of operation is the same. Diagram (*b*) shows a panemone with flat surfaces as paddles, the back pressure on them, as they move against the wind, being removed by means of a semi-cylindrical screen which must, however, be moved round by means of a wind vane as the wind direction changes.

In Fig. 70 the operation of two other forms of panemone with movable paddles is shown. The first (*a*) has hinged flat paddles (shown in plan) which swing, about a vertical axis up against a stop as they begin to move with the wind. An alternative arrangement is to mount the paddles on horizontal rods which are carried on the central axis of rotation. The paddles then turn into the horizontal plane as they move against the wind presenting only their edge to it until they have completed the half revolution when they fall to present their full surface. Both of these have the disadvantage of involving maintenance for the additional axes carrying the paddles and are

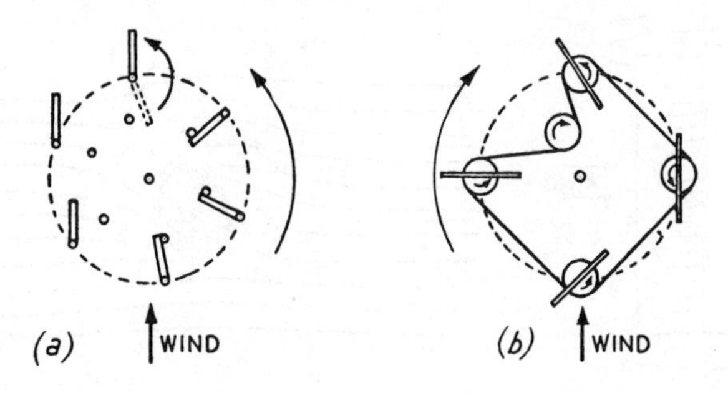

FIG. 70. *Panemones with moveable paddles*

also subjected to continual shocks as the paddles bump up against the stops.

Diagram (*b*) shows a less crude, though inevitably more complicated, form in which the inclination of the blades, or paddles, to the wind is controlled by a system of chains or belts which rotate them, about their vertical axes, through half a revolution for each complete revolution of the whole rotor. This form clearly departs considerably from the simple panemones and thereby loses one of their most important advantages—the possibility of cheap construction.

Fig. 1 on p. 7 shows an example of an ancient windmill of the panemone type while an illustration of a more modern one is given in Fig. 71. It is due to a Finnish inventor S. J. Savonius (Refs. 16 and 17) and is a modification of the simple form shown in Fig. 69(*c*).

The Savonius rotor has the two halves of the bent sheet displaced so that the wind can pass between them. Thus, in addition to the pressure on the concave surface of the half facing the wind, there is additional pressure, assisting rotation, on the back side of the other half of the rotor, thus improving the torque. A number of Savonius type machines were built some years ago but only in small sizes. The construction is still used for ventilators.

A rotor built by a Czechoslovakian inventor F. Stastik (Ref. 18, p. 264) is, in effect, a modification of the simple form shown in Fig. 69(*b*) but the axis is horizontal and the whole machine, with its screen, must orientate to face the wind. A generator of this type, one model of which gave an output of 12·3 h.p. in a wind speed of 14 m/sec, was erected near Benghazi some years before World War II but was destroyed during the war.

Although they may prove useful, and of economic construction,

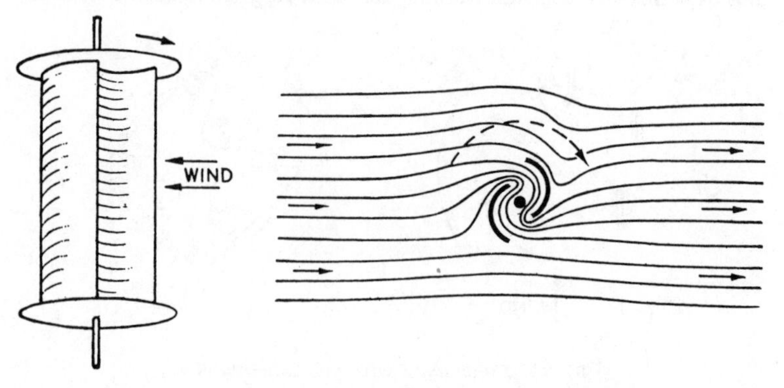

FIG. 71. *The Savonius rotor*

in small sizes, panemones are not likely to be applicable to large-scale power production. Their low power coefficient would necessitate the use of very large surfaces to obtain the required power. Not only would this be expensive in material for the blades but the machines would be difficult to construct with sufficient strength to withstand the high wind pressures encountered on very windy sites.

Class (*b*) *machines*. In these machines the active surfaces are placed at a very small angle to the relative wind instead of being perpendicular to it, and the driving force, instead of being displaced in the direction of the relative velocity $V - v$, makes an angle with it.

There are several sub-divisions of this class but the general principle of operation is outlined in the simple treatment given below.

Referring to Fig. 72, suppose that the active surface, which may be flat or, more usually, may have an aerofoil section, is placed so that it makes a large angle $(\phi + \alpha)$ with the direction of the wind. V is the velocity of the wind when it approaches the surface. (As shown on p. 192 this wind speed is, ideally, two-thirds of that at a considerable distance upstream from the windmill.) The forces brought into play result in the surface moving, with velocity v, in a direction perpendicular to V. The velocity of the relative wind is V_R which is the vector difference $V - v$. This relative velocity makes an angle α (the 'angle of attack') with the surface. A force F acts on the surface and this has two components—a 'lift' force L perpendicular to the relative velocity and a 'drag' force D parallel to this velocity. The drag causes resistance to the motion of the surface in the air with a consequent dissipation of energy in the form of eddies in the track.

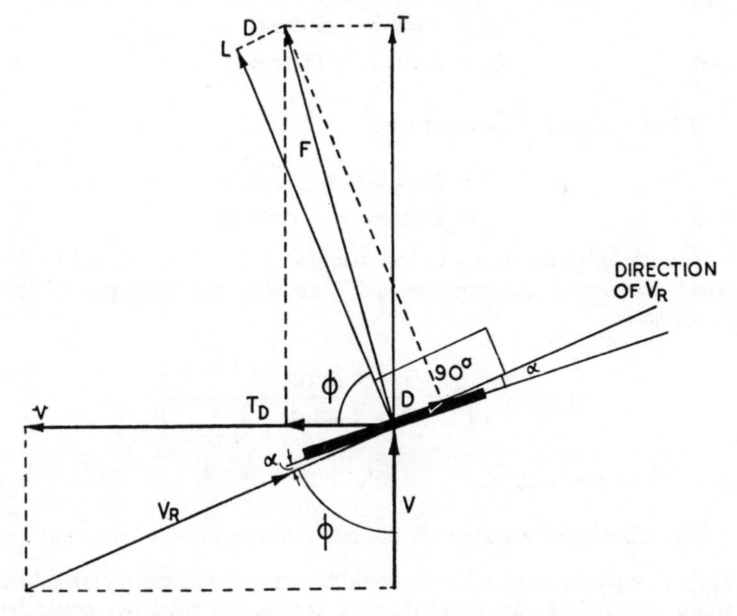

FIG. 72. *Wind velocities at a blade element*

The lift has a component which provides the driving force on the surface. Thus, profiles for the moving surfaces used in practice are designed to have $\frac{\text{lift}}{\text{drag}}$ ratios which are as large as possible and which may lie between 50 and 100.

Introducing the pressure $\frac{1}{2}\rho A V_R{}^2$ of the relative wind on the surface, of area A, (ρ = air density), these two forces can be written

$$L = C_L \cdot \tfrac{1}{2}\rho A V_R{}^2$$

and
$$D = C_D \cdot \tfrac{1}{2}\rho A V_R{}^2$$

where C_L and C_D are the 'lift coefficient' and 'drag coefficient' respectively. These coefficients can be determined experimentally for any profile and they vary with the angle of attack α.

Now, alternatively, F can be split up into two components T and T_D. The first, in the direction of the wind V, is the thrust component while T_D, in the direction of motion of the surface, is the driving component.

It can be seen that these forces are related to the lift and drag forces by the equations

$$T = L \sin\phi + D \cos\phi$$

and

$$T_D = L \cos\phi - D \sin\phi$$

If k is the ratio $\frac{D}{L}$, we can write

$$T = L \sin\phi\,[1 + k \cot\phi]$$

and

$$T_D = L \cos\phi\,[1 - k \tan\phi]$$

The useful power given to the machine is $T_D \,.\, v$ while that in the wind *arriving at the machine* is TV so that the efficiency of the machine is

$$\eta = \frac{T_D \,.\, v}{T \,.\, V} = \frac{1 - k \tan\phi}{1 + k \cot\phi} = \frac{1 - k\dfrac{v}{V}}{1 + k\dfrac{V}{v}}$$

since $V \sin\phi = v \cos\phi$.

The efficiency thus depends upon k and the value of the ratio $\frac{v}{V}$. If there were no drag, when k would be zero, the efficiency would be unity. Actually k can be made very small. The efficiency could be low if the ratio $\frac{v}{V}$ were very large or, again, if this ratio were very small—since then $\frac{V}{v}$ would be great. In practice the speed of movement, v, of the surface is large compared with the wind speed, V, at the machine, so that it is only the first of these possibilities which need be considered.

It must be appreciated that, apart from the simplifying assumptions concerning the flow of the wind which underly the above discussion, there are other factors which cause the full theory of operation in practice to be more complex than that developed here. Thus, for example, instead of the whole of the active surface moving laterally at speed v as supposed above, in any practical machine its movement must be rotational. This will introduce complications due to changing values of the relative wind speed V_R and angle of attack α at different points on its path. Again, with rotors having sails or blades mounted radially on a horizontal axis, the surface velocity v increases with the radius along the blade so that each radial element of the blade must be considered in calculating the total driving torque or thrust; the blades may, or may not, be given a slight twist to allow for this increasing velocity. Further, operation with an optimum value for the ratio $\frac{v}{V}$, to obtain maximum efficiency, implies that the two speeds, v and V, shall vary together, the rotational speed of the rotor keeping step with changes in wind speed. In practice this is very difficult, if not impossible, to achieve. If the rotor speed is constant, as it is likely to be when driving a constant-frequency alternator, clearly no attempt can be made in this way to maintain maximum efficiency in changing wind speeds and other means, such as variation of the blade pitch, must be adopted. Propeller-type machines, which are by far the most important for wind power generation, merit the fuller discussion given in the next chapter.

In spite, however, of the overwhelming importance of the propeller for machines in this general class there are some other types of machines which deserve mention.

Vertical-axis turbine rotors. During the last few decades many inventors have designed, in one form or another, wind-driven machines having a vertical axis and bearing some superficial resemblance to a water turbine. Claims have been made that the simple construction, avoidance of speed-control mechanisms, and acceptance of winds from any direction without orientation, make it superior to the propeller type with a horizontal axis.

These machines can be distinguished from the panemones because the operating forces are produced by wind which strikes the moving surfaces, not perpendicularly, but at a small angle.

Lacroix (Ref. 1) suggests that, while the precise theory of operation for such machines has not been established, they are driven by forces set up by a wind flow, through them, which may be as shown in Fig. 73. He considers that the wind flow is distorted by the Magnus effect (see p. 203) the rotor behaving like a complete cylinder. Then the wind over the arc *ab* produces only a little driving torque; the main torque is provided by the outgoing wind over the arc *bc* while the blades over arc *ca* give a braking torque since they move against the wind, though with their convex surfaces meeting it. There is thus an obvious disadvantage in low efficiency since the driving torque is due only to the difference between opposing torques. This calls for a larger swept area than that needed with the propeller type for a given power output but there is a partly counter-balancing advantage—with increasing speed of rotation the braking torque increases more rapidly than the gross driving torque so that there is inherent speed control which prevents the rotor from running at excessive speeds even in very high winds. Hence the lack of need for expensive speed-controlling devices. On the other hand the large area required for

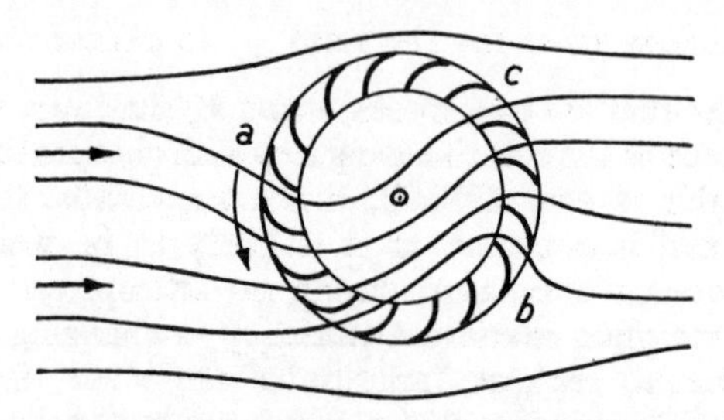

FIG. 73. *Wind flow diagram for turbine-type rotor*

the production of high power output introduces heavy thrusts in high winds thus necessitating very strong supports.

To improve the efficiency of such machines, some inventors have added a ring of fixed blades, surrounding the rotor, these blades being arranged to direct the wind on to the moving blades at the best angles. R. Vezzani (Ref. 19) has described an arrangement for the projected construction of a 500 kW wind-driven unit in Italy. In this, an attempt is made to obtain a highout put from a relatively small rotor by employing a number of fixed ducts to increase the wind speed striking it from different wind directions. The rotor is to take the form of a three-bladed propeller, of 10 m diameter, running on a vertical axis and located at the entrance to a subterranean duct, or diffuser, leading away the outgoing air with fixed or rotating aspirators to increase the effect.

Ducts with a diminishing cross section to accelerate the wind have been suggested by other inventors. They may certainly be worth while if they can be used to reduce the size of the rotor required for a given power output without themselves introducing operating difficulties and high construction costs. The last question needs full investigation before any serious consideration is given to the adoption of this form of machine.

It has been suggested by Lacroix (loc. cit.) that a machine of the vertical-axis type might have a high efficiency and be suitable for large power production if, by using blades of good aerodynamic forms mounted on a rotor running at high speed, driving torque can be produced over the whole of the circle of rotation. This may be done either by changing the blade pitch slightly as it rotates or even by using a small number of fixed blades, of suitable profile, and running at a sufficiently high speed so that the tangential velocity is much greater than that of the wind.

Other types. A. Flettner developed the 'Flettner rotors' and succeeded in crossing the Atlantic in a rotor-driven ship in 1925. He also suggested that these rotors, which are actually cylinders spinning about their axis, could be used as blades for large windmills.

Flettner rotors depend for their operation upon the 'Magnus effect' (Ref. 20). When a cylinder spins at a sufficiently high speed in a transverse wind the wind flow round the cylinder is unsymmetrical and a pressure is exerted on the cylinder in a direction perpendicular to the wind; thus the cylinder acts like an aerofoil.

J. Madaras (Ref. 21) planned to build at West Burlington, New Jersey, a very large generating unit consisting of a number of vertical Flettner rotors, each 90 ft high and 28 ft in diameter, made of aluminium and mounted on wagons moving on a circular track of about 1 km diameter. The rotors, spun by electric motors and reversed in rotation twice per circuit of the track, would be driven round the track by the wind pressures and the power would be generated through electric generators driven by the wagon axles. In 1931 the estimated cost of the plant, based on 1000 kW from each rotor, was forty dollars per kilowatt for a total capacity of 40,000 kW.

The project was advanced to the stage of a test upon a single full-scale rotor in October 1933. Although the test results confirmed the calculations on the forces brought into play, the project was abandoned, probably because of the large power losses inherent in such a system and its unsuitability for construction on windy, hill-top sites.

For full discussion of the Magnus effect and both Flettner and Savonius rotors the reader is referred to Refs. 20, 18 (pp. 302–335), 22 (pp. 60–61 and pp. 132–139) and 16.

Another interesting experimental machine, designed to work without orientation at sites such as those in the Rhône valley where the wind is predominantly uni-directional, was the Morel 'Barrage' of J. B. Morel of Grenoble. In its 100-kW prototype form it had three vertical 'Venetian blinds', each consisting of hinged slats, of aerofoil section, carried between two endless belts running over wheels at the top and bottom. All three drove a single generator mounted on a shaft at ground level. The front slats were driven upwards by the wind and the back ones driven downwards. The swept area for each 'blind' was about 30 square metres. After a series of tests over three years, beginning in 1947, the machine was abandoned, mainly because its construction cost was likely to prove excessive. Morel has since worked on the development of a wind-driven machine of a new type.

BIBLIOGRAPHY

(1) LACROIX, G. L'énergie du vent. *La Technique Moderne*, T. XLI, Nos. 5 et 6, pp. 77–83 (1er et 15, Mars 1949), et Nos. 7 et 8, pp. 105–111 (1er et 15, Avril 1949).

(2) LACROIX, G. Les problèmes électriques soulevés par l'utilisation de l'énergie du vent. *Bulletin de la Société Française des Electriciens*, Vol. V, No. 103, pp. 211–215 (Avril 1950).

(3) BETZ, A. Das Maximum der theoretisch moglichen Ausnutzung des Windes durch Windmotoren. *Zeitschrift für das gesamte Turbinen wesen*, V. 17 (20 September 1920).

(4) BETZ, A. Windmills in the light of modern research. (U.S. National Advisory Committee for Aeronautics, *Technical Memorandum* No. 474, p. 27, 2 plates (August 1928)). From *Die Naturwissenschaften*, Vol. XV, N. 46 (18 November 1927).

(5) FROUDE, R. E. On the part played in propulsion by differences of fluid pressure. *Trans. Inst. Naval Architects*, 30, p. 390 (1889).

(6) GOLDSTEIN, S. On the vortex theory of screw propellers. *Proceedings of the Royal Society*, V. 123 (1929).

(7) LOCK, C. N. H., PANKHURST, R. C. and CONN, J. F. C. Strip theory method of calculation for airscrews on high speed aeroplanes. *R. & M.* 2035 (October 1945).

(8) PRANDTL, L. and BETZ, A. Ergebuisse der aerodynamischen Versuchsanstalt zu Gottingen, 111, Lfg., R. Oldenbourg (München und Berlin, 1935).

(9) GLAUERT, H. An aerodynamic theory of the airscrew. *A.R.C.R. & M.* 786 (January 1922).

(10) SABININ, G. KH. The theory of an ideal windmill. *Transactions of the Central Aero-Hydrodynamical Institute*, Moscow. (Lock, C. N. H. A discussion of *The Theory of an Ideal Windmill*, by G. Sabinin, *A.R.C.* No. T2615 (May 1928).)

(11) FATEEV, E. M. *Windmotors and wind power stations* (Moscow, 1948).

(12) Final report on the wind turbine. *Research Report* PB25370, Office of Production, Research and Development, War Production Board, Washington, D.C. (1946).

(13) THOMAS, P. H. *Aerodynamics of the wind turbine*. Federal Power Commission (1948).

(14) THOMAS, P. H. *The wind power aerogenerator, twin wheel type*. Federal Power Commission (1946).

(15) ROSENBROCK, H. H. An extension of the momentum theory of wind turbines. Electrical Research Association, *Technical Report*, C/T105 (1951).

(16) SAVONIUS, S. J. *The wing rotor in theory and practice*. Savonius & Co. (Helsingfors).

(17) SAVONIUS, S. J. S-rotor and its applications. *Mechanical Engineering*, V. 53, p. 333–8 (May 1931).

(18) Kašpar, F. Větrné motory a elektrárny. Vol. 1, *Elektrotechnický Svaz Československý* (Praha, 1948).
(19) Vezzani, R. Study of a project of a wind power generating station of medium power driving a pumping station for hydraulic accumulation. *Elettrotecnica*, Vol. 37, pp. 398–419 (September 1950).
(20) Flettner, A. *The story of the rotor*. F. O. Willhoft (New York, 1926).
(21) Madaras, J. Possibilidades economicas do aproveitamento da forca rotora do vento no Brasil. Club de Engenharia, *Revista*, V. 1, p. 523–7 (August 1935).
(22) Witte, H. *Windkraftwerke*. Rudolf A. Lang (Possneck, 1950).
(23) Stein, D. Modern wind power generators for rural applications. *Reichsarbeitgemeinschaft Windkraft*, Berlin-Steglitz (December 1944).
(24) Stein, D. Statistics of the power production of wind power plant and their evaluation. *Elektrizitätswirtschaft*, Vol. 50, pp. 279–285 (October 1951).
(25) Parallelbetrieb einer 8,8 kW-Windkraftanlage mit einem Stromversorgungsnetz. Studiengesellschaft Windkraft e.V., Stuttgart. *Windkraft Mitteilungen*, Nr. 1 (1 April 1953).
(26) Lanoy, H. *Les aéromoteurs modernes*. Girardot (Paris, 1947).
(27) Heys, J. W. van. *Wind und Windkraftanlagen*. Georg Siemens (Berlin, 1947).
(28) Deparis, G. Générateur éolien d'énergie électrique à axe vertical. *Le Génie Civil* (15 September 1947).
(29) Champly, R. *Les moteurs à vent*. Dunod (Paris, 1933).
(30) Vezzani, R. Il problema Italiano dell'utilizzazione del vento. *Annali dei Lavori Pubblici*, XX, Fasc., 3 (Anno. 1942).
(31) Batrla, J. New designs of wind-power generators. *Elektrotech. Obz.*, Vol. 38, pp. 582–585 (November 1949).
(32) Dettmar, E. h. G. Die Elektrotechnik in China. *E.T.Z.*, Vol. 7, pp. 199–204 (12 February, 1931).

CHAPTER 13

PROPELLER TYPE WINDMILLS

Development

WHILE there may exist, under some circumstances, economic possibilities for the use of wind-driven machines of the other types described in the previous chapter, almost all the successful modern machines of small and medium sizes have been of the propeller type, and it is generally accepted that this type should be used in developing large units.

Much has been written already about the design of machines of this type and the literature will doubtless be extended when the projects for large sets, which are now being developed in several countries, have achieved practical realization. All that can be done in the limited space available here is to describe, and explain the necessity for, their main features and to give a brief survey of the most important designs which have been published, leaving the reader to consult the references given at the end of the chapter for fuller discussions of the design problems involved.

Design

Reduced to its simplest terms, the propeller-type windmill consists of a number of blades disposed radially round a shaft, to which they are attached, and which lies parallel to the wind direction so that the blades rotate in a plane approximately normal to this direction. The rotor is carried aloft by a supporting structure and provision is made for it to orientate, or 'yaw', so that it is held into the wind, and for its rotational speed to be controlled. The power developed by the rotor has also to be transmitted to the machine to be driven.

These components are dealt with separately below.

The rotor. The blades of the rotor, usually shaped to follow one of the conventional aerofoil designs whose aerodynamic characteristics are known, may be solid or fabricated, i.e. consisting of a thin skin covering a framework. They may vary in number from two to twelve or more, may be tapered or of the same chord width throughout, and, again, may be of plane form or twisted. Their pitch may be fixed or variable and they may be rigidly mounted or allowed to 'cone' or 'drag' to relieve the stresses set up by rapidly changing wind speeds.

Blade construction. The material used for the blades must be strong and yet light and must not be subject to serious deterioration in bad climatic conditions.

Wood is used for the solid blades of small rotors and, in the form of plywood skins, for the built-up blades of medium-size machines.

Plastics, being weather resistant, may be used later but have not yet been used for rotors of significant size.

Stainless steel, or aluminium-alloy, blades are the most likely alternatives for large rotors. The former, which was used for the 65·6 ft long blades of the Grandpa's Knob generator (Ref. 1), resists corrosion but presents some difficulties in fabrication, and an Alclad aluminium alloy, with a pure aluminium coating, has been recommended in the War Production Board report (Ref. 2).

Blade characteristics and rotational speed. In Ch. 12 the underlying theory of windmills was discussed but it was assumed that the whole of the active surface was moving at the same speed v when met by a wind of speed V. The combination of the two speeds resulted in a relative wind speed V_R making an angle of attack α with the surface and producing lift (L) and drag (D) forces perpendicular and parallel, respectively, to the direction of V_R. The relative wind speed has actually a third component—the local air movement associated with the aerodynamic forces on the blades. This was mentioned in the previous chapter when considering tip losses.

With a propeller-type rotor the whole of the active surface of the blade is not, of course, moving with the same speed. For a given speed of rotation N (r.p.m.) of the rotor, the speed v for an elemental section of the blade at radius r is given by $v = 2\pi \, r.N.$ so that v increases with the radius along the blade to its extremity at radius R when $v = 2\pi \, RN$. Thus, for a wind speed V, uniformly distributed over the rotor surface, both the magnitude and direction of the relative wind velocity will vary with the radius r. This means that the useful lift force L, per unit of blade surface, will vary with r.

Detailed calculations of the forces on the blades are based upon elemental sections treated as parts of an aerofoil. The lift force, for a given relative wind speed, increases with the angle of attack α until this reaches the stalling value, which may be about 15°, after which the lift decreases. Information on lift coefficients for conventional aerofoils such as would be used for windmill blades exists to enable these calculations to be made, and for tip-loss coefficients to be applied, but drag data, which depend so much upon the finish of the blade surface, are the subject of some uncertainty.

As was indicated in the discussion in Ch. 12, for high efficiency the blade sections must be shaped to have the greatest possible lift and the smallest possible drag. To extract optimum power at each succeeding section along the blade, it is thus necessary that both its shape, and the blade angle which its principal axis makes with the plane of rotation, shall be varied to suit the changing magnitudes and directions of the relative wind. The smaller the peripheral speed $2\pi rN$, the greater the angle which, for any given wind speed, the relative wind will make with the plane of rotation. It follows, therefore, that to maintain the best angle of attack, the blade angle should vary continuously along the blade and should be greatest at the root and least at the tip. Again, with a constant lift coefficient, the width of the blade should diminish continuously from its greatest value at the hub to a minimum at the tip.

The implications, therefore, are that the blades should have both a twist and a taper but, in fact, the inner portions of the blades, near the hub, contribute very little to the power extracted by the windmill. The higher cost of construction of such blades as compared with the cheaper rectangular plan-form blades, untwisted, must be considered in relation to the possible gain in annual energy which would result from their use (see Ref. 2). If the rotor has to run at a constant rotational speed, as is probable if it drives an alternator for electricity supply to a constant frequency network, a further complication is introduced. The blade angles for optimum efficiency can only be calculated on the assumption that the ratio $\frac{2\pi rN}{V}$, and hence the angle of attack, remains constant; this cannot be so if N is constant while the wind speed V varies.

Thus, as V increases, the angle of attack α will increase and so will the lift force until the stalling angle is reached, after which it will slowly fall. The power coefficient of the rotor thus varies with the relationship between rotational speed and wind speed and is most conveniently expressed, for any given blade design, in terms of the 'tip-speed ratio' $\mu_o = \frac{2\pi RN}{V_1}$ (V_1 being the undisturbed up-wind speed of the wind). Fig. 74 gives curves of power coefficients for different blades used in low and high-speed rotors and these curves exhibit optimum values of μ_o which are seen to vary from about 1 for slow-speed multibladed rotors to 2·5 for the old fashioned four-bladed windmill and 6, or more, for the high-speed rotors. If, therefore, the rotational speed of the rotor can be kept proportional to the wind speed, optimum output is obtained, but this is impossible to achieve in practice because, first, the inertia of the rotor is high and, second, the wind speed varies over the swept area.

Another difference between the forces brought into play in a multi-bladed rotor as compared with the single surface considered in Ch. 12

arises from the interference between the blades. Thus, each blade moves in an air flow which has been disturbed by its predecessor and this fact influences the number and width of the blades to be used in relation to the speed at which they are to rotate.

It has already been shown that for maximum power extraction the retardation of the wind by the rotor should be such that $\frac{\text{downwind speed}}{\text{upwind speed}} = \frac{1}{3}$. While it is not possible, in practice, to produce this retardation in all the wind passing through the rotor, the

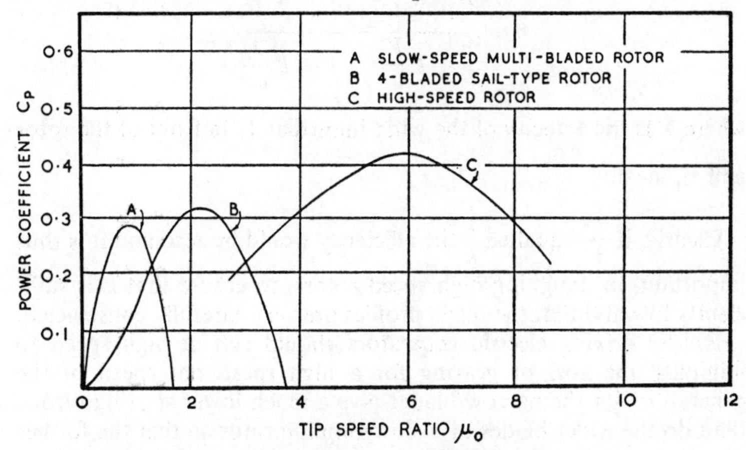

FIG. 74. *Power coefficients for different types of rotor*

design must aim at doing so. In spite of the gaps between the blades the rotor can be made to extract power from all the air passing through it. This may be done by using many wide blades rotating slowly or only two or three thin blades rotating rapidly.

Betz (Ref. 3) showed that, for optimum power extraction, the 'solidity ratio' at any radius r—i.e. the ratio $\frac{\text{blade width}}{\text{distance between blades}}$ —is

$$\frac{1}{C_L} \cdot \frac{16}{9} \cdot \left(\frac{V_1}{\omega r}\right)^2 \frac{1}{\sqrt{1 + \frac{4}{9}\left(\frac{V_1}{\omega r}\right)^2}\left(1 - \frac{3}{2}\,k\,\frac{\omega r}{V_1}\right)}$$

where C_L = lift coefficient
V_1 = undisturbed wind velocity
ω = angular velocity of the rotor

and k = the ratio $\frac{\text{drag}}{\text{lift}}$.

Since the last term does not usually differ greatly from unity, it follows that the solidity ratio depends very much upon $\left(\frac{V_1}{\omega r}\right)^2$; the greater the rotational speed of the rotor the smaller the solidity ratio. A high speed windmill, therefore, has a small number (often two) of narrow blades.

In Ch. 12 the efficiency of such a rotor was shown to be

$$\eta = \frac{1 - k\dfrac{v}{V}}{1 + k\dfrac{V}{v}} = \frac{1 - k\dfrac{\omega r}{V}}{1 + k\dfrac{V}{\omega r}}$$

where V is the velocity of the wind immediately in front of the rotor and is, ideally, $\frac{2V_1}{3}$.

Clearly, if $\frac{\omega r}{V}$ equalled $\frac{1}{k}$ the efficiency would be zero and it is thus important, in designing high speed rotors, to ensure that k is sufficiently low and that the blade profiles are very carefully constructed.

Rotors driving electric generators should run at high speed to minimize the cost of gearing for a high rotational speed of the generator. But the narrow blades give a much lower starting torque than do the wider blades of a slow-running rotor so that the former may not be self-starting. With large rotors for wind-driven electric generators, the rotational speed is limited by the need to keep the tip speed of the blades well below the speed of sound.

The War Production Board Report (Ref. 2) in its Chapter III discusses the aerodynamics of high-speed windmills in detail and reaches conclusions, based on the analysis of a number of blade designs, some of which are stated briefly below:

(i) The best number of blades is two.

(ii) The best possible power coefficient C_p with laminar-flow blades (Drag/Lift ratio = 0·005 approx.) is about 0·51 with an open rotor hub and about 0·535 if the hub is 'faired', i.e. enclosed by a streamline surface of revolution which enables power to be extracted from the wind meeting the hub area. This maximum C_p occurs at a tip-speed ratio of about 11·5 although the loss of efficiency from increasing the rotational speed above that for optimum C_p is slight.

(iii) Low-drag blade sections and optimum plan-form are of little advantage at tip-speed ratios below 3.

(iv) Constant C_p, or maximum efficiency, can be maintained with a variable-speed rotor which always rotates at a speed proportional to the wind speed, i.e. with a constant tip-speed ratio.

(v) The proper average chord width for the blade over the outer half-radius is usually more important than the detail plan-form or section. Incurring the extra cost of a tapered blade is almost certainly economic due to its greater efficiency and lower weight than a rectangular blade for equal power.

(vi) With a tapered blade a hub fairing over the inner quarter radius is worth while.

It is further concluded that more experimental data are required on the performance of windmill rotors.

The cost of construction of the blades, in the form needed to give the optimum power coefficient, must be compared with that of a rather longer blade of cheaper, though less efficient, design. The power is given by

$$P = C_p \cdot \frac{\rho}{2} AV^3$$

(where A = area swept by the blades)
and if, for example, C_p is reduced to $\frac{9}{10}$ of its value the same power can be produced by increasing A to $\frac{10}{9} A$. The increment of area when the radius R is increased by a small amount $\triangle R$ is, approximately, $2\pi R \cdot \triangle R$ and to make this increment equal to $\frac{1}{9} A$, or $\frac{1}{9} \pi R^2$, the increase in the radius $\triangle R$ must be $\frac{1}{18} R$.

Thus, in a rotor of radius 50 ft, an increase to about 53 ft radius would counterbalance a reduction of the power coefficient from 0·5 to 0·45.

Control of speed and output

Although, as already shown, maximum rotor efficiency could be maintained during operation if the rotational speed of the rotor were always proportional to the wind speed, the attainment of this condition in practice is well-nigh impossible. In variable-speed machines, such as the small sets for supply to individual premises, much depends upon matching the characteristics of the d.c. generator and battery to those of the rotor as well as upon inherent speed control of the rotor itself. This matter is discussed in the next chapter.

In larger machines directly connected to a.c. networks, when the rotational speed must be constant (unless some special type of generator is used (see p. 219)) constant efficiency, through constant tip-speed ratio, cannot be maintained though the optimum output for any particular tip-speed ratio can be obtained through variable pitch of the blades if this complication proves to be worth while.

It is, however, necessary also to limit the power output at high wind speeds to that corresponding to the maximum rated output of the generator. This may call for variation of blade pitch to bring about a reduction of efficiency and so to 'spill' surplus power. There are other ways of limiting the output. In the old fashioned sail-type mills this was done by reducing the area of frame covered by canvas when the wind was high or, as in the Danish Lykkegaard mill (see Plate VI) by the use of sails consisting of a number of hinged shutters which can move about an axis perpendicular to that of the sail and so allowing strong winds to pass through it. The rotor of a small machine may be mounted off-centre from the axis of the supporting pole or tower so that it turns out of wind at excessive wind speeds. This is done in the Lucas 'Freelite' generator (see p. 241). Another method of controlling the speed and output of small machines is to use auxiliary air-brake vanes rotating with the rotor but carried on separate arms. These are spring-controlled but fly outwards, to give a braking effect, when the rotational speed of the rotor rises above a given value (see Ch. 14). The Danish F. L. Smidth windmill generators employ a small auxiliary wing situated near the tip of the blade, on the trailing side, and normally lying flush with the blade, which is of fixed pitch. This small wing is spring-controlled and at high wind pressures it rotates through an angle from the blade surface so that it exerts a braking action. Bilau's Ventimotor (Ref. 5) also had small spring-controlled flaps near the blade tips.

In other small machines the centrifugal force on the blades, as the rotational speed rises, is made to change their pitch. The blade fixing at the hub may have a spiral socket so that the blade changes pitch as it is pulled radially outwards by this force or, alternatively, the blade spindle may be bent near the hub so that the centre of gravity of the blade lies off the axis of rotation. Pitch changing with a fly-ball governor is used in the small Winpower machines.

The controls just mentioned have been mainly for speed limitation in high winds. To return to the question of changing the blade pitch in large, constant-speed machines for a.c. generation, this pitch must be set to suit the wind speed and the load being supplied by the generator driven by the rotor. Since, in practice, the wind speed may change very rapidly, suiting the pitch to the wind speed will call for a pitch-control mechanism having an equally rapid response and actuated by the changes in wind speed and in the output. Servo mechanisms will probably be used for pitch change in such machines and the controlling element may be a small pilot windmill, which will also initiate the starting and stopping sequence of the machine, or it may take the form of flaps on the blades.

The Smith-Putnam (Grandpa's Knob) aerogenerator used a centrifugal fly-ball speed governor of great sensitivity with a hydraulic coupling which could slip and so allow a slight change of rotor speed actuating the governor. In considering, however, modifications to the Smith-Putnam unit to reduce the cost of construction, Putnam

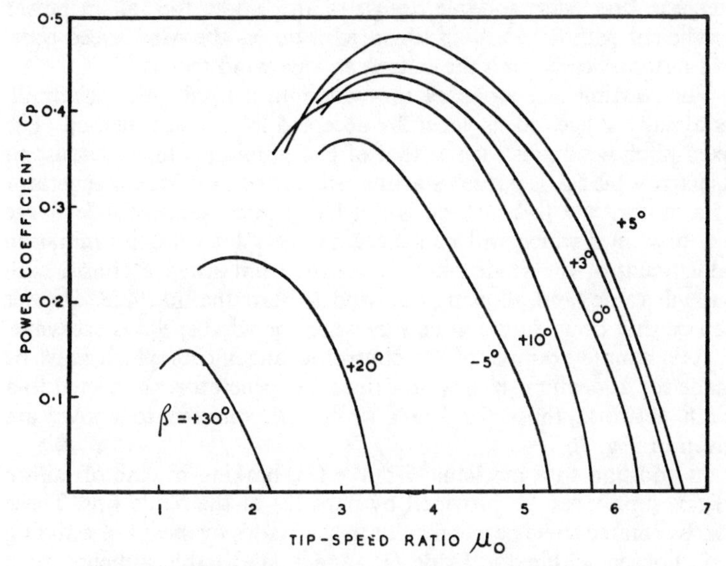

FIG. 75. *Variation of power coefficient with blade angle*

(Ref. 1) suggests that control of output by flaps, rather than by pitch control, would be cheaper.

Opinions differ about the desirability of attempting to change the pitch over the operating range between cut-in and full rated wind speed. This is probably not worth while and the trend in the most recent designs is rather to limit the pitch-changing to control outside this range.

U. Hütter (Ref. 6), in considering the influence of the frequency of various wind speeds occurring at a given site upon the choice of the most economic rotational speed for a windmill located there, shows curves similar to those reproduced in Fig. 75 for variation of power coefficient with tip-speed ratio and with blade angle. The conclusion is that a fixed blade setting of 5° is the most favourable, for the design and wind régime considered, for all tip-speed ratios. Change of blade pitch, while it might bring the operating point on to the peak of another curve, would not improve the output so that the pitch should only be changed to limit the output at high wind speeds or to facilitate starting.

J. Juul (Refs. 7, 8, 9 and 10), in describing his recent experiments involving the construction of a new prototype windmill generator for a.c. working in Denmark, follows similar lines of reasoning and also suggests that, with suitable design of the blade, the fall in power coefficient with diminishing tip-speed ratio, as the wind speed rises, will automatically limit the output at high wind speeds.

For starting and stopping the windmill, a small pilot windmill, as already suggested, is generally accepted as the best method. The fixed-pitch windmills such as that of F. L. Smidth which, because of its narrow blades, is not self starting, are started by using the generator as a motor. But this method is not likely to be successful for large a.c. machines which will be started by the pilot windmill initiating adjustment of the blade pitch, when the wind speed attains a high enough value, and allowing the wind to start the machine. The set can be shut down in the same way when the wind speed is excessive.

As a simple example of the control arrangements which must be made for a medium sized wind-driven a.c. generator, connected to a main network, those for Juul's prototype, referred to above, are given in Fig. 76.

In addition to a mechanical brake (3), braking in case of failure of the supply can be provided by flaps (2) at the blade tips. These can be rotated to lie perpendicular to the blades by means of a sliding bar, housed within the blade (1), and a steel cable attached to a lever system on the main shaft. The shaft is hollow and through it runs a movable rod connected to a piston in a compressed air cylinder supplied by an air pressure system (6) through an electro-magnetically operated valve.

When the main circuit breaker is closed the air valve is opened and sets the flaps (2) parallel to the blades ready for the rotor to run. An induction generator (5) is driven from the main shaft through a gearbox (4) and supplies its output to the network through an electro-magnetically operated switch. A centrifugally operated regulator (7) engages the generator when synchronous speed is reached while a reverse-current relay (8) disengages the generator when the wind speed falls below the cut-in point.

Blade mounting. To relieve the stresses at their roots under high and suddenly occurring wind speeds, the blades of large rotors may be flexibly mounted so that they can 'cone', i.e. bend backwards under the wind pressure instead of continuing to rotate in a plane normal to the wind. The position which they finally take up is governed by the moment of the wind pressure opposing that of the centrifugal forces although, instantaneously, the angular momentum of coning and the coning damping will influence this position.

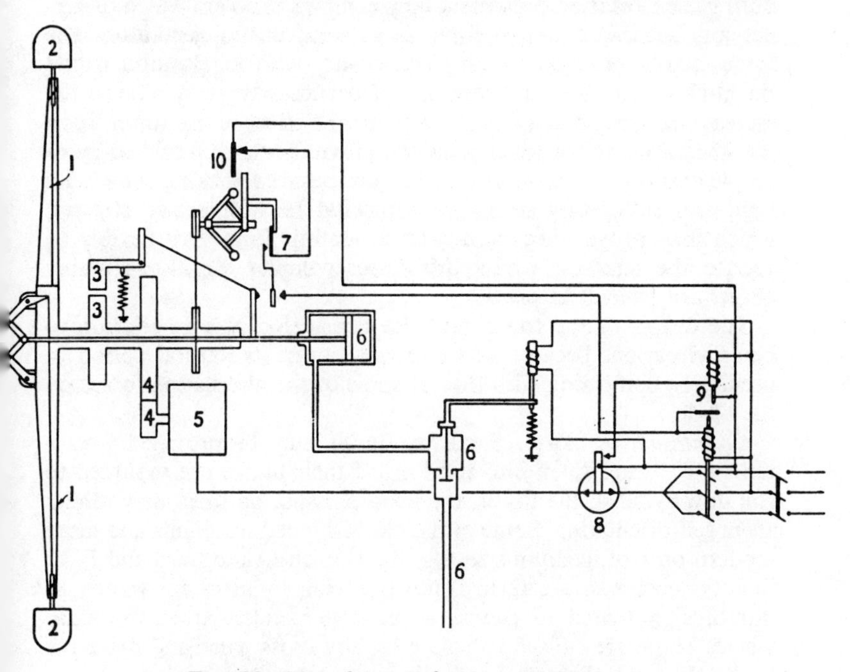

FIG. 76. *Control circuit for a.c. aerogenerator*

1. Main blades; 2. Brake flaps; 3. Mechanical brake; 4. Gearbox; 5. Generator; 6. Air pressure system; 7. Centrifugally-operated cut-in switch; 8. Reverse-current relay; 9. Auxiliary breaker switch; 10. Overspeed safety switch

When variable coning is embodied in the design, the blades are mounted down-wind of the tower to avoid the excessive shaft overhang which would otherwise be needed to prevent the blades from fouling the tower.

In medium-sized rotors, with rigidly mounted blades, there may be (as in the F. L. Smidth generators) some built-in coning which has the effect of counterbalancing the wind-pressure moment by that due to centrifugal force. Both moments increase together, with a variable speed machine, as the wind speed rises.

In addition to provision for variable coning, drag hinges may be used to give more flexibility in the mounting by permitting the blades momentarily to lag slightly behind the hub when the latter suddenly accelerates.

There is no doubt that the calculation of the blade stresses set up during their rotation presents a very complex problem which underlies any successful design for a large wind-driven generator. The forces acting on each rotating blade vary with its position round the circle, even when the wind is not particularly gusty, due to the increase in wind speed with altitude, to the effect of the tower upon the wind flow, to the reversal of the gravitational pull and to gyroscopic and other effects. The choice lies between making the blades rigid and sufficiently strong to withstand the alternating stresses, which may prove uneconomic, or mounting them very flexibly to reduce the stresses, which introduces complex vibrations whose effects are difficult to predict.

The design of the tower must be considered also in relation to blade vibrations because of the possibility of its natural period of oscillation coinciding with that of some of the alternating forces on the blades.

Orientation or 'Yaw'. Small windmills may be provided with a tail vane to turn them into wind or, if their blades are mounted to run downwind of the tower, the wind pressure on them may render them self-orientating. Some of the old-fashioned windmills and more modern ones of medium size, e.g. the Danish Lykkegaard and F. L. Smidth machines, are turned into the wind by auxiliary wheels or 'fan-tails' mounted to run in a plane perpendicular to the main wheel. These are caused to rotate by any cross wind and drive the head of the mill through a reduction gear giving a very slow orientation (the Danish mills take something approaching 1 hr to make a complete revolution of the head).

For larger machines an electrically-driven yawing motor, controlled by a small yaw-vane or by a small pilot windmill, and rotating the head of the main machine through a bull gear, might be preferable, especially in producing much quicker orientation. An objection to this is that the electric supply to the yaw motor may fail, so that the fan-tail drive might indeed be more reliable. In the Smith-Putnam machine the yawing motor was hydraulic.

Power Transmission. In all but the very small, fast-running, machines, gearing is generally necessary to provide the high rotational speed required by electric generators of normal design when driven by a slow-running rotor. (For a rotor tip speed of 500 ft/sec the rotational speed would be 95·4 r.p.m. at 100 ft diameter or 47·7 r.p.m.

at 200 ft diameter.) Such gearing is expensive and several novel means of eliminating it have been suggested. Thus, for example, in the Honnef design (Refs. 11, 12, 13, 14, 15, 16 and 17) the rotor and stator of a generator of large diameter and having many poles, were built into contra-rotating propellers to give a relative speed of rotation equal to the sum of the speeds of the two propellers. In an earlier form of the MAN-Kleinhenz design (Refs. 15 and 17) the generator rotor was built into the propeller hub and the stator was mounted in the streamlined housing. An obvious difficulty with such designs is that of obtaining satisfactory operation with any economically short air gap in the generator.

The latest design, to eliminate gearing, which has now reached the stage of actual construction, (with a rotor of 8 m diameter in a French prototype and of 24 m diameter in a 100 kW British experimental set) is the pneumatic system of the French inventor J. Andreau (Refs. 20, 21 and 22). It has hollow blades, open at the tip, and when these are driven round by the wind, air is thrown out centrifugally; the depression created draws up air through a tubular supporting tower at the base of which is a turbine driven by this ascending air. The turbine drives the generator which can thus be placed at ground level instead of aloft (see Plate XXVII). The scheme has the advantages of reducing the weight to be carried at the top of the tower, of avoiding the need for any specific relationships between the rotational speeds of the rotor and alternator and those of the wind, and of damping-out rapid changes in torque due to gusts. It may also have reduced maintenance costs through avoiding those associated with the more conventional form of drive using a gearbox and flexible coupling. On the other hand, two conversions of energy from aerodynamic to mechanical form are involved—one aloft and one at ground level—so that a rather lower overall power coefficient than that for the more conventional design must be expected.

The efficiencies which must be taken into account in this system are: (*a*) the aerodynamic efficiency of the rotor, (*b*) its internal efficiency acting as a suction pump, (*c*) the efficiency of air discharge, (*d*) the efficiency of the air turbine and (*e*) the efficiency of the generator. The overall efficiency is the product of these five and it must be compared with the product of those of the rotor, gears and generator in the more conventional type of machine. Reasonable comparative estimates for these overall efficiencies may be about 30 per cent for the Andreau and about 40 per cent for the conventional machine. In the Andreau machine it may be more easily possible to utilize economically the energy from higher wind speeds. The main limiting factor will be the tip speed of the blades.

The Danish Lykkegaard machine also has the generator at ground level, but the drive is through gearing. There is a vertical driving-shaft down the centre of the tower. This method is not, however, likely to be economic for much larger machines with high towers and long driving shafts.

Other methods of power transmission, such as by a hydraulic circuit, or by the Constantinesco system using vibrations in liquids contained in a pipe circuit, have been suggested but do not yet appear to have been adopted.

When the electrical generator is driven through gearing, a coupling which can slip should be used. Two alternatives are the hydraulic coupling and the electric induction type of coupling. Putnam (Ref. 1) suggests that the second of these should be used because of the ease with which the maximum torque transmitted through it can be controlled by the designer to any specified value.

Supporting structure. To support small wind-driven generators, either a wooden pole or light lattice tower is used. For medium sizes, lattice steel towers are commonest; concrete was used by F. L. Smidth in Denmark, but mainly because of the shortage of steel when their machines were built during the World War II. Concrete towers have the advantages of handsome appearance and a well-protected access to the machinery aloft but they are almost certainly more expensive, and more liable to damage by vibration, than steel towers.

For large units a self-supporting lattice steel tower may still be best but other forms of structure have been proposed. Both the MAN-Kleinhenz (Refs. 15 and 17) and Andreau machines are designed to have the supporting structure—in the second case tubular in form—maintained vertical by steel wire guys. The Russian Balaclava 100 kW set (Refs. 17, 23, 24 and 25) had a fixed lattice supporting tower but an inclined strut, also of lattice construction, was used to take the thrust on the windmill head. This strut, of course, rotated with the head. Its lower end ran on a circular track on the ground. Another suggestion (see also p. 260) has been to use an orientating tripod with two identical inclined legs, carrying the major portion of the weight, having their feet running on a circular track. The lower end of the third leg would remain at the centre of the track circle where the slip rings would be located. The supply cables from the generator aloft would run down this third leg which, being tubular, also provides a convenient means of ascent to the head.

The designers of several very large German wind power projects, in the decade before World War II, all appeared to have been impressed with the need to build a very tall tower, some 200 or more metres high, to ensure high and steady wind speeds for their wind turbines (Refs. 15, 16, 17, 18 and 19). To make full use of these expensive structures they proposed to mount several rotors on an orientating framework carried by a bearing at the top of the tower. P. H. Thomas, of the American Federal Power Commission has described (Ref. 26) designs for 6500 and 7500 kW wind-turbines consisting of twin wheels mounted at the two ends of a double cantilever which could revolve on a turntable to orientate the wheels. His 475 ft high tower is also intended to enable high-speed winds to be tapped. A capacity of between 5000 and 10,000 kW is considered by Thomas to be the minimum significant size for an aerogenerator connected to an electricity supply network.

The Russians, in a recent project (Refs. 27 and 28) have gone even further; A. G. Ufimtsev and V. P. Vetchinkin have proposed a large, diamond shaped, trellis-like framework mounted, with its longer axis vertical, on a narrow lattice tower with cross-arms. The structure rests on a pedestal bearing and is supported by guys attached to a bearing at the top. The trellis, which can thus orientate to face the wind, carries twelve wind turbines with rotors of 20 m diameter to give a total output of about 500 h.p. (373 kW) at a wind speed of 8 m/sec (18 m.p.h.).

Generators

Small wind-driven machines usually have d.c. generators whose characteristics are discussed in Ch. 14.

In medium and large-sized units for a.c. supply there are three main alternatives:

(*a*) an induction generator
(*b*) a synchronous alternator
(*c*) a d.c. generator used with a converter.

A fourth possibility which, so far as the author knows, has not yet been used in a practical machine, is an a.c. commutator generator which can give a constant-frequency output while running at a variable speed. This machine could perhaps be used for outputs up to 400–500 kW but commutation difficulties would probably rule it out for larger capacities. M. P. Kostenko (Ref. 29) has proposed the use of the a.c. commutator generator in this way.

It has been shown earlier that maximum energy can be extracted from a given wind régime by a wind turbine which runs at a variable speed so as to maintain a fixed design ratio between this speed and that of the wind. V. R. Sektorov (Ref. 24) has considered the relative merits, for a.c. supply to small rural networks, of the two operating

TABLE XXXII

Designs for wind-driven generators

Project or machine	Ref.*	Rotor diameter (feet)	Rated wind speed (m.p.h.)	Output (kW)	Form of rotor	Speed or output control	Optimum tip-speed ratio	Rotational speed of rotor (r.p.m.)	Tower height (feet)	Method of weather-cocking
Honnef (1945)	M	525	—	20,000	One 5-bladed, one 6-bladed with counter rotation	—	—	—	785	—
MAN-Kleinhenz (1947)	M 57	425	36	14,000 a.c. Four synch. generators	Four blades, aerofoil section	Variable pitch: servo motor	4·4	10·5	820	Servo motor and impulse mechanism
P. H. Thomas (*a*)	26	200	34	7500 d.c. to a.c.	Twin, 3-bladed	Speed control through generator field	9	Max. 42·75	475	Electric drive
designs (*b*)	26	200	28	6500 d.c. to a.c.	Twin, 2-bladed	Speed control through generator field	12	Max. 47·5	475	Electric drive
War Production Board design	2	200	30	1500 a.c. induction generator	Two blades	Blade pitch control: electro-mechanical system	12	50	150	—
Smith-Putnam	1	175	30	1250 a.c. synch. generator	Two blades rectangular	Hydraulic pitch control through flyball governor	6	29	110	Yaw vane servo mechanism hydraulic yaw motor
Russian design	23 24	164	24·6	1000 a.c. synch. generator	Three blades	Blade tip rotation by stabilizers	6	25	—	—
Balaclava machine	25	98	24·6	100 a.c. induction generator	Three blades	Variable pitch by flaps	4·75	30	76	Tail vane and electric drive

Russian VIM-GUSMP D-18 and TsAGI. 1-D-18	24	59	18	30 induction generator	Three blades	Blade tip rotation by stabilizers	4·75	40	76	Tail vane or fantail
F. L. Smidth machines (*a*)	37	79	24·6	70 d.c.	Two or three wood blades of aerofoil section	Fixed coning and pitch; speed control by flaps	9	variable	78	Lateral auxiliary wheels
F. L. Smidth machines (*b*)	37	57	24·6	50 d.c.			9	variable	78	
Lykkegaard machine	M 62	59	24·6	30 d.c.	Four wood sails hinged flaps	Flap springs adjustable for speed control	2·5	variable	66	Lateral auxiliary wheels
Darrieus (1929)	45 51	65	13·5	15 d.c.	Three blades aerofoil section	Fixed pitch	8·7	Variable Max. 50	66	—
Design and costing study†	Ch. 16	225	35	3670 a.c. induction generator	Two blades aerofoil, 2-stage taper $11\frac{1}{2}°$ twist, fixed coning	Blade pitch control by aileron fly-governors. Start/stop by pilot windmill	9·74	42·5	135 Tripod design	By fantail coupled to bogey wheels through centrifugal clutch on fluid flywheel
Enfield (Andreau type)	59	80	30	100 a.c. synch. generator	Two hollow blades	Automatic pitch control by hydraulic system. Variable coning	—	Variable. Max. 95·4	100	Self orientating but assisted by wind-sensitive power control system
John Brown	60	50	35	100 a.c. induction generator	Three tapered untwisted blades free to cone and drag	Hydraulic control of blade pitch	6·5	130	78 to hub centre	Automatic electrical control
SEAS (Denmark)	61	43	28	45 induction generator	Three blades	Blade tip rotation spring controlled	5·4	56	66	Yawing vane and electric motor drive

* M = Manufacturer's or designer's data given in descriptive brochures: other references are to the bibliography.
† This study has been made by Folland Aircraft Ltd., under the sponsorship of the Ministry of Fuel and Power.

methods—constant rotational speed, with or without variable blade pitch, and variable rotational speed with some means of obtaining constant frequency. He is particularly concerned with the gain in annual output achieved by the second method and concludes that, although it may give better utilization of low wind speeds, it does not show a considerable advantage for sites with mean wind speeds between 4 and 8 m/sec.

P. H. Thomas (Refs. 26, 30 and 31) adopts for his large aero-generator project a d.c. generator, driven through gearing, feeding a twelve-phase synchronous converter running in synchronism with the network. The conclusion in the War Production Board Report (Ref. 2) (which discusses the question of choice of generating equipment very fully in its Chapter VII) and in a design study recently made in Great Britain, is that the higher capital cost of such a scheme, as compared with either of the alternating-current alternatives, outweighs the advantage gained in enhanced output of energy the amount of which varies with the wind régime at the site. In both of these studies the induction generator is preferred to the synchronous generator mainly because of its greater stability and robustness and the ease with which it can be made fully automatic in operation—as the generating plant will need to be when it is in one of many wind-power stations located in remote districts. In the Smith-Putnam project (Ref. 1) a synchronous generator was selected on the grounds that this type was more acceptable to supply authorities and because, taking into account the cost of power-factor improvement when an induction generator is used, there would be little or no economic advantage to be gained. There is not much difference between the costs of the two machines. The induction machine may be rather more expensive than the synchronous alternator; it will give rather less energy from a given wind régime and has also the disadvantage of requiring power-factor correction for the magnetizing current which it takes from the network. On the other hand, there are justifiable doubts about the electrical stability of a synchronous generator when subjected to the rapidly-varying driving torques which will occur in a gusty wind. A sudden increase of (say) 50 per cent in the wind velocity would increase the torque—approximately proportional to (wind speed)^3—more than threefold so that the machine might well be thrown out of synchronism unless the automatic pitch-control mechanism for the blades can be made to respond fast enough to prevent this happening.

This argument applies, of course, especially to the conventional form of plant in which the rotor drives the generator through gearing. With the pneumatic system of power transmission used in the

Andreau machine the rotor and generator speeds are much less inter-dependent so that there is then a stronger case for the synchronous generator.

Main features of some recent designs. It is, of course, impossible to deal in detail with all the designs which have emerged in different countries during the last two or three decades but, for purposes of comparison, Table XXXII gives some of the most important features of a number of those which have been published.

Plates XXIV, XXV, XXVI, XXVII and XXVIII show some modern machines which have actually been built.

BIBLIOGRAPHY

(1) Putnam, P. C. *Power from the wind.* Van Nostrand (1948).

(2) Final report on the wind turbine. *Research Report*, PB25370, Office of Production, Research and Development, War Production Board, Washington, D.C. (1946).

(3) Betz, A. Windmills in the light of modern research. (U.S. National Advisory Committee for Aeronautics, *Technical Memorandum* No. 474 (August 1928), p. 27, 2 plates.) From *Die Naturwissenschaften*, Vol. XV, N. 46 (18 November 1927).

(4) Betz, A. Das Maximum der theoretisch moglichen Ausnutzung der Windes durch Windmotoren. *Zeitschrift für das gesamte Turbinen wesen.* V. 17 (20 September 1920).

(5) Lacroix, G. L'énergie du vent. *La Technique Moderne* (March–April 1949).

(6) Hütter, U. Der Einfluss der Windhaufigkeit auf die Drehzahlabstimmung von Windkraftanlagen. *Zeitschrift für Elektrotechnik*, Heft 6, 1948 und Heft 1, 1949.

(6*a*) Aéromoteurs Allgaier, système Hütter. *Le Génie Rural* (July–August 1950).

(7) Juul, J. Report of results obtained with the SEAS experimental wind power generator. *Elektroteknikeren*, Vol. 47, pp. 5–12 (7th January 1951). (In Danish.)

(8) Juul, J. Supplement til beretning om resultater opnået med Seas' forsøgsmolle. *Elektroteknikeren*, Vol. 48, No. 4, pp. 65–79 (22nd February 1952).

(9) Juul, J. Investigation of the possibilities of utilization of wind-power. *Elektroteknikeren*, Vol. 45, pp. 607–635, No. 20 (22nd October 1949). (In Danish.)

(10) Juul, J. Application of wind power to rational generation of electricity. *Elektroteknikeren*, Vol. 43, pp. 137–148 (7th August 1947). (In Danish.)

(11) Honnef, H. *Windkraftwerke.* Vieweg & Sohn (1932).

(12) Honnef, H. Hohenwindkraftwerke. *Elektrotechnik und Maschinenbau*, Heft 41/42, Wien (1929).

(13) Honnef, H. *Honnefs Kohlenfreie Elektrizitätserzeugung*. Berlin, 1947.

(14) Honnef, H. Wind power generation—new design of high-speed turbine. *Electrical Review* (17th September 1948).

(15) Heys, J. W. van. *Wind und Windkraftanlagen.* Georg Siemens (Berlin, 1947).

(16) Hamm, H. W. German wind-turbine projects planned during the Hitler era. *FIAT Final Report* No. 1111 (May 1947).

(17) Witte, Hans. *Windkraftwerke.* Rudolf A. Lang (Possneck, 1950).

(18) KLEINHENZ, F. Projekt eines Grosswindkraftwerkes. *Der Bauingenieur* (5th June 1942).

(19) KLEINHENZ, F. Gewichts und Kostenvergleich von Grosswind-Kraftwerken verschiedener Höhe bei gleichem Windrad-durchmesser. *Archiv für Warmewirtschaft und Dampfkesselwesen* (November 1943).

(20) ANDREAU, J. L'éolienne nouvelle formule n'a ni mécanisme ni engrenage. *Science et vie* (Paris, April 1950).

(21) ANDREAU, J. Utilisation de l'énergie du vent. Société des Agriculteurs de France. *Les journées d'études sur l'utilisation de la force motrice dans l'enterprise agricole*, p. 48 (21 et 22 May 1947).

(22) BOUDINEAU, A. Une solution nouvelle Andreau pour l'utilisation de l'énergie éolienne. *Revue générale de l'électricité* (May 1951).

(23) SECTOROV, V. R. Report on the operating characteristics of the initial 100 kW aero-electric unit at Balaclava. *Elektrichestvo*, No. 2 (1933) (in Russian).

(24) SECTOROV, V. R. Operating conditions and types of wind power installations for rural districts. *Elektrichestvo*, No. 10, pp. 33–37 (October 1949) (in Russian).

(25) SEKTOROV, V. R. The present state of planning and erection of large experimental wind-power stations. *Elektrichestvo*, No. 2, pp. 9–13 (1933). (In Russian.)

(26) THOMAS, P. H. *The wind power aerogenerator, twin wheel type*. Federal Power Commission (1946).

(27) FATEEV, E. M. *Windmotors and wind power stations*. (Moscow, 1948).

(28) KAZHINSKY, B. and KARMISHIN, A. Wind power multi-wheel plants. *Technika Molodezhi*, No. 12 (December 1951) (in Russian).

(29) KOSTENKO, M. P. A.C. commutator generation with frequency regulation independent of rotational speed. *Elektrichestvo*, No. 2 (1948) (in Russian).

(30) THOMAS, P. H. *Aerodynamics of the wind turbine*. Federal Power Commission (1948).

(31) THOMAS, P. H. *Electric power from the wind*. Federal Power Commission, (1945).

(32) FATEEV, E. M. and ROZHDESTVENSKII, I. V. Achievements of soviet wind power engineering. *Vestnik Mashinostroeniya*, No. 9, pp. 24–27 (1952) (in Russian).

(33) ASTA, ANTONINA. Esperienze sull'utilizzazione dell'energia del vento per produzione d'energia elettrica. *La Ricerca Scientifica*, Anno 23, No. 4 (April 1953).

(34) VEZZANI, R. Il Calcolo delle pale degli aeromotori veloci. *Annali dei Lavori Pubblici* (1944).

(35) BILAU, K. Elektrizitätserzeugung durch Riesen-Windkraftanlagen. *E.T.Z.* (2 May 1935).

(36) SERRAGLI, G. Observations on modern wind-driven electric plant. *l'Elettrotecnica* (December 10/25 1947).

(37) MÖRCH, O. V. Moderne Windkraftwerke. *Ingenioren*, No. 40 (1941).

(38) HUTTER, U. Die Entwicklung von Windrädern hoher Leistung. *Bericht* No. 4 der Ventimotor G.m.b.H. (Weimar, 1941).

(39) SCHENFER, K. and IVANOV, A. Lines of development of rural wind-power plants. *Elektrichestvo*, No. 5 (May 1941) (in Russian).

(40) COLAMARINO, G. Possibilità di utilizzazione dell'energia eolica mediante generatori a corrente alternata. *Rendiconti della XLVIII, Riuniune annuale dell'AEI*, Rapporto No. 134 (Torino, 1947).

(41) CARRER, A. Utilizzazione dell'energia del vento con schema tipo Ward-Leonard a funzionamento invertito. *L'Elettrotecnica* (August 1949).
(42) CARRER, A. Generatori a corrente continua par l'utilizzazione dell'energia del vento. *L'elettrotecnica* (August 1949).
(43) BAIRSTOW, L. *Applied aerodynamics*. Longmans Green & Co. (1939).
(44) PIERCY, N. A. V. *Aerodynamics*. English Universities Press, 2nd Edition (1947).
(45) LACROIX, G. Les éoliennes électriques Darrieus. *La Nature* (Paris, 15 December 1929).
(46) MONNEY, C. R. Le problème des aéromoteurs. Son importance économique en France. Principes concernant l'étude et l'emploi des aéromoteurs. *Bulletin de la Société des Ingénieurs civils de France*, No. 1/2 (1943).
(47) MONNEY, C. R. *Les aéromoteurs. Mémoires de la Société des Ingénieurs civils de France* (March 1943).
(48) Congrès du Vent. (Carcassonne, 1946).
(49) BASIAUX, P. L'énergie du vent dans l'Aude. *La Nature* (Paris, November 1945).
(50) LACROIX, G. Les problèmes électriques soulevés par l'utilisation de l'énergie du vent. *Bulletin de la Société française des électriciens* (April 1950).
(51) DARRIEUS, G. Utilisation de l'énergie éolienne. Ses perspectives, principes et modes de réalisation. Conférence. *Bulletin de la Société d'encouragement pour l'Industrie nationale* (7 May 1953).
(52) CARLEVARO, E. Recenti progressi dei motori a vento e futuri aviluppi. *Energia* (February 1946). *L'Elettrotecnica* (December 1946).
(53) VEZZANI, R. Le caratteristiche costruttive delle grandi centrali aero-elettriche. XLVIII Riunione Annuale dell'A.E.I. *Memoria* No. 132 (1947).
(54) ACKERET, J. and CAILLE, C. Untersuchungen am Modell eines Windkraftwerkes. *Schweizerische Bauzeitung* (22 July 1939).
(55) STAMBACH, E. Ueber Windkraftwerke. *Schweizerische Bauzeitung* (22 March 1952).
(56) CHRISTALLER, H. Utilization of the wind energy.* *Elektrizitätswirtschaft*, Vol. 50, No. 11, pp. 320–322 (1951).
(57) KLEINHENZ, F. The utilization of wind power by high altitude wind power stations.* *Technik*, Vol. 2, No. 12, pp. 517–523 (1947).
(58) KLOSS, M. Problems of the electrical design of wind power stations.* *Technik*, Vol. 2, No. 11, pp. 471–479 (1947).
(59) Wind-Generated Electricity. Prototype 100-kW Plant. *Engineering*, Vol. 179, No. 4652, p. 371 (25th March, 1955).
(60) VENTERS, J. The Orkney Windmill and Wind Power in Scotland. *The Engineer*, 27th January, 1950.
(61) JUUL, J. Wind machines. Paper given at the UNESCO Symposium on Wind and Solar Energy (New Delhi, 1954).
(62) STEIN, D. Developments in the utilization of wind power in Denmark. *Elektrizitätswirtschaft*, Vol. 41, pp. 346–349, 370–374 and 390–392, 5th and 20th August and 5th September, 1942.
(63) GOLDING, E. W. and STODHART, A. H. The use of wind power in Denmark. *Electrical Research Association, Technical Report* C/T112 (1954).

* Translated title.

CHAPTER 14

PLANTS FOR ISOLATED PREMISES AND SMALL COMMUNITIES

RECENT interest in wind-power development in Great Britain and some other industrialized countries has centred mainly on the practicability of using large units connected to electricity supply networks and some designs for these have been described in the preceding chapter. But there is considerable interest also in the economic possibilities of medium-sized sets of capacity between (say) 10 and 100 kW which could be used either alone or in conjunction with some other sources of power to supply small communities on islands or in thinly populated districts for which an electricity distribution system would be prohibitively expensive.

Again, there are in existence many thousands of small wind-driven generators, of up to about 3 kW capacity, which have been introduced and used successfully during the last two or three decades for electricity supply to isolated premises.

This chapter is concerned with such medium and small machines, which have rather different characteristics from the proposed large units, particularly because some degree of energy storage, to cover calm periods, is involved. Such storage frequently calls for accumulators and, hence, for a direct-current generator. Variable speed of rotation for the generator is thus quite practicable, and is commonly adopted, instead of the constant speed which is demanded by an alternator giving a constant-frequency supply.

Siting and installation of small sets

Another distinguishing feature of these small installations is that the generator voltage is low. This, combined with the low power capacity, implies that they must be located near to their load so that the choice of site for them is more limited than that for large machines.

To illustrate this point, consider a 1500 W generator with a terminal voltage of 32 volts. The full output current is $\frac{1500}{32}$ = 46·9 A. If a 7/·064 cable, of cross section 0·0225 in.^{2} is used, the voltage drop per yard (double) run of cable would be 0·102 v. Thus, even a 50 yd cable from the generator terminals to the load point would mean a voltage drop of 5·1 V with a consequent power loss of nearly 16 per cent of that generated. The loss could be reduced by using a heavier cable but this might add an appreciable fraction to the cost of the installation. (The cost would be of the order of five shillings per yard run, or £12 for 50 yd.) Incidentally, the internal wiring of the premises must also be done in rather heavy cable on low voltage systems such as 32 V or less.

The wind-driven generator should be mounted on a rigid pole or lattice-steel mast, well guyed to resist high wind pressures and to minimize vibration, and should be high enough (usually at least 30–40 ft) to be well clear of buildings, trees or other obstructions which may interfere with the wind flow. Choice of the highest, and best exposed, site close to the premises is very important in getting a good output. The machine should *not* be mounted on a roof top: the wind flow over the roof, at low height, is usually turbulent and the thrust on the machine may be sufficient to cause damage to the roof in high winds.

Capital and annual costs for small machines

It is somewhat difficult to state, with any precision, comparable capital costs, per kilowatt capacity, for small sets of various designs made in different countries. Apart from the facts that prices, when quoted, are often f.o.b. at some foreign port and that the equipment included, additional to the wind rotor and generator, is not always clearly stated in the advertisements, the rated wind speed is often indefinite. Thus an unfair comparison can be made in considering the costs of machines claiming the same capacity: one may produce its full output in a wind of 15 m.p.h. and another at 20 m.p.h. or even more. The second, having a smaller rotor, may be cheaper but will give much less annual output than the first. Some manufacturers try to avoid this difficulty by stating the annual output as well as, or even instead of, the capacity but this is unsatisfactory because the output varies so much with the site chosen. It would be helpful if the rotor diameter, rated wind speed and generator capacity were always clearly stated, and even more helpful if the cost for an alternative rated wind speed and generator capacity could be quoted though, admittedly, some changes in the design might be needed to fit a different size of generator with the same rotor.

From some recent quotations (1954) it can be concluded that the capital costs vary from about £150 per kilowatt for machines of a few hundred watts capacity, including batteries and control gear, down to £100–£120 per kilowatt for machines up to 10 kW capacity including tower and control gear but not batteries.

To include transport, erection, and wiring, a cost of about £150 per kilowatt is probably a fair estimate for a complete installation over this range of capacity. Provision for storage in the larger sizes may increase the kilowatt cost but the tendency is rather to install only limited storage, to cover essentials such as lighting, using the machine for power purposes when there is sufficient wind.

The annual charges on the plant for depreciation, interest and maintenance should be split up according to the different lengths of life which might apply to the various components. Thus, on £150 capital, a reasonable distribution of the depreciation charges (taken as flat rates since sinking funds are scarcely applicable to such small capital sums) may be as follows:

	£	*s.*	*d.*
£30 with life 20 years	1	10	0
£80 with life 15 years	5	6	8
£40 with life 6 years	6	13	4
Total depreciation	£13	10	0

To this must be added

	£	*s.*	*d.*
Interest charges (at 4 per cent per annum)	6	–	–
Maintenance charges (1 per cent)	1	10	0

The total annual charges may thus be £21 per kilowatt or 14 per cent of the initial cost.

Outputs of small and medium-sized machines

C. A. Cameron Brown (Ref. 1) in tests upon nine types of wind driven generators, of capacities from 250 W to 10 kW, at Harpenden in 1924–5 found that specific outputs varied from 686 kWh per annum per kilowatt to 1710 kWh per annum per kilowatt.

With A. H. Stodhart, the author has investigated (Ref. 2) the question of what outputs could be expected from wind-driven generators designed for a given range of wind speeds when these are located in different wind régimes. A later report (Ref. 3) by the same authors gives curves for annual outputs, plotted against annual mean wind speeds, for three rated wind speeds. These curves, constructed from wind data for British sites, have already been reproduced in Ch. 3. In Fig. 77 they are shown with some check points added to compare British output figures with some from other countries.

The specific outputs (kilowatt-hours per annum per kilowatt) for these check points, derived from the publications referred to, are based upon test results on small machines in the countries of their origin. These are listed, with notes, below:

Canada. Reports (Refs. 4, 5 and 6) of the University of Saskatchewan give the results of tests on 1000 W, 32 V. Wincharger machines in operation in Saskatchewan at sites with annual mean wind speeds of 11·4 m.p.h. With a rated wind speed around 30 m.p.h. the annual output was about 1320 kWh/kW.

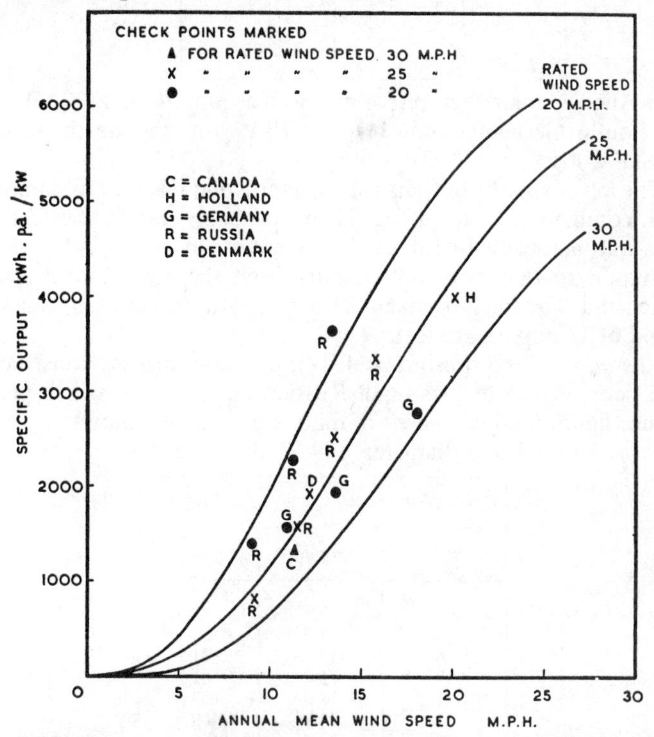

FIG. 77. *Specific output/wind speed curves for different parts of the world*

Holland. A report (Ref. 7) of the Prinsenmolen Commissie gives the results of wind measurements at Noordwijkerhout, near the Dutch coast, and estimates from these the annual output which would be obtained from a 50 kW, F. L. Smidth, wind-driven generator if it were located there. This output is 4000 kWh/kW from a machine having a rated wind speed of 24·6 m.p.h. The mean wind speed for the year at the place and measuring height (70 m) is not stated but, from results for a period of two and a half months, this may be estimated as 9 m/sec or 20 m.p.h.

Germany. Two German plants, (*a*) Nordwind (18 kW at 18 m.p.h.) and (*b*) Allgaier (8 kW at 22 m.p.h.) are now commercially available. The manufacturers of the former give the annual outputs for three annual mean wind speeds as follows:

Annual mean wind speed (m.p.h.)	Annual output (kWh)	kWh/kW
11·2	28000	1550
13·5	35000	1930
18	50000	2790

The Allgaier manufacturers give specific outputs of 1200 kWh/kW for south Germany and 2670 kWh/kW for the north German coastal districts.

The lower specific outputs for the Nordwind machine as compared with comparable ones in Fig. 77 are probably due to the fact that it is automatically furled at the exceptionally low wind speed of 27 m.p.h. so that potential outputs from the higher wind speeds, which must be fairly frequent at a site with an annual mean wind speed of 18 m.p.h., are lost.

Russia. A. V. Karmishin (Ref. 10), has described the construction and performance of two small Russian machines for which performance figures and estimated outputs are given as follows:

(*a*) Rotor 1·5 m diameter, 109 W at 8 m/sec.

Wind speed		Output
m/sec	*m.p.h.*	*watts*
4	9	2·5
5	11·2	13
6	13·5	31
7	15·7	61
8	18·0	109
9	20·1	109
10	22·4	109

Annual mean wind speed		Annual output	Annual specific output
(m/sec)	*(m.p.h.)*	*(kWh)*	*(kWh/kW)*
4	9	154	1410
5	11·2	250	2290
6	13·5	395	3620

(*b*) Rotor 3·5 m diameter, 1000 W at 11 m/sec.

Wind speed		Output
m/sec	*m.p.h.*	*watts*
5	11·2	120
6	13·5	200
7	15·7	340
8	18	500
9	20·1	700
11	24·6	1000

Annual mean wind speed		Annual output	Annual specific output
m/sec	*m.p.h.*	*kWh*	*kWh/kW*
4	9	855	855
5	11·2	1610	1610
6	13·5	2500	2500
7	15·7	3315	3315

Denmark. J. Juul (Ref. 11), in estimating the annual outputs obtainable from 17·5 m diameter, 50 kW, and 24 m diameter, 70 kW, F. L. Smidth machines if located at a good Danish site for which the detailed wind data is available, gives figures of 1892 kWh/kW and 1910 kWh/kW. The annual mean wind speed, from Juul's figures, for a height of 20 m, is around 5·4 m/sec (12·2 m.p.h.).

Juul's figures for the measured output of a 13 kW wind-driven generator during eight months of operation connected to a network (Ref. 12) have already been given in Ch. 11. The corresponding annual output would be 616 kWh/m^2 of swept area. On the basis of these tests the design of a 24 m diameter, three bladed, 100 kW generator was outlined which, with a rated wind speed of about 11 m/sec (24·6 m.p.h.), would give an annual output of 280,000 kWh, or 2800 kWh/kW.

Usable outputs

Records of annual output for the Danish 30, 50 and 70 kW windmill generators during the years of the World War II (Ref. 13) show that the maximum annual outputs of these machines were about

1800 kWh/kW but, since these supplied local networks with a very small night load, so that the windmills were usually shut down during the night, this figure does not represent the maximum output which could be obtained.

This introduces a point which is most important in connection with small and medium-sized machines, as distinct from large sets connected to extensive networks which can absorb the wind-generated power at all times, namely, that the loading schedule must be carefully planned if full advantage is to be taken of the annual energy available. Haphazard loading can result in only half—or even less—of the energy being used; the remainder is dissipated in overcharging the batteries when these are used for storage. This fact is clearly recognized in one of the Saskatchewan reports (Ref. 4) which divides the measured output of a windmill generator between periods during the day and the night. It is especially important to plan the loading if a wind-driven plant is to supply a small network with only limited storage to cover essentials such as lighting. The power requirements of the community must be studied in relation to the probable occurrence of wind which points to the need for careful analysis of the hourly wind speeds over several years.

Costs per unit of energy: battery storage

The two factors, annual charges and annual output, determining the cost per kilowatt-hour, have already been discussed.

The conclusions to be drawn are that, for small sets (under 10 kW) the annual charges may be about £21 per kilowatt while the annual specific output (for sites with annual average wind speeds around 11 m.p.h.) may be about 1000 kWh/kW. Thus, 1000 kWh costs £21 which means a cost per kilowatt-hour of $\frac{21 \times 240}{1000} = 5{\cdot}04$d. per kilowatt-hour. Unless the site is exceptionally windy, this cost should probably be regarded as a minimum since it implies that all the available energy is utilized. Especially with small sets (of only a few hundred watts capacity) when full battery storage, to cater for calm periods, is provided, the efficiency of the battery is very important.

A Saskatchewan report (Ref. 4) gives performance figures for a 1000 W, 32 V Wincharger machine under test conditions with the load adjusted to suit the state of charge of the battery. In twelve months during which the average wind speed was 10·2 m.p.h. the energy produced by the machine was 1105 kWh and that used from the battery was 956 kWh, i.e. the average battery efficiency was 86·5 per cent.

If, however, a small machine is at a site with a lower average wind speed, say only 6 or 7 m.p.h., the wind speeds at which charging of the battery begins and at which full output is reached should be correspondingly low—about 4 m.p.h. and 15–20 m.p.h. respectively (as compared with the more usual limits of about 6–8 m.p.h. for cut-in to between 20 and 25 m.p.h. for full output). Otherwise, long periods with no charging may cause loss of charge in the battery, even with no load being taken from it, so that its average efficiency may fall to around 50 per cent. This doubles the cost per kilowatt based on the annual energy produced by the generator.

Since the cost of the battery may sometimes approach that of the machine itself and, if not properly maintained, its life may be only two or three years, it is important that the type of battery to be used should be decided with care and that adequate attention be given to it in service. The main causes of rapid deterioration are over-charging, over-discharging and being left without charging.

There are two types of lead-acid battery which could be used. The Planté type is less liable to lose capacity due to the shedding of active material by its plates than the Faure type but it may suffer more easily from sulphation and requires good conditions of service. Probably the Faure type, constructed for heavy duty with a generous size of plates, will give a longer life when used with a wind-driven generator.

The Nickel-iron alkaline battery is usually more expensive than the lead-acid type and has the disadvantages of an inferior charge–discharge voltage characteristic and higher internal resistance. On the other hand it is mechanically robust and less liable to damage from the irregular operating conditions which may be met with in wind-power work. It is not damaged by over charging, nor by occasional over-discharging, has no self-discharge and is less liable than the lead-acid type to damage by frost.

The capacity of the battery installed depends upon the capacity of the generator and upon its voltage, as well as upon the number of consecutive hours of calm which may have to be covered by supply from the battery. The generator voltage may be 6–12 V for sizes up to 500 W and 24, 32 or 110 V for the larger machines. The battery capacity may vary between 130 and 450 Ah.

To summarize, the factors mainly influencing the cost of the energy are (*a*) the annual mean wind speed at the site (*b*) the operating range of wind speeds for which the machine is designed (*c*) the loading schedule adopted, including the storage capacity provided (*d*) the initial cost per kilowatt of the plant and (*e*) the percentage allowed for annual charges, including maintenance.

Generating costs for medium-sized sets are discussed in Ch. 15.

Design features of small wind-power plants

To minimize the size, and cost, of the generator for a given output, it should be driven at high speed. In the smaller machines, with propeller diameters of only 6–8 ft, which can be designed for high rotational speeds, gearing can be dispensed with, but it becomes necessary with larger rotor diameters. The rotor tip-speed is a limiting factor. Thus, for example, the tip-speed for a 6 ft diameter rotor running at 1200 r.p.m. is 377 ft/sec or 257 m.p.h.

The set runs at a speed varying with the wind speed up to the point of full rated output of the generator. For higher wind speeds the rotational speed must be controlled by some form of governor. The generator voltage rises with speed until, at the cut-in point, it is slightly above that of the battery so that a charging current is supplied. Thereafter, rising speed causes a gradual rise in the generated voltage with consequent higher current output but the terminal voltage is held approximately constant at that of the battery to which the generator is connected. There are thus three main characteristics to be considered in studying the performance of the set:

(i) the rotor characteristic, the power produced increasing with wind speed in a way which depends upon the rotational speed of the rotor and upon the type and design of governor;

(ii) the generator-load characteristic which varies with rotational speed;

(iii) the battery-load characteristic.

The energy extracted from the wind, and eventually supplied to the load circuit, is influenced by all three of these characteristics so that the derivation of the overall efficiency of the rotor-generator-battery combination is rather complex.

Rotor characteristics. The family of curves of Fig. 78, given by G. Lacroix (Ref. 14) and obtained from tests upon a 10 m diameter fixed-bladed rotor, show the typical variation of the power developed by the rotor for different rotational speeds and wind speeds. Their peaks, as would be expected, occur at the same tip-speed ratio, which has the value 8 in this particular design. Lacroix draws attention to the way in which the curves coincide on their left-hand side and suggests that, if a machine of this design could be operated over this coinciding range, a number of different wind speeds could be covered with only a small change in rotational speed. But the rotor would not then be worked so as to give optimum output. For this, the power output curve must pass through the peaks of the family of curves.

With direct drive (or neglecting losses in the intermediate gearing) the power output from the rotor is identical with the power input to the generator. Again if, for ease of consideration, one thinks of the generator as 100 per cent efficient, its output to the battery and load circuits for different rotational speeds should follow this optimum

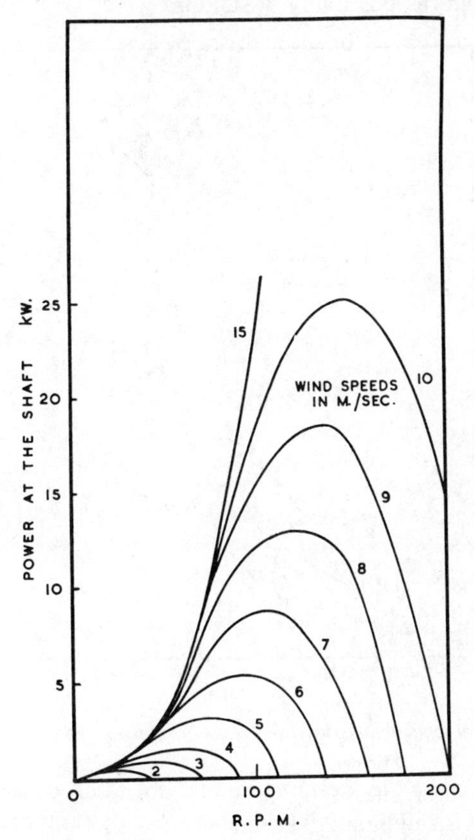

FIG. 78. *Power/rotational speed curves for different wind speeds*

curve through the peaks. The power in the wind being proportional to the cube of the wind velocity, the output of the rotor will be proportional to this cube if its efficiency remains constant. This implies that the power requirement of the generator driven by the rotor should be proportional to the cube of its rotational speed.

Clearly the rotor and generator characteristics must be matched to produce the best effect.

Fig. 79 shows the power characteristics for the 6 ft, fixed-pitch, wooden propeller used in the Lucas 'Freelite' set and also the curves of power input and output for the generator used with it. The curve of maximum power for the rotor is shown dotted and the set operates with a generator curve well to the right of the curve through the peaks to avoid the possibility of stalling.

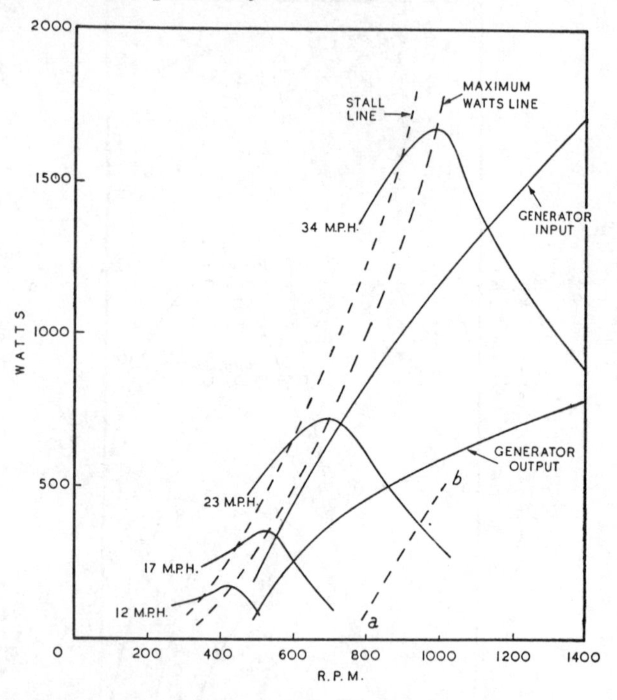

FIG. 79. *Power characteristics for Lucas "Freelite" wind-driven generator*

Generator and battery characteristics. If a shunt generator is used connected to a battery, a great increase in output current will occur for a small increase in rotational speed. This can be seen from the shunt generator characteristics shown in Fig. 80 on which the battery characteristics, for three states of charge, are shown. Taking the 50 per cent charged condition, a 25 per cent increase in the speed of the generator from 0·8 N to N increases the charging current from $I_{.8N}$ to I_N, i.e. approximately five-fold in this instance. Although the current variation differs a little with the state of charge of the battery, the effect persists. Since the battery terminal voltage rises only slightly with charging current, it is close enough to say that the power given out by the generator increases about five-fold for this relatively small rise in rotational speed. Thus a plain shunt generator would result in a steeply rising curve as *ab* in Fig. 79. If the generator characteristics in Fig. 80 fell more steeply with increasing current output, or if the voltage rise with the speed were smaller, this curve of power output would rise less steeply. Instead of a plain shunt

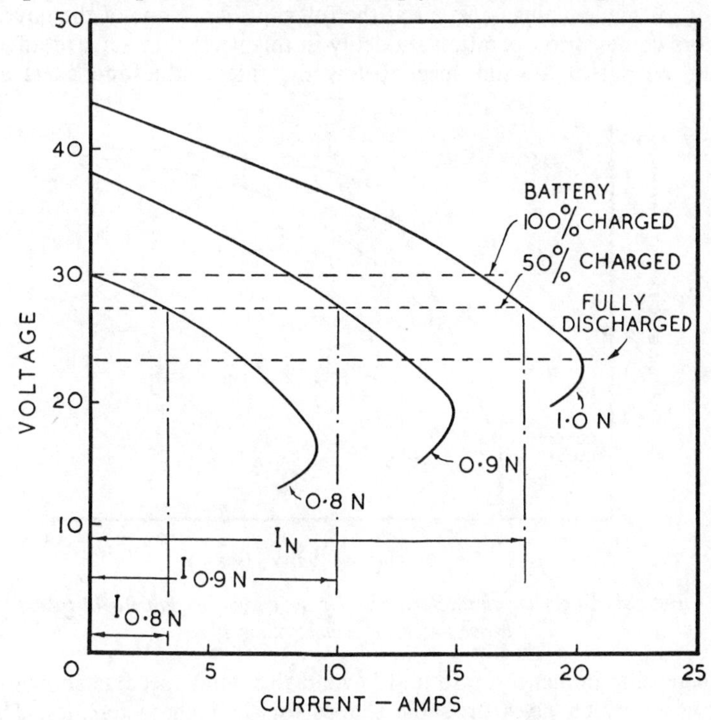

FIG. 80. *Generator and battery characteristics*

machine it may, therefore, be better to use either one which is compound wound or a three-brush generator. The former can be given a characteristic which falls at a rate depending on the number of series demagnetizing ampere-turns on the field. In the three-brush machine, the voltage applied to the field winding is reduced as armature reaction, rising with the output current, diminishes the magnetic field. Lacroix (loc. cit.) discusses this question and describes an automatic regulator, used by G. Darrieus of Compagnie Electro-Mécanique with the machine referred to on p. 234, for variation of the generator excitation.

Governor characteristics. One of the three forms of governor—(i) wind spilling, (ii) air brake or (iii) centrifugallv-operated pitch change—described in Ch. 13, is used to limit the power output to the full rated capacity of the generator. Ideally the governor should come into operation only at the rated wind speed and should operate by reducing the efficiency of the rotor so that, at all higher wind speeds, the output remains constant at the full capacity. None of the governors comes into operation suddenly in this way; they start to affect the output at a wind speed below the rated value and exert an

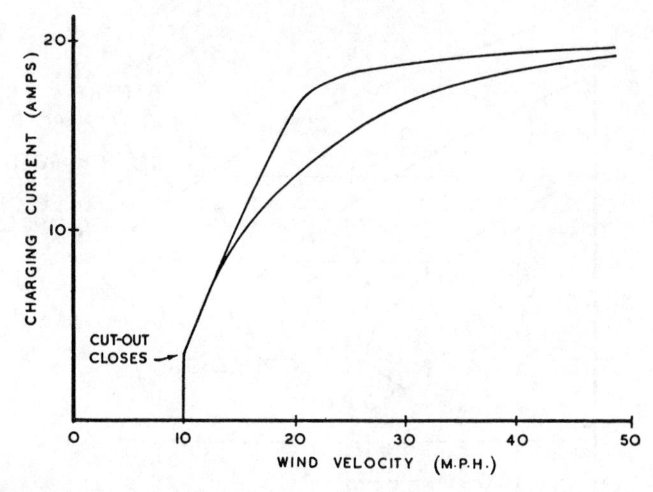

FIG. 81. *Governor characteristics: Upper curve for centrifugal types; lower curve for wind-spilling type*

increasing influence up to and beyond this wind speed as shown in Fig. 81 which gives probable shapes for their characteristics. The wind-spilling type of governor is more gradual in action than the other two types so that it gives a rather low aerodynamic efficiency over the upper part of the output curve.

In Fig. 82 typical power/speed curves, such as those in Figs. 78 and 79, are converted to torque/speed curves and constant efficiency curves are shown dotted: the curve of maximum efficiency corresponds to that passing through the peaks of the power/speed curves. In this figure no governing is assumed.

Fig. 83 and 84 (for which the author is indebted to an unpublished report, on the efficiency of small wind-driven generators, prepared some years ago by Messrs. Merz and McLellan) show the effects of the two forms of governor (wind spilling and centrifugally-operated)

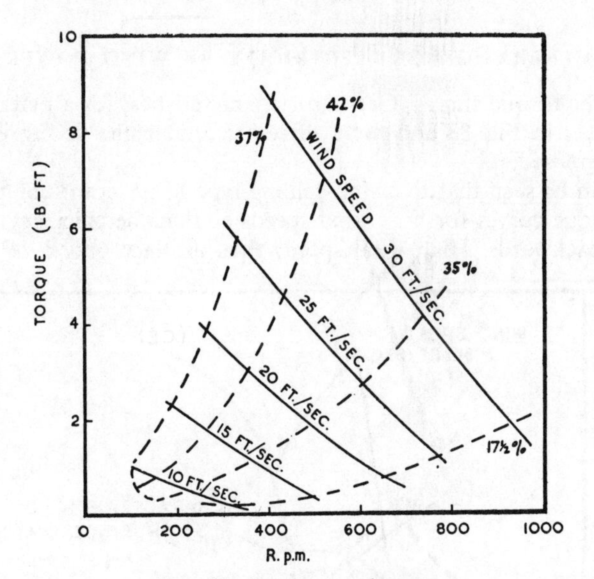

FIG. 82. *Torque-speed curves*

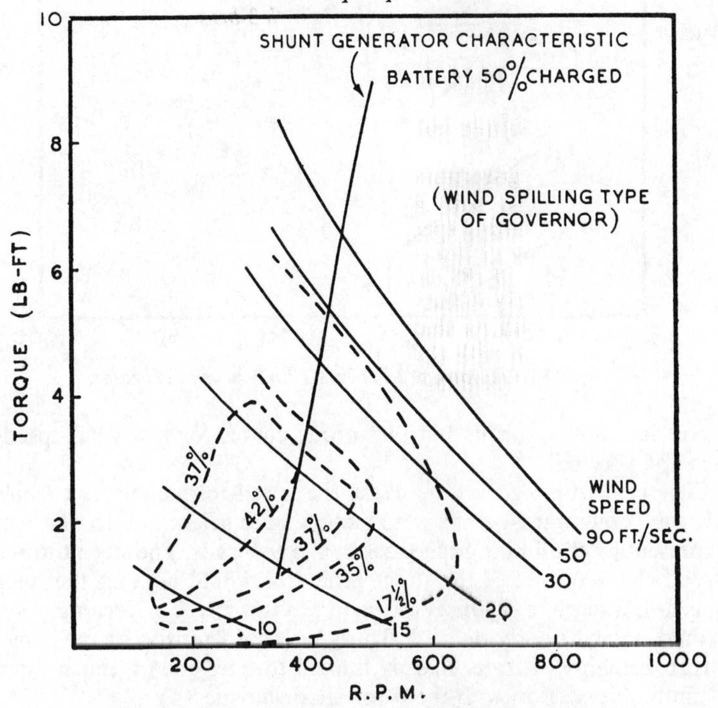

FIG. 83. *Torque-speed curves with shunt generator*

upon the torque curves. Generator characteristics, for a plain shunt generator in Fig. 83 and for a three-brush machine in Fig. 84, are superimposed.

It can be seen that the wind-spilling type of governor compresses the torque curves for high wind speeds so that the efficiency curves bend backwards. High wind speeds thus produce only a relatively

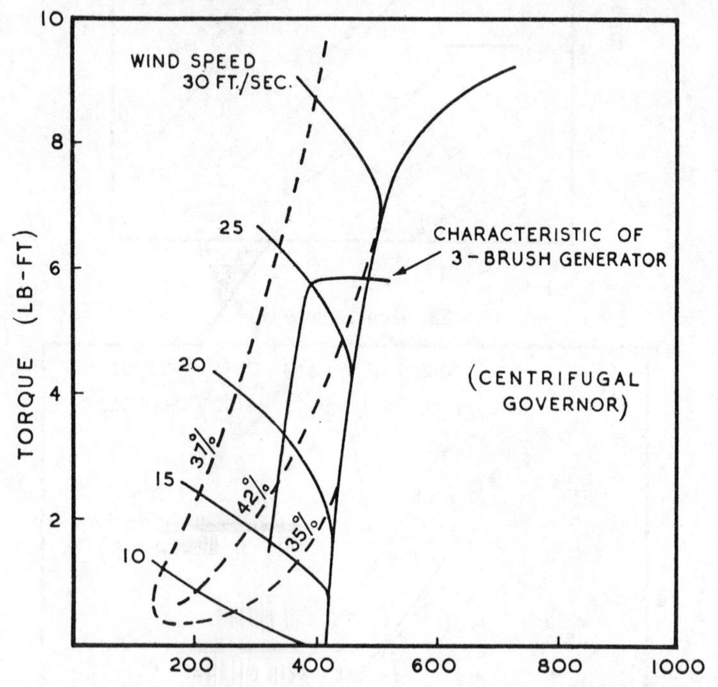

FIG. 84. *Torque-speed curves with 3-brush generator*

small increase in torque but the torque curves for low wind speeds are not affected.

The centrifugal governors cause the speed-torque curves to turn sharply downwards, with a reduction of efficiency, into the line representing the limiting speed set by the governor. The steep torque-speed characteristic of the shunt generator would not suit the steep speed-limiting line of the centrifugal governor: the governing point would not be clearly defined. Thus, the characteristic of the three-brush machine, with its sharply limited torque, gives a much more definite intersection with the rotor characteristic.

Wind-spilling governing, which turns the rotor out of wind by rotating it about either a vertical or horizontal axis, may be extended to a rotation of 90° so as to stop the machine when the wind speed becomes dangerously high, or at other times. It can be so operated by a furling handle at the base of the tower (see Fig. 85). In some small

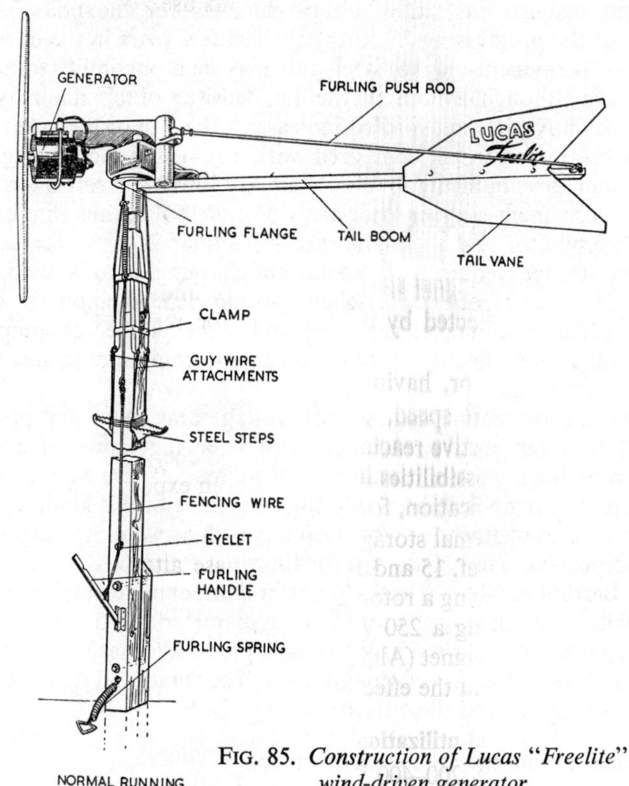

FIG. 85. *Construction of Lucas "Freelite" wind-driven generator*

sets a mechanical brake, similarly operated from ground level, is used.

Overall efficiency of small sets. The aerodynamic efficiency of some of the small, fixed-bladed rotors may have an optimum value of between 40 and 45 per cent. But since, as already shown, they cannot be operated so as to attain this over the working range, a value of about 35 per cent, as a maximum before governing begins to decrease the efficiency, is probable.

The efficiency of small-power, low-voltage, direct current generators is low and may be only about 55 or 60 per cent even at full load. The overall efficiency of a small set of a few hundred watts capacity may thus be a little under 20 per cent.

The necessary exciting current for the generator is an important source of loss and several suggestions have been made to use permanent magnets instead of electro-magnets for the poles. As a result of the progress made during the last few years in the development of permanent-magnet steels this may be a possibility for small machines although, even now, the flux densities obtainable may not be sufficiently high to avoid an increase in the size of the generator, for a given power, as compared with one using electromagnets. A further development of this theme is to use a small alternator with a permanent-magnet rotor of six or more poles, thus eliminating the commutator and its auxiliaries. A rectifier is then necessary if battery storage (requiring direct-current charging) is to be used. The economic success of such a scheme would depend upon the costs of the permanent-magnet steel used, and of the rectifier, as compared with the saving effected by omission of the commutator and field-magnet windings.

Such an alternator, having both voltage and frequency proportional to rotational speed, offers scope for regulation of current output by the inductive reactance of the circuit. There would appear to be interesting possibilities in using a machine of this kind, without a battery and rectification, for purposes such as water heating which would provide thermal storage for the wind energy.

N. Bertholon (Ref. 15 and 16) has built an experimental machine, the 'Altervent', having a rotor of 2·5 m diameter and three, variable-pitch, blades driving a 250 VA, single-phase alternator with a six-pole, permanent magnet (Alnico) rotor. The rotational speed is from 600–1000 r.p.m. and the effective voltage 24 V.

The load circuit and utilization of energy

With sets of only 200–400 W capacity, having low-voltage battery storage, the function is merely to supply lighting and a small amount of power for such purposes as a battery-operated radio set. Thus, for example, the Lucas 'Freelite' is intended for use with up to six lighting points—three 25 V, 40 W bulbs and three 25 V, 25 W bulbs. The wiring circuit for this set is given in Fig. 86 which shows the cut-out to bring in, and cut-out, the generator as its voltage exceeds, or falls below, the charging voltage. The provisions for eliminating radio interference are also shown.

In larger sets, with only partial storage, arrangements must be made to change over from battery charging to power supply when the wind-speed is high enough to provide power surplus to battery requirements. Thus, for example, the Allgaier machine has a three-phase alternator (380/220 V) of 6–8 kW capacity and a 220 V d.c. exciter of about 2 kW capacity which can be used to charge a battery serving lighting and small power needs. The circuit arrangements suggested by the manufacturers are as shown in Fig. 87. The

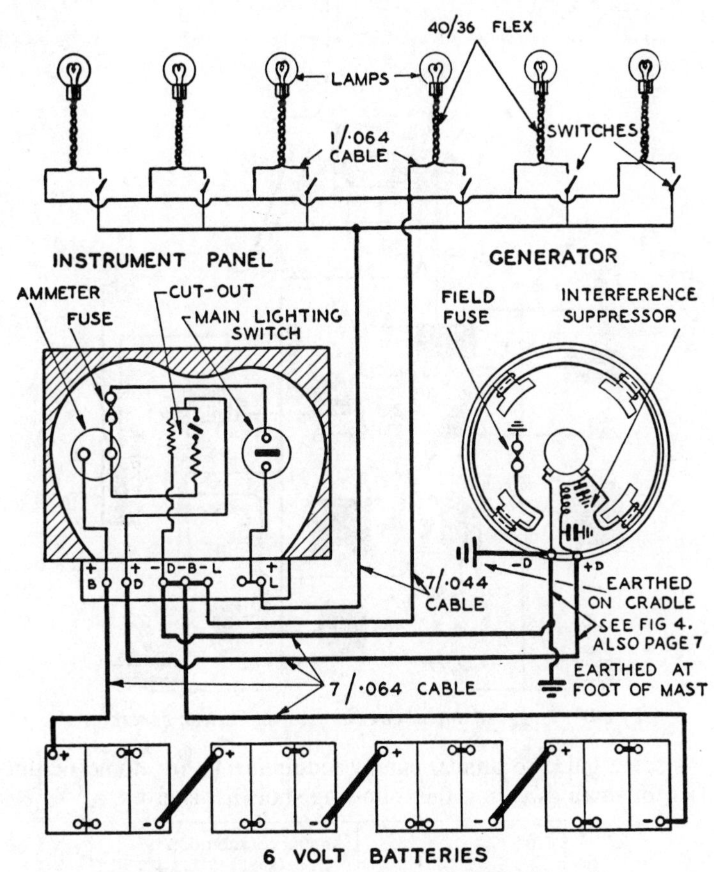

FIG. 86. *Wiring diagram for Lucas "Freelite" machine*

propeller drives the alternator (3) and exciter (4) through gearing with a speed regulator (2) which, it is claimed, keeps the speed so constant that directly-connected incandescent lamps do not flicker. The exciter is connected, through a reverse-current circuit-breaker (5), to the battery (6) supplying a circuit containing lamps, radio receiver, electric iron, vacuum cleaner and washing machine.

The alternator supplies pumps for water supply, crop-spraying equipment, a water heater, thermal-storage cooker, a refrigerator and, for use when there is sufficient wind, a fodder chopper, a threshing machine and a soil warming unit for horticultural use.

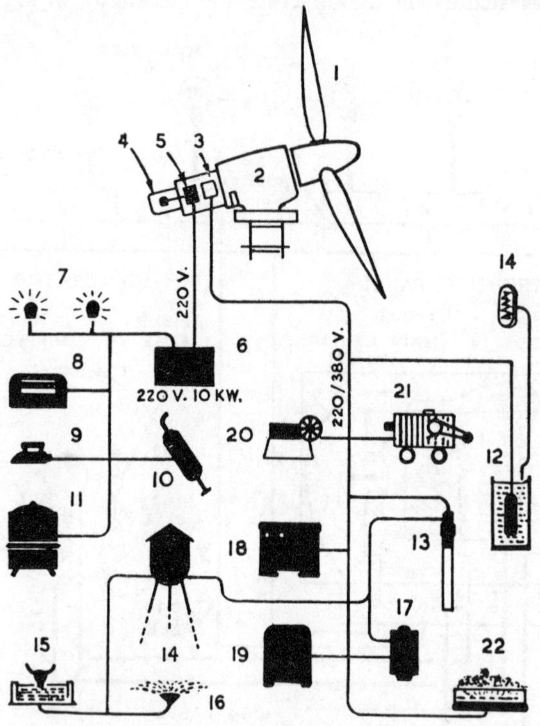

FIG. 87. *Suggested load circuits for wind-driven generator*

As some guide to annual energy requirements, for domestic purposes, of small dwellings the following short list is given.

Purpose	Annual consumption (kWh)
Lighting	100 to 200
Lighting and small domestic equipment	200 to 350
Cooking	1500 to 3000
Water heating	500 to 1500
Refrigerator	250 to 750
Iron	100
Water pumping	200

For the annual energy requirements for various purposes on farms see Refs. 17 and 18.

It must, of course, be borne in mind that low voltage (24 or 32 V) sets may demand specially-constructed equipment if they are to be used for power purposes.

The question of utilizing, to the best advantage, the annual output of energy from medium-sized wind-driven machines is discussed in Ch. 17.

BIBLIOGRAPHY

(1) CAMERON BROWN, C. A. *Windmills for the generation of electricity*. Institute for Research in Agricultural Engineering. (Oxford, 1933).

(2) GOLDING, E. W. and STODHART, A. H. The potentialities of wind power for electricity generation (with special reference to small-scale operation). Electrical Research Association, *Technical Report*, Ref. W/T16 (1949).

(3) GOLDING, E. W. and STODHART, A. H. The selection and characteristics of wind-power sites. Electrical Research Association, *Technical Report*, Ref. C/T108 (1952).

(4) *Wind electric research report*. University of Saskatchewan, Department of Agricultural Engineering (December 1st 1939 to November 30th 1940). Summary of 12 months' data. No. 499/40.

(5) TRAPP, D. L. *Comparative costs of producing electrical power from farm lighting plant units*, No. 44/47. University of Saskatchewan, Department of Agricultural Engineering.

(6) YOUNG, J. R. W. *The selection and installation of a farm lighting plant*, No. 114/45. University of Saskatchewan, Department of Agricultural Engineering.

(7) Wind-energie. Mededeling Prinsenmolen-Commissie. Overdruk uit het Weekblad *De Ingenieur*, No. 45, 1951. Algemeen (Gedeelte, 45).

(8) *Het Prinsenmolenboek*. H. Veenman En Zonen (Wageningen, 1942).

(9) Experimenten met de Benthuizer Bovenmolen. (Uitgegeven Door De Vereeniging *De Hollandsche Molen* Te Amsterdam In Samenwerking Met De Nijverheidsorganisatie T.N.O. Te s-Gravenhage) (January 1952).

(10) KARMISHIN, A. V. Small power wind-electric generating units. Priroda (*Nature*), No. 11, pp. 24–31 (November 1949).

(11) JUUL, J. Undersøgelse af muligheder for vindkraftens udnyttelse. *Elektroteknikeren*, Vol. 45, No. 20, pp. 607–635 (22nd October 1949).

(12) JUUL, J. Report of results obtained with the SEAS experimental wind power generator. *Elektroteknikeren*, Vol. 47, pp. 5–12 (7th January (1951). (In Danish.)

(13) PEDERSEN, M. Oversigt over Undelektricitetsproduktionen fra 1940 til 1948. Maaneds—Meddelelse (Marts 1948).

(14) LACROIX, G. Les problèmes électriques soulevés par l'utilisation de l'énergie du vent. *Bulletin de la Société Française des électriciens*, Vol. X, No. 103 pp. 211–215 (Avril 1950).

(15) BERTHOLON, N. *Groupe aéro-moteur 'Altervent' à génératrice courant alternatif monophasé*. Congrès du Vent (Carcassonne, 1946).

(16) LANOY, H. *Les aéromoteurs modernes*. Girardot et Cie (Paris, 1947).

(17) GOLDING, E. W. *The electrification of agriculture and rural districts*. English Universities Press (1937).

(18) GOLDING, E. W. Wind-generated electricity—and its possible use on the farm. *Farm Mechanization* (March 1953).
(19) SCHENFER, K. and IVANOV, A. Lines of development of rural wind-power plants. *Elektrichestvo*, No. 5, pp. 21–22 (May 1941). (In Russian.)
(20) MOGILNITSKII, I. D. On the regulation of slow-running wind-power generators. *Papers of the Lenin Academy of Agricultural Sciences*, No. 5, pp. 36–40 (1950).
(21) ANDRIANOV, V. N. and POKATEAV, A. I. Regulation of the output of a wind power station. *Elektrichestvo*, No. 6, pp. 19–24 (1952). (In Russian.)
(22) A farm electrification programme. *Report of Manitoba Electrification Enquiry Commission* (Winnipeg, Manitoba, 1942).
(23) SEIDEL, G. R. Kleine windelektrische Anlagen für den Export. *E.T.Z.*, Bd. 70, S. 158 (1949).
(24) CARRER, A. Generatori a corrente continua per l'utilizzazione dell'energia del vento. *L'Elettrotecnica*, Vol. 36, p. 376 (1949).
(25) CARRER, A. Utilizzazione dell'energia del vento con schema tipo Ward-Leonard a funzionamento invertito. *L'Elettrotecnica*, Vol. 36, p. 383 (1949).
(26) HÜTTER, U. Der Einfluss der Windhäufigkeit auf die Drehzahlabstimmung von Windkraftanlagen. *Zeitschrift für Elektrotechnik*, Heft 6, pp. 117–122 (December 1948) und Heft 1, pp. 10–14 (January 1949).

CHAPTER 15

THE ECONOMY OF WIND POWER GENERATION

THE source being free, it is obvious that the cost of energy generation by the wind depends upon (*a*) the annual output of energy and (*b*) the annual charges for interest, depreciation and maintenance of the plant. When these are known the economy of wind power utilization can be assessed by comparison of its generating costs with those applying to alternative sources of energy.

Let p = Percentage annual charges including interest, depreciation and maintenance.

C = Cost of the plant in £ per kilowatt.

T_s = Specific output in kilowatt-hours per annum per kilowatt.

Then the generating cost, in pence per kilowatt-hour $= 2{\cdot}4\dfrac{p\,.\,C}{T_s}$.

So that, for example, if $p = 8$ per cent, $C = £50$ and $T_s = 4000$, the cost per kilowatt-hour $= 0{\cdot}24$d.

Annual charges (p). The figure to be used for the percentage annual charge to include interest, depreciation and maintenance depends upon the rate at which capital can be borrowed, upon the life of the plant, upon its design and its place of installation (which affects maintenance costs).

The rate of interest will vary with time and with the status of the borrower. Large-scale generation would generally be undertaken by a large national authority, or public utility company, when an interest rate of about 4 per cent would apply. For small or medium scale generation, by private individuals or small communities, the rate will be higher.

The life of the plant cannot well be expressed as a definite number of years because, in fact, the number varies for different parts. Thus, for example, assuming that a large propeller-type aerogenerator is used, the blades may need replacing after a period between five and ten years while the supporting tower may last for thirty or forty years. Small direct-current plant (see Ch. 14) will include batteries for storage and these will need fairly frequent replacement.

The maintenance cost will naturally depend upon the size, type and design of the plant and on its location. Large machines will be designed to operate automatically and may be installed in groups so that only a small maintenance staff will be needed to attend to them. Medium-sized plants would normally be installed as auxiliaries supplementing the main supply at small power stations and here again maintenance would be low. For example, in Denmark, the only country which has used such sets on any appreciable scale and for a sufficient length of time for a reliable maintenance figure to be obtained, this has been found to average about 1 per cent of the capital cost.

The maintenance of small equipments, installed to supply individual premises, is usually undertaken by the occupier who does not account for the maintenance required, but this will be mainly for periodic attention to the batteries and, very occasionally, to the machine and supporting structure.

It will be appreciated that, especially for medium- and small-scale utilization, annual charges are difficult to state precisely but the following figures are probably representative:

For large-scale plants $p =$ 8–9 per cent
For medium-scale plants $p =$ 10–12 per cent
For small-scale plants $p =$ 12–15 per cent

Cost of plant (C). This subject will be discussed in more detail in Ch. 16. It is perhaps sufficient here to note that the cost, per kilowatt of installed capacity, will vary with:

(*a*) the type of machine
(*b*) the actual magnitude of the installed capacity
(*c*) rated wind speed.

The type of plant cannot be considered independently of the size because very small sets will be for direct current and will run at variable speed, medium-size plants may be either d.c. or a.c. and large plants will be almost invariably alternating current without any provision for storage. Again, it is possible that cheaply constructed sets, of perhaps a vertical-axis type, may have advantages in small-scale utilization while being out of the question for large-scale use.

In this chapter we are concerned more with the economic choice of the main design features, particularly the rated wind speed, in relation to the wind régimes applying to regions having varying costs of energy generation by alternative sources.

Specific output (T_s). Sufficient has been said already in earlier chapters for it to be realized that the specific output obtainable from a wind power installation depends upon the power-duration curve for its site and upon the rated wind speed chosen in its design. For a given diameter of propeller in a wind turbine, the higher the rated wind speed the lower the cost per kilowatt of capacity. This is not surprising since the capacity increases with the cube of the rated wind speed. On the other hand, the specific output decreases rapidly with increase in rated wind speed. To obtain the lowest cost per kilowatt-hour generated is thus a matter of striking a balance between these two effects: it cannot be concluded immediately that the highest specific output corresponds to the lowest energy cost.

The cost of generation by large aerogenerators

While construction costs for large aerogenerators cannot be known with any precision until a number have been built, sufficient information exists, both from design and costing studies and from the American experience at Grandpa's Knob (Ref. 1), for reasonable estimates to be made and for conclusions to be drawn concerning the variation of these costs with propeller diameter and with rated wind speed. Thus it is clear that, certain costs remaining fixed independently of the size of the machine, the cost per kilowatt will continuously decrease as the diameter is increased from about 80 or 100 ft to the maximum which is at present considered practical, namely about 200–240 ft. Again, for a given diameter, the cost per kilowatt will fall with increasing rated wind speed.

To show the influence, upon the choice of rated wind speed and upon generating cost, of the annual mean wind speed at the site of a large generator, Table XXXIII, has been drawn up. It refers to an aerogenerator with a propeller diameter of 210 ft and thus having about the lowest costs per kilowatt which might at present be practicable. The costs shown are representative of recent estimates from independent costing studies applying to series production of large units (see Ch. 16). The figures for specific outputs are taken from Fig. 57 (Ch. 10) and so refer, at the different annual mean wind speeds, to sites having similar wind régimes to those in Great Britain. Annual capital charges have been taken as 8½ per cent.

The values of kilowatt capacity have been included, these being calculated on the assumption of an overall power coefficient of 40 per cent.

In the table, the minimum generating costs are shown in heavy type and it must be noted that, except when the annual mean wind speed is as low as 10 m.p.h., the rated wind speed for minimum energy cost does not correspond to maximum specific output but rather to the reverse. The reason is, of course, that an increase in rated wind speed increases the kilowatt capacity at a relatively low

TABLE XXXIII

Variation of output and energy cost with rated wind speed at sites with different annual mean wind speeds

Wind turbine diameter 210 ft. Overall power coefficient 40 per cent. Annual capital charges 8½ per cent.

Annual mean wind speed (m.p.h.)	Rated wind speed (m.p.h.)	Capacity (kW)	Cost of plant (£/kW)	Specific output (kWh p.a./kW)	Energy cost (pence per kWh)	Total annual output (kWh)
25	35	3220	40	3420	**0·24**	11,000,000
	30	2030	55	4400	0·26	8,830,000
	25	1160	90	5400	0·33	6,500,000
	20	592	160	6200	0·52	3,730,000
20	35	3220	40	2100	0·39	6,760,000
	30	2030	55	3100	**0·36**	6,300,000
	25	1160	90	4150	0·45	4,810,000
	20	592	160	5350	0·65	2,960,000
15	35	3220	40	1000	0·82	3,220,000
	30	2030	55	1800	**0·63**	3,660 000
	25	1160	90	2700	0·68	3,130,000
	20	592	160	3750	0·87	2,220,000
10	35	3220	40	200	4·1	644,000
	30	2030	55	650	1·74	1,320,000
	25	1160	90	1200	**1·53**	1,390,000
	20	592	960	1900	1·55	1,245,000

additional capital cost. Thus, the fall in the cost per kilowatt more than compensates for the fall in specific output.

From the last column it is seen that, although, in each group, the total annual outputs do not differ greatly for rated wind speeds close to the optimum, they increase very rapidly as the annual mean wind speed rises. The ratio between the total for the optimum condition in the first group and for the optimum in the last group is 7·9:1. Since the ratio between the cubes of the mean wind speeds is $\frac{25^3}{10^3} = 15{\cdot}6$ the usable energy pattern factor (see p. 30) at an annual mean wind speed of 10 m.p.h. is almost exactly twice that at 25 m.p.h.

If minimum energy cost is to be the sole criterion, then the rated wind speed should be 35 m.p.h. when the mean speed is 25 m.p.h.; 30 m.p.h. when the mean speed is either 20 m.p.h. or 15 m.p.h.; and 25 m.p.h. (or perhaps 20 m.p.h.) when the mean speed is 10 m.p.h.

It is, however, of some interest to observe that, at sites having an annual mean wind speed of 25 m.p.h. or over, a 592 kW unit might be installed which would give the equivalent of full load for well over 6000 hr a year with an energy cost of 0·52 pence per kilowatt. Such an aerogenerator, having a rated wind speed of 20 m.p.h. and a cut-in speed of 10 m.p.h., would actually be in operation for about

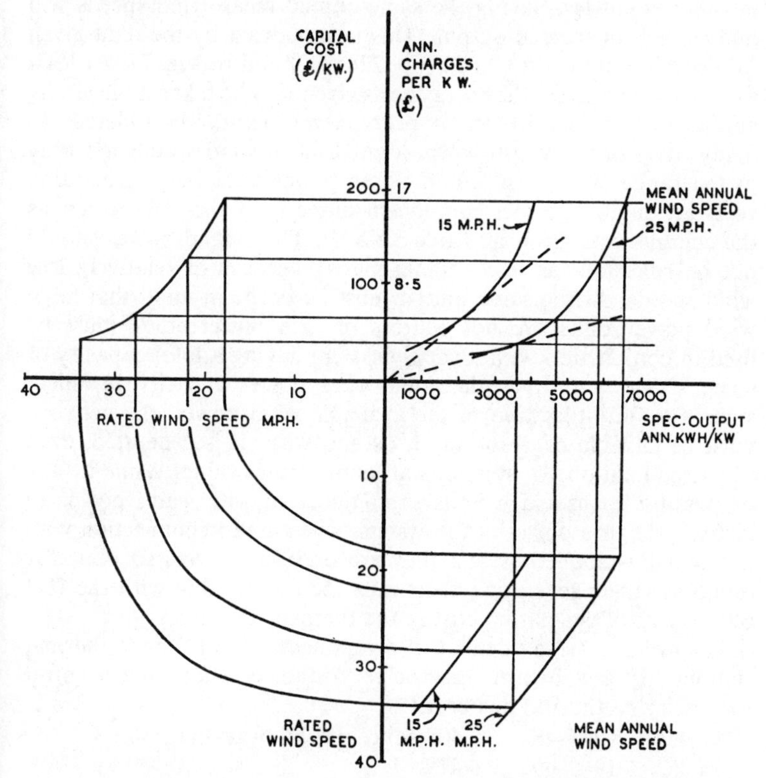

FIG. 88. *Relationships between generating costs and rated wind speed*

7500 hours in the year. During this period some 5500 hours would be at full output and the remainder at a reduced output.

Fig. 88 exhibits some of the data from Table XXXIII in graphical form and shows a graphical construction to obtain curves of annual charges per kilowatt for different specific outputs. The costs per kilowatt-hour are proportional to the slopes of the lines joining any point on the curves to the origin and are clearly minimum at the points of contact of the tangents. By projection downwards from these points of contact the optimum rated wind speed for each annual mean wind speed is obtained.

The figures for energy cost in Table XXXIII are based on specific outputs for wind régimes in Great Britain but, although this must be checked by hourly wind speed records, it is unlikely that sites, in other countries, having the same annual mean wind speeds will differ much in specific output. This is supported by the data given for specific outputs in Ch. 14 (pp. 228–231) and in Fig. 77. At least as an approximation, the energy costs given may be taken as generally applicable for the size and type of aerogenerator considered. In many parts of the world where annual mean wind speeds are only in the range 10–15 m.p.h. and water power is lacking, generating costs are high. The fuel component alone is of the same order as the energy costs given in Table XXXIII. Thus, wind power should not be ruled out as uneconomic merely because of relatively low wind speeds. At the same time it must be borne in mind that large wind power units are not sources of firm power; they must be used in conjunction with a supply system having a total capacity of some six to ten times the total wind power capacity installed, since, for full utilization of the available wind power, the network must be capable of absorbing it as and when it is generated, even when the load on the system is at its minimum value. While it may be possible to use such units in conjunction with water power to increase the firm power of the system, when used in connection with a thermally-supplied system they are only fuel savers so that, for economy, their generating costs must be comparable with the fuel component of generating cost in the thermal stations.

The average value of this fuel component of cost for the thermal stations of the British Electricity Authority (now the Central Electricity Authority) has been:

1947–48	0·405 pence per kilowatt-hour
1948–49	0·419 ,, ,, ,,
1949–50	0·408 ,, ,, ,,
1950–51	0·416 ,, ,, ,,
1951–52	0·451 ,, ,, ,,
1952–53	0·463 ,, ,, ,,
1953–54	0·480 ,, ,, ,,

Taking a general value of 0·4 pence per kilowatt-hour it is seen from Table XXXIII that sites having annual mean wind speeds of 20 m.p.h. or above could be used economically for large-scale wind power generation in Great Britain the energy costs at such sites being less than 0·4 pence. The fuel component must be 0·6 pence or more for sites having mean annual wind speeds of only 15 m.p.h. to be used economically.

The cost of generation by medium-scale aerogenerators

The costs of construction for medium-sized wind-driven plants, up to about 100 kW capacity, can be given with more confidence than those for units of the largest size because there is some supporting evidence from actual construction as, for example, in Denmark where they have been used in appreciable numbers. While the cost per kilowatt increases with reduction in rated wind speed, as in the case of large-scale plant this cost is almost inversely proportional to the kilowatt capacity, i.e. the total cost of the plant falls only slightly as the rated wind speed is reduced.

Following the same method as in Table XXXIII, Table XXXIV gives estimated energy costs, with different rated wind speeds, for a wind turbine of 50 ft diameter. Again the specific outputs quoted

TABLE XXXIV

Variation of output and energy cost with rated wind speed at sites with different annual mean wind speeds

Wind turbine diameter 50 ft. Overall power coefficient 37·5 per cent. Annual capital charges 12 per cent

Annual mean wind speed (m.p.h.)	Rated wind speed (m.p.h.)	Capacity (kW)	Cost of plant (£/kW)	Specific output (kWh p.a./kW)	Energy cost (pence per kWh)	Total annual output (kWh)
25	35	169	75	3420	**0·64**	578,000
	30	106	118	4400	0·77	466,000
	25	61·5	190	5400	0·98	343,000
	20	31·4	360	6200	1·65	198,000
20	35	169	75	2100	**1·02**	355,000
	30	106	118	3100	1·09	329,000
	25	61·5	190	4150	1·32	255,000
	20	31·4	360	5350	1·98	165,000
15	35	169	75	1000	2·16	169,000
	30	106	118	1800	**1·89**	191,000
	25	61·5	190	2700	2·03	166,000
	20	31·4	360	3750	2·76	118,000
10	35	169	75	200	10·8	33,800
	30	106	118	650	5·2	68,900
	25	61·5	190	1200	**4·6**	73,800
	20	31·4	360	1900	4·9	65,900

are based on wind régimes at British sites. Annual charges of 12 per cent, as against $8\frac{1}{2}$ per cent for large plant, have been taken to allow for the possibly different financial status of the users of such medium-size units and for a somewhat higher rate for maintenance costs. A lower overall power coefficient of 37·5 per cent has also been assumed in calculating the kilowatt capacity.

The energy costs for the optimum rated wind speeds exhibit the same trend as with large-scale plant but are, of course, higher. Again the lowest costs are obtained with high rated wind speeds and low specific outputs except when the mean wind speed is as low as 10 m.p.h.

BIBLIOGRAPHY

(1) Putnam, P. C. *Power from the wind.* Van Nostrand (1948).

(2) Kleinhenz, F. Projekt eines Grosswindkraftwerkes. *Der Bauingenieur*, S. 173 (1942).

(3) Kleinhenz, F. Gewichts und Kostenvergleech von Grosswindkraftwerken verschiedener Hoke bei gleichem Windradurchmesser. Denkschriften der Reichsarbeitsgemeinschaft, *Windkraft*, Nr. 1–7 (Berlin, 1943).

(4) Kleinhenz, F. Die Ausnutzung der Windenergie durch Hohen—Windkraftwerke. *Technik, Bln.*, 2, S. 517 (1947).

(5) Kleinhenz, F. Das Grosswindkraftwerke M.A.N.-Kleinhenz. Denkschriften der Reichsarbeitsgemeinschaft, *Windkraft*, Nr. 1–7 (Berlin 1943).

(6) Juul, J. Application of wind power to rational generation of electricity. *Elektroteknikeren*, 43, 137–148 (7th August 1947). (In Danish.)

(7) Golding, E. W. *The economic utilization of wind energy in arid areas.* Paper given at the UNESCO Symposium on Wind and Solar Energy (New Delhi, 1954).

(8) Golding, E. W. *The utilization of wind power in desert areas.* Proceedings, International Symposium on Desert Research (Jerusalem, 7th–14th May, 1952).

(9) Golding, E. W. *Economic aspects of the utilization and design of wind power plants.* World Power Conference, Brazilian Sectional Meeting (1954).

(10) Golding, E. W. *The economic and practical aspects of utilizing wind energy in arid areas.* UNESCO/NS/AZ/139 (Paris, August 1953).

(11) Haldane, T. G. N. and Golding, E. W. G. *Recent Developments in Large-Scale Wind Power Generation in Great Britain.* Paper No. 1, Section K, Fourth World Power Conference (London, 1950).

CHAPTER 16

CONSTRUCTION COSTS FOR LARGE WIND-DRIVEN GENERATORS

TAKING the two main factors in the economy of wind power generation—annual wind energy and construction cost—enough is known about the former to encourage the belief that its magnitude is ample in many parts of the world, but evidence on costs for large plants is rather meagre. Hence it became obvious, at the start of the wind power investigations now proceeding in Great Britain, that a comprehensive design and costing study should be undertaken to determine the main design features for the optimum form of wind-driven generator to operate connected to the supply network and being located at a typical windy site. Such a study was sponsored by the Ministry of Fuel and Power as part of their contribution to the investigations and was made by Folland Aircraft Ltd. With the permission of the Ministry, some of the conclusions reached are given in the following paragraphs. The design finally suggested is outlined and its details and probable component costs are compared with those for three other designs, of comparable size.

British design and costing study.

This study, which had the widest possible terms of reference, was based on the wind régime as measured at Costa Hill, Orkney, for the year 1949, a rather sub-normally windy year. This régime is represented by the velocity frequency curve of Fig. 89. Certain calculations on the cost of work on the site were also based on the topography of this hill as being reasonably typical.

The conclusions reached and the main optimum design features which emerged can be summarized as follows:

(*a*) Type of rotor: horizontal axis, two bladed propeller.

(*b*) Rotor diameter: 225 ft.

(*c*) Rotor speed of rotation 42·5 r.p.m. Tip speed 500 ft/sec.

(*d*) Blades: section NACA 0015; two stage taper; $11\frac{1}{2}°$ twist; 10° included coning.

(*e*) Capacity: 3670 kW at rated wind speed 35 m.p.h.

(*f*) Generator: a.c. induction type.

(*g*) Tower: Rotatable tripod; hub height 135 ft, two legs carried on bogeys running on a circular rail track, the base of the third leg being at the centre of the track circle.

(*h*) Transmission: spur gears, direct drive without clutch or hydraulic coupling.

(*i*) Controls: blade pitch change by ailerons. Start and stop by pilot windmill. Yaw by fantail coupled to bogey wheels through centrifugal clutch or fluid flywheel.

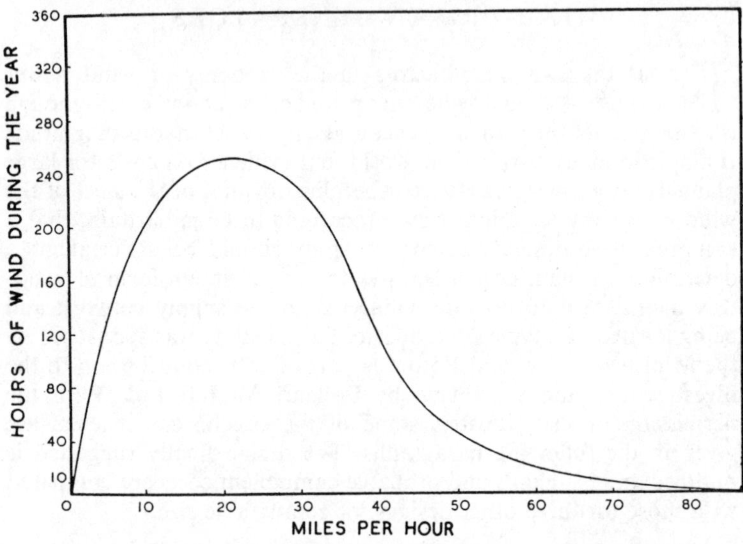

FIG. 89. *Velocity frequency curve, Costa Hill, Orkney*, 1949

(*j*) Capital cost of completed installation: £150,462 or £41 per kilowatt (January, 1951) on a basis of forty off.

(*k*) Energy cost: 0·18 pence per kilowatt-hour.

Notes

1. The conventional propeller type with a horizontal axis was chosen after a preliminary examination of the possibilities of the most likely alternative, the vertical-axis type. This showed that a vertical-axis machine with a rotor 140 ft diameter and 70 ft high might, at a somewhat optimistic estimate, produce 450 kW with a rated wind speed of 35 m.p.h. the cost per kilowatt being about £206, i.e. rather more than twice the cost of a propeller-type machine of the same capacity. The corresponding energy costs were 1·075 pence against 0·45 pence.

There was little to choose between two and three blades for the propeller type from the points of view of performance, but two were more convenient for the method of erection suggested.

2. Rotor diameters up to 300 ft were considered and the construction

costs for different sizes were calculated. Blades for 300 ft diameter would present manufacturing problems but could be made; with a bigger machine than this the gearbox would be difficult to make because of the very low rotational speed, resulting from tip speed limitations, and the high torque. Fig. 90 shows how the construction cost falls with increases in diameter and in rated wind

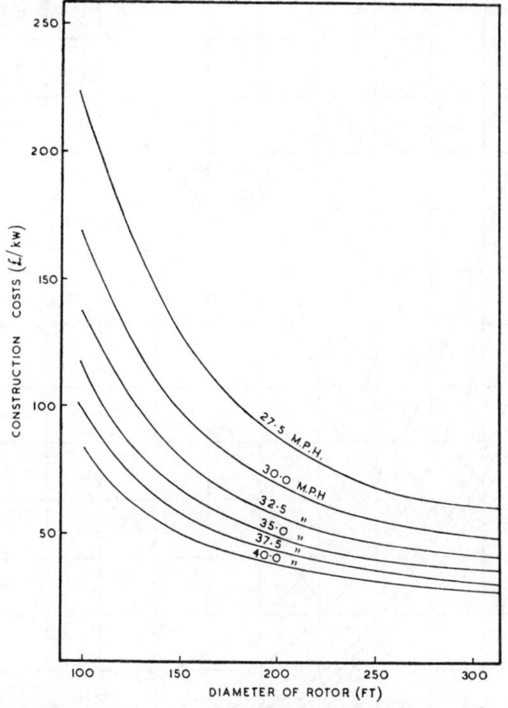

FIG. 90. *Variation of constructional cost with rotor diameter*

speed while Fig. 91 gives corresponding curves of energy cost with the given wind régime and for the percentage annual charges adopted in this study. A diameter of 225 ft, and rated wind speed of 35 m.p.h. were chosen because these values give costs which are little above the minimum while they are unlikely to involve excessive gearbox difficulties which would result from the higher torques and lower rotational speeds corresponding to larger diameters or increased rated wind speeds. For example, at 225 ft diameter, increasing the rated wind speed from 35–40 m.p.h. increases the power capacity from 3670 kW to 5400 kW. Incidentally, it also reduces the specific output from 3474 to 2777 kWh per annum per kilowatt.

The rotor tip speed should be as high as practicable and, since the curve of annual output against tip speed was found to be almost flat between 400 and 500 ft/sec, the latter tip speed was chosen. The

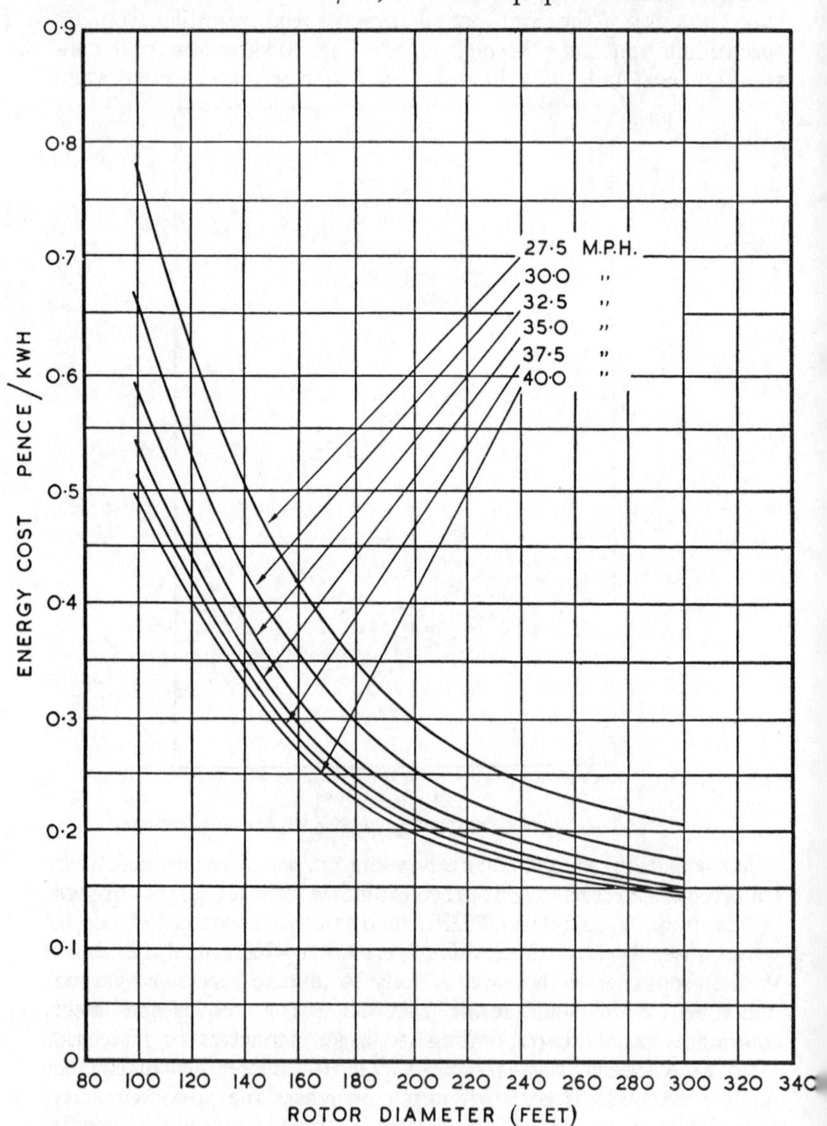

FIG. 91. *Variation of energy cost with rotor diameter*

higher the tip speed the greater the proportion of the total energy generated by wind speeds around, and a little above, the rated wind speed.

3. The increased output from a twisted blade with two stage taper was found to be worth the manufacturing costs involved. Coning was also found to be advantageous.

4. Alternative methods of electrical generation, including d.c. generation, with variable rotational speed, and static conversion, were investigated, and the induction generator was chosen as the most economic. Power factor correction by static capacitors would be used if necessary. Optimum output would be achieved by employing the highest practicable tip speed instead of by choosing the rated wind speed to give maximum power coefficient. Thus the tip-speed ratio varies with the rated wind speed considered and is 9·74, with a power coefficient of 0·475, at a rated wind speed of 35 m.p.h. and a tip speed of 500 ft/sec.

The alternator efficiency was taken as 0·95 at full load, falling to 0·90 at 33⅓ per cent of full load.

5. The tower, or supporting structure, takes the form of a rotatable tripod as shown in Fig. 92.

Its advantages are that: (*a*) it is almost, if not entirely, self orientating; (*b*) it gives a reduced effect of 'tower shadow' on the blades as they rotate and (*c*) it eases the problem of erection, particularly by reducing the time to be spent in work in an exposed position aloft. Most of the work of assembling the rotor blades, shaft, gearbox and generator could be done at ground level and the whole thereafter raised at one lift by using the two outer legs of the structure as shear legs. This is a very important matter since the bad weather which often persists for a period of considerable length at an especially windy site may greatly retard work at the top of the tower. Such delays are certainly expensive when erection staff have to be located at an isolated site remote from their main headquarters. The facilities for raising and lowering the rotor and associated machinery from and to ground level are also likely to reduce maintenance costs through easing the work required at a major overhaul or for replacement of any main components.

A possible disadvantage of the tripod form of construction is the need for a fairly large circular track which might be difficult to locate on the summit of a very sharply peaked summit.

6. To determine the possibilities of reducing the construction cost some alternative constructions were considered.

The first of these was a non-orientating machine. It was concluded that the reduction of cost by building the tripod on fixed foundations

FIG. 92. *Suggested construction for large windmill*

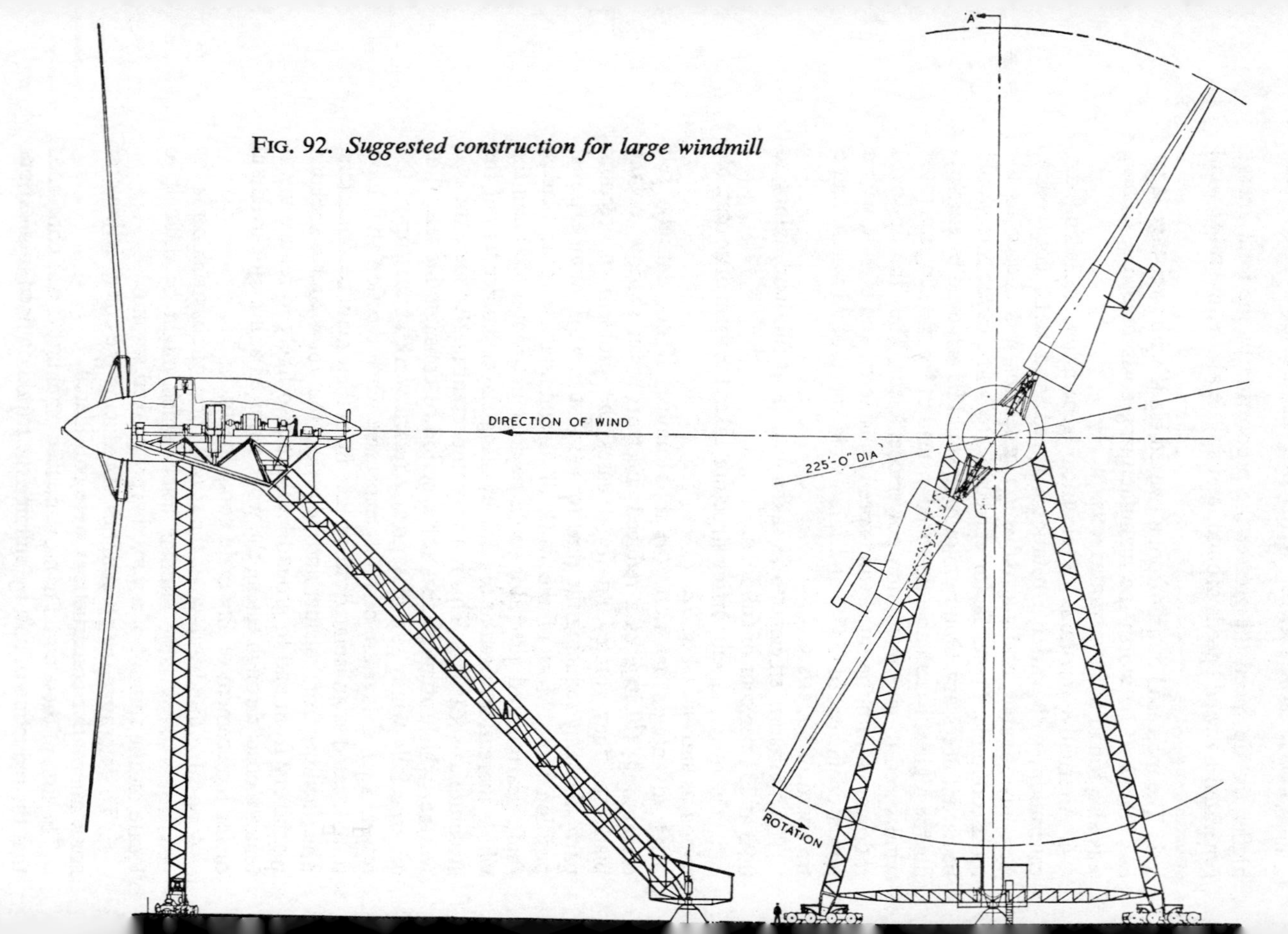

would be very small. The maintenance costs would probably not be reduced; the elimination of maintenance costs for the track might be counterbalanced by higher replacement costs due to increased wear and fatigue on the blades and shaft as a result of vibrations induced by winds blowing at an angle to the disc swept by the blades. On the other hand, non-orientation inevitably implies some loss of energy production, the annual magnitude of which will depend on the wind rose at the site (see Ch. 5). Another scheme, using modified non-orientating windmill having three rotors running in vertical planes at 120° to one another, was considered but, even neglecting difficulties from mutual interference between the three rotors, the trebled rotor cost makes this uneconomic.

A machine with contra-rotating rotors was also studied but the expected aerodynamic performance would not make it attractive. Fuller consideration was given to the possibilities of tandem rotors placed on opposite arms of the tower. Although this would have some advantages with a conventional type of tower, the estimates of energy cost proved to be slightly higher than the rotatable tripod and a single rotor.

The hub height of 135 ft was chosen as the minimum to give a reasonable ground clearance (about 22½ ft). With a machine located on a hilltop, when the gain in wind speed with height above ground up to two or three hundred feet is not great, the extra cost for a higher tower is not justifiable.

7. For power transmission from the rotor to the generator both mechanical transmission by a shaft running down to ground level and also hydraulic transmission to a generator on the ground were considered. The first would have some advantages over the spur gear drive chosen—the main cables would not pass down the tower, slip rings for the main output circuit would not be needed and more rigid foundation would be provided for the generator—but it would be more expensive. The overall efficiency of the hydraulic transmission, including generator losses, was estimated as only 60 per cent against 90 per cent for the spur gear and, in addition, expensive, specially developed components would be needed for such a drive.

8. Control of blade pitch by ailerons offers the advantages of lightness and cheapness but feed-back from the hub is essential for governor stability.

For yaw control a fantail rotor which is set at right angles to the main rotor, and which will rotate at a speed which is dependent upon the amount of yaw, has the advantage of immunity from trouble caused by an electrical failure.

9. The effect, upon the total capital cost, of the number of machines

built was investigated for the 225 ft. diameter size. This cost was found to fall as shown in Fig. 93.

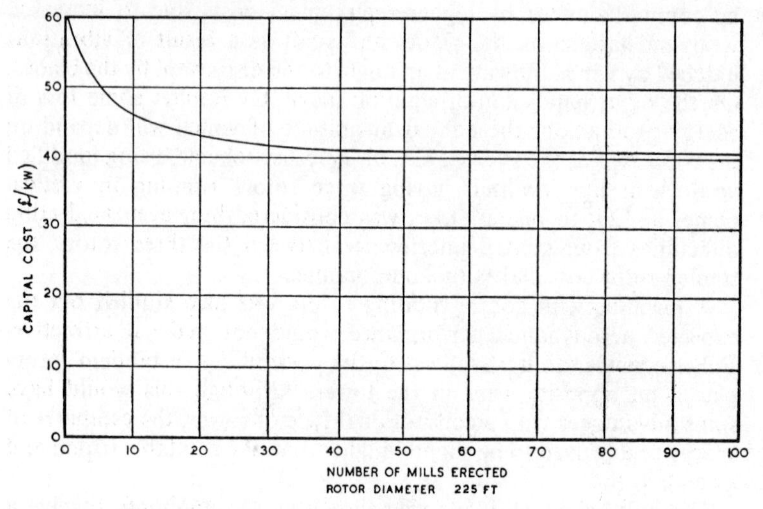

FIG. 93. *Variation of capital cost with number built*

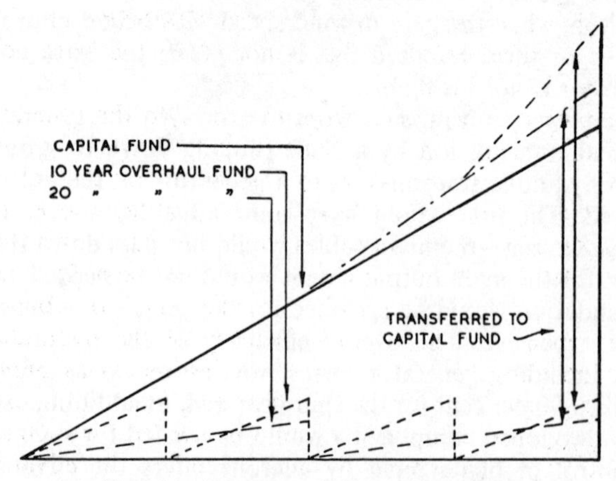

FIG. 94. *Diagram of depreciation costs*

10. The derivation of the cost of the energy produced involves basic assumptions concerning interest, depreciation and maintenance. These assumptions may be briefly stated:

(i) The machine will be scrapped at the end of forty years during which time there will be:

(*a*) Every ten years a general overhaul with replacement of the blades.

(*b*) Every twenty years, in addition to the 'ten-year overhaul' there will be replacement of blade, windshaft, and gearbox bearings.

(*c*) Minor repairs, painting and other maintenance at the annual charge of 0·8 per cent of the capital cost.

(*d*) Wages and expenses of an attendant, at the rate of one man for five machines, $= \frac{£1000}{5} = £200$ per annum.

(ii) The necessary capital will be obtained by Government borrowing, the rate of interest on the capital and for the sinking funds being $3\frac{1}{4}$ per cent.

The sinking fund accumulation and expenditure can be shown diagrammatically as in Fig. 94 in which the funds for the second twenty-year overhaul and for the fourth ten-year overhaul are added to the capital repayment, or obsolescence, fund at the end of the forty years.

TABLE XXXV

Annual running costs for 225 ft diameter aerogenerator with rated wind speed 35 m.p.h.

Item	Annual charge rate per cent		Cost	Annual charge
	Sinking fund	*Direct*	(£)	(£)
10-year overhaul, blades, transport and labour	8·624		16090	1387·6
20-year overhaul, bearings and labour, etc.	3·628		7550	273·9
Capital cost less 10- and 20-year overhaul	1·253		1·253 % of (150,462 −16,090 −7550)	1589·1
Interest on capital		3·25		4890
General maintenance		0·80		1203·7
Attendance				200
Total				£9544·3

With an interest rate of 3¼ per cent the sinking fund charges are:

For the ten-year overhaul	8·624 per cent
For the twenty-year overhaul	3·628 per cent
For the obsolescence (forty-year)	1·253 per cent

For the 225 ft diameter machine with a rated wind speed of 35 m.p.h. the running costs are thus as shown in Table XXXV.

The annual energy production from the wind régime at Costa Hill is estimated as in Table XXXVI.

The energy/wind speed curve, plotted from the figures in the first and last columns of Table XXXVI is shown in Fig. 95 from which

TABLE XXXVI

Estimation of annual energy production by 225 ft diameter aerogenerator with a rated wind speed of 35 m.p.h.

Gross kilowatts $= \frac{16 \cdot 1}{10^6} C_P \, . \, R^2 V^3$ (R = radius in feet; V = wind speed in m.p.h.)

Gear-box efficiency, 95 per cent

Wind speed (m.p.h.)	C_p	Gross kW	% Full load	Alternator efficiency (%)	Nett kW	Annual Duration of wind speed (hours)	Energy output (kWh)
10	0					206	
12·5	0					224	
15	0·043	29·3	0·72			237	
17·5	0·216	235·5	5·76	19·5	43·6	244	10650
20	0·325	522	12·8	42·8	213	248	52800
22·5	0·396	935	22·9	72	640	249	159500
25	0·441	1392	34·2	90·5	1200	245	294000
27·5	0·464	1910	46·7	93	1690	237	401000
30	0·472	2580	63·2	94·8	2330	225	525000
32·5	0·475	3280	80·2	95	2960	208	616000
35	0·474	4080	100	95	3670	183	671000
37·5	0·383	4080	100	95	3670	149	546000
40	0·315	4080	100	95	3670	110	403500
45	0·222	4080	100	95	3670	67·5	248000
50	0·1613	4080	100	95	3670	44	161500
55	0·1214	4080	100	95	3670	30	110000
60	0·0936	4080	100	95	3670	19	69700
65	0·0735	4080	100	95	3670	12	44000
70	0·0588	4080	100	95	3670	7	25700
						Annual total	4338350

Table XXXVII

Analysis of construction costs

Part of installation	British design study	Smith-Putnam	P. H. Thomas	Kleinhenz
	Percentage of total cost	Percentage of total cost	Percentage of total cost	Percentage of total cost
Blades	7·4	11·2	3·9	9·4
Hub, blade supports, blade shanks and bearings	7·4			
Main shaft and bearings	4·7	41·5	5·9	15·3
Nacelle or head, structure and auxiliary equipment	7·4			
Tower or other support	8·1	7·7	11·2	31·3
Gearbox	16·7	9·5	2·3	5·5
Electric generator and installation	12·5	3·4	33·6	15·8
Control equipment for speed, yaw and load	4·4	6·5	8·3	10·4
Foundations and site work	31·4	16·6	20·3	9·0
Engineering		3·6	14·5	
Notes on project	225 ft dia., 3670 kW at 35 m.p.h. Cost £150,462 (1951). Cost per kW, £41. for 40 off. Energy cost 0·18d.	175 ft dia., 1500 kW at 30 m.p.h. Cost 286,673 dollars (1945). Cost per kW, 191 dollars 20 off. Energy cost 6·9 Mils per kWh.	200 ft dia. (twin wheels), 7500 kW at 34 m.p.h. Cost 5,095,770 dollars (1945). Cost per kW, 68 dollars. 10 off. Energy cost 2·04 Mils per kWh.	425 ft dia., 14000 kW at 37 m.p.h. Cost 6,900,000RM (1947). Cost per kW, 495RM. Energy cost 3·25 pfg per kWh.

the total annual output of energy, proportional to the area under the curve, is $12{\cdot}75 \times 10^6$ kWh.

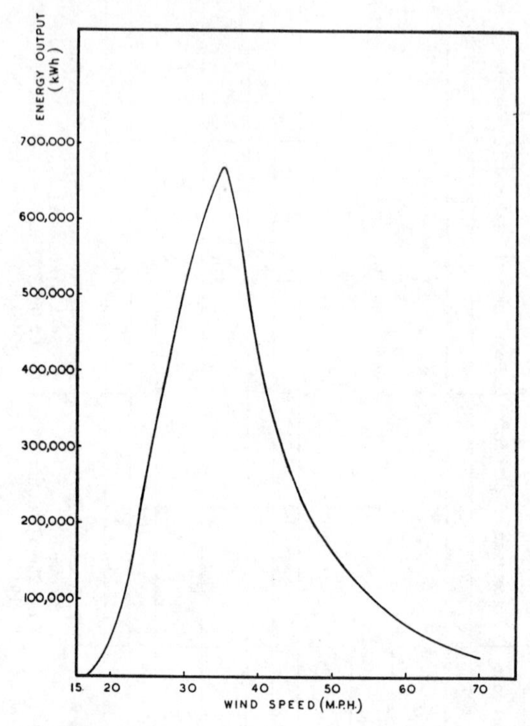

FIG. 95. *Relationship between energy output and wind speed*

Since this output is achieved for an annual expenditure of £9544·3 (from Table XXXV) the cost per kilowatt-hour

$$= \frac{9544{\cdot}3 \times 240}{12{\cdot}75 \times 10^6} = 0{\cdot}18 \text{ pence}$$

The specific output is

$$\frac{12{\cdot}75 \times 10^6}{3670} = 3470 \text{ kWh per annum per kilowatt}$$

Comparison of construction costs for large wind-power plants.

Table XXXVII shows the results of analysis of the construction costs for four large projects varying in rotor diameter from 225 ft to 425 ft.

BIBLIOGRAPHY

(1) KLEINHENZ, F. Projekt eines Grosswindkraftwerkes. *Der Bauingenieur*, S. 173 (1942).

(2) KLEINHENZ, F. Gewicht und Kostenvergleech von Grosswindkraftwerken verschiedener Hoke bei gleichem Windraddurchmesser. Denkschriften der Reichsarbeitsgemeinschaft, *Windkraft*, Nr. 1–7 (Berlin, 1943).

(3) KLEINHENZ, F. Die Ausnutzung der Windenergie durch Hohen—Windkraftwerke. *Technik. Bln.*, 2, S. 517 (1947).

(4) KLEINHENZ, F. Das Grosswindkraftwerke M.A.N.—Kleinhenz. Denkschriften der Reichsarbeitsgemeinschaft *Windkraft*, Nr. 1–7 (Berlin, 1943).

(5) JUUL, J. Application of wind power to rational generation of electricity. *Elektroteknikeren*, 43, 137–148 (7th August 1947). (In Danish.)

CHAPTER 17

THE RELATIONSHIP OF WIND POWER TO OTHER POWER SOURCES

IN the preceding three chapters the cost of wind power, in its different scales of use, has been discussed. Here we are concerned particularly with placing it in its right perspective in relation to other methods of generating power.

In Fig. 96 an attempt is made to show this relationship diagrammatically. An examination of the diagram will show that the utilization of wind-generated energy is greatly influenced, in the three scales—large, medium and small—by the question of energy storage. Thus, for large-scale power generation, no direct storage is involved; small machines will usually operate with full storage by an electric battery while, in the medium scale, there are various possibilities with or without storage. Let us consider separately the several methods of utilization indicated in this diagram.

Large-scale utilization

When a wind-driven generator operates in connection with a main network its principal function is that of a fuel saver, if the network is supplied from thermal stations; or perhaps as a 'water saver' if the stations are hydroelectric. In the diagram of Fig. 96 various sources of energy are shown as supplying the main network. Some of these—nuclear energy, marine heat (i.e. power production by utilization of temperature differences in the sea at different depths), tidal energy and wave energy—are under development so that their ultimate economy for electricity production is still unknown with any precision. Indeed, it is unrealistic to attempt to give energy costs for any of the methods of large-scale production because so much depends upon the circumstances in an individual case: the cost of thermal generation must depend upon the availability and cost, of coal, oil, peat, natural gas, etc. as influenced by the distance of the generating station from its source of supply. Again, the constructional costs of conventional hydro-electric schemes vary greatly according to their location while, for tidal and (possible) wave-power schemes, location has an even greater importance. The same applies to wind power, the generating cost for which depends largely upon the mean wind speed.

The possibilities for economic utilization of wind power on a large scale must, therefore, be assessed by comparison of its ascertainable costs, as discussed in Ch. 15, with those of the power sources with which it will be in competition.

There is also the much broader, though not unimportant, question of the part which wind power might play in relieving the rate of expenditure of the world's reserves of solid and liquid fuels, and of

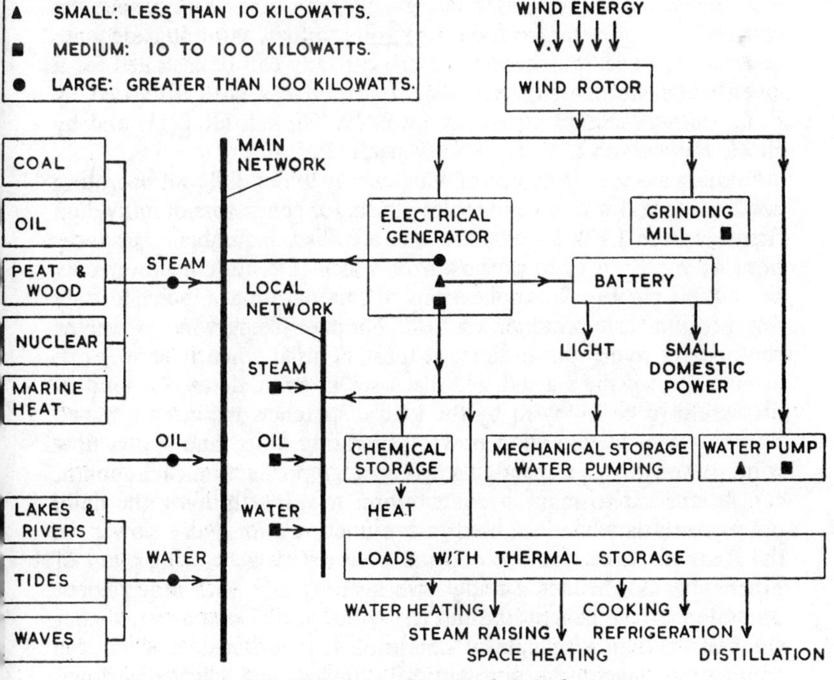

FIG. 96. *Utilization diagram for wind power*

adding to the total of power production by inexhaustible energy sources. This becomes more important as developed hydro-power sources approach the limit of economic exploitation (see Refs. 1–9).

If used in conjunction with a system fed from thermal power stations, wind power cannot be given any credit for capacity value; the capacity of the thermal generating plant cannot be reduced by the introduction of unfirm wind power capacity. The worth of the wind generated energy, per kilowatt-hour, is then only the incremental cost of generating this energy by the thermal plant—little greater than the cost of the fuel burnt. With an extensive network there is some possibility of obtaining firm wind power capacity through diversity in the occurrence of wind at widely separated sites, but the results of such tentative studies as have been made to determine this possibility are not very encouraging. There are times when calm weather exists over the whole area covered by even a large system.

In conjunction with hydro-power, or with a mixed thermal and hydro system, there is, however, some capacity value attaching to wind power. Wind-generated energy can then be used to reduce the consumption of water from the reservoirs so that, assuming adequate reservoir storage, an increased hydro-capacity can be installed for a given water storage capacity. This question has been discussed by P. C. Putnam (Ref. 10, Ch. 11), by P. H. Thomas (Ref. 11) and by E. M. Fateev and I. V. Rozhdestvenskii (Ref. 12).

Putnam assesses the value of wind energy under different operating conditions, and with several scales of use, for generators of individual capacity from 1 kW to 3000 kW and also lists a number of applications of wind power to purposes for which intermittent power may be suitable. In Fig. 96 applications of this nature are included only for medium-scale wind power use. Thomas likens wind power to run-of-river hydro-power since it must be used when it happens to occur or it will be wasted. He discusses in some detail the kind of procedure to be followed by the load dispatcher operating a mixed power system having wind power generating units, and shows how wind power, with a dependable total energy production each month, can be applied to make a contribution towards fulfilling the daily power demand when it is used in conjunction with hydro power. In the Russian paper, the use of wind power with generating plant of other types is considered under two headings: (i) with independent operation of the generators and (ii) with parallel operation. Under the first heading a scheme of operation is mentioned in which the wind-driven machine is automatically loaded, and relieved of load, in accordance with its instantaneous output, with frequent switching over of the consumers from the wind power bus bars to those of the other generators. The object is to ensure maximum utilization of the wind energy as it becomes available but there are disadvantages in the switching relays needed and in the complication introduced into the operation of the station. Although treated as large-scale utilization because of the methods of operation suggested in the paper, the wind-driven machines in the Russian paper are individually only some 30 kW capacity. One type has an 'energy accumulator' in the form of a flywheel.

Pumped storage, i.e. the pumping of water into a reservoir at high level so that it can later be let down through water turbines to generate power, is one possibility for firming power which fluctuates in quantity (see Refs. 12, 13 and 14), but its provision as an adjunct to a wind power station is not likely to be generally economic. It may, however, be feasible in some special circumstances.

H. Duquennois, (Ref. 30), in a valuable paper presented to the Sectional meeting of the World Power Conference in Brazil (July, 1954) considers in some detail the effects of operating, in Algeria, new wind power capacity with an established network fed mainly by hydro-electric stations.

He shows that the worth of the wind power depends upon the effects which it will have on the annual periods of running for the hydro-electric plants, upon the existing capacities of the dams, the allowable water flow at any time as well as its annual average volume, and on the wind régime in relation to the annual rainfall. The introduction of wind power plant has the effect of increasing the annual volume of water and it is important to determine the cost of providing extra storage capacity in the dams before considering the installation of new power plant. It happens that, in Algeria, strong winds commonly occur in the dry season so that wind power could be complementary to hydro-power in an especially favourable way.

Another recent paper by A. Kroms (Ref. 36) follows rather similar lines to that of Duquennois. He analyses the operation of wind, hydro-electric and thermal power stations in combination to feed a large network and discusses the usefulness of wind power under different conditions of load on the network. Kroms points out that wind power stations improve the energy balance by reducing the contributions required from thermal stations and may, under some circumstances, reduce the capacity of the compensating thermal plant needed to fulfil the power demand when water supplies are inadequate. He stresses the advantage to be gained from combining, through one interconnecting network, energy sources with varied characteristics.

Both of these papers give estimates of the construction cost which can be allowed for wind power plant to be economic under different assumed conditions of operation.

Full utilization of the wind energy which becomes available annually, requires that the network into which it is fed must be able to receive the power from the wind-driven machines at any time. This implies a limit to the wind power capacity which can be connected to a network. The limit is approximately that of the minimum load on the system—about 15 per cent of the total generating capacity—but is sufficiently high to be neglected at the present stage of development of wind power.

Particularly if they are to be installed, not in groups, but on separate hilltops, large wind-driven generators must be run unattended and with automatic operation. There must be controls which depend on the wind velocity. Thus the machine may run, unloaded, in light winds with its speed regulated by a governor which adjusts the blade angle. When the wind speed is high enough for the generator to be able to produce some output, the rotational speed must be matched to the synchronous speed corresponding to the network frequency and the switching-in sequence initiated. This process, in detail, will depend on whether the generator is synchronous or asynchronous. For wind speeds between the cut-in value and the rated speed, the blade angle will remain constant, and the output will depend entirely on the wind speed. At higher wind speeds the governor will adjust this angle to maintain constant speed and output up to the furling point when the machine may be shut down or, at least, the generator be disconnected from the line. The question of automatic control systems for wind-driven plant is very fully treated in the War Production Board Report (Ref. 15), Chapters V and VII.

Putnam (Ref. 10, p. 121) gives wattmeter chart records showing the output obtained with the Smith Putnam (Grandpa's Knob) generator during starting in a low or a high wind and also during periods of steady and gusty winds of speeds above the rated value.

L. Linner (Ref. 16) has given a theoretical treatment of the problem of running a wind-driven synchronous generator in connection with an infinitely stiff network which holds the machine in synchronism in varying wind speeds. He considers also the possibility of the generator becoming overheated through overloads caused by gusty winds. T. F. Wall (Ref. 17) has also discussed the possibilities and economy of large wind-driven synchronous generators.

V. N. Andrianov and D. N. Bystritskii (Ref. 18) have described some Russian experiments with a 40 kVA synchronous generator operated in parallel with a powerful grid. They give oscillograms showing the changes in armature and field currents which occur during the self-synchronization of the machine. Their record of power output, taken by a high-speed recording wattmeter (see Fig. 97) shows pulsations in the output of a frequency almost coinciding with the rotational speed of the wind rotor. They suggest that the continuous fluctuations of wind speed under gusty conditions may lead to the occurrence of beats in the angular velocity of the wind power plant with consequent pulsations in the output.

Slower speed fluctuations in power output for a wind of varying speed are referred to by V. R. Sektorov (Ref. 19) in a paper describing the experimental operation, in 1931, of the 100 kW wind-driven asynchronous generator at Balaclava and outlining a 5000 kW wind power project. He stresses the importance of utilizing a number of wind-driven machines to give a much steadier output than is obtainable from a single unit. The average daily load diagrams of the Balaclava generator during March and August, 1931, as given by Sektorov, are shown in Fig. 98.

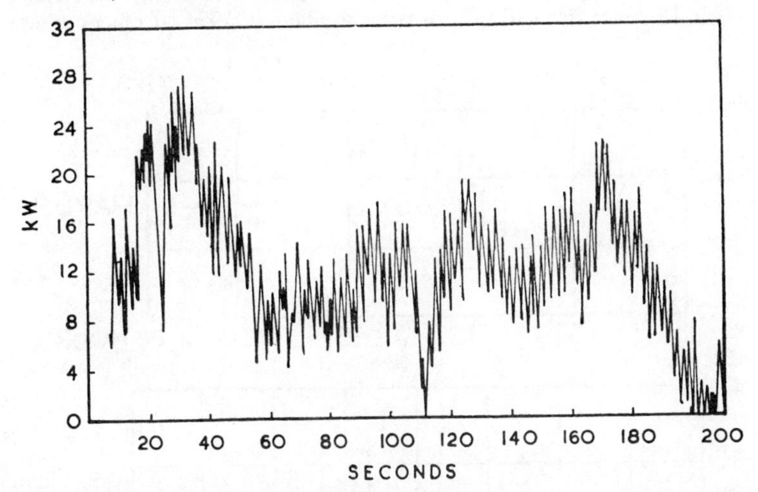

FIG. 97. *Power output record*

A thorough discussion of the operating characteristics of wind-driven electric generators, with calculations relating to their stability, is to be found in Ref. 15, Ch. 7.

A useful paper by L. D. Anscombe and A. J. Ellison (Ref. 20), though it does not deal specifically with wind-driven machines, discusses technical aspects of operating both induction and synchronous generators on industrial premises with the grid network and gives notes on the protection of small machines used in this way.

Medium-scale utilization

Referring again to Fig. 96, the energy output from a medium-sized wind-driven generator is shown as being fed into a local network or, alternatively, into various loads which have inherent storage. The first use implies competition with the source of power supplying the network, the wind generator then acting as a fuel saver as it would in large-scale use with a main network. The object of the introduction of loads with inherent storage is to avoid the need for expensive equipment, usually a battery, serving no purpose other than the storage of energy.

Most commonly, local networks would be supplied by diesel-driven generators so that the competition is then with the fuel component of the cost of generating by this means. Taking a fuel consumption of 0·5 pint per kilowatt-hour as the minimum, this fuel component is just under 1 penny per kilowatt-hour when diesel fuel costs 15 pence per gallon (this price applied, in 1954, at places where

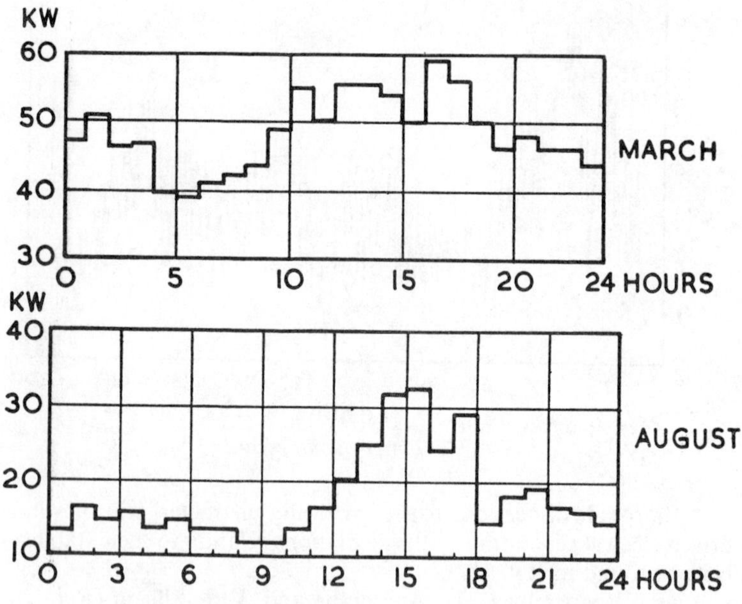

FIG. 98. *Daily load diagrams for Balaclava generator*

transport charges for the fuel are not appreciable). On the other hand, at some remote places where diesel engines may be used to drive electric generators, the price per gallon may be increased, by high transport charges, to between 3 and 4 shillings per gallon. Combined with a rather higher consumption rate of perhaps 0·6–0·7 pint per kilowatt-hour, the fuel component of generating cost is then between 3·5 and 4·0 pence per kilowatt-hour. In Fig. 99 curves of generating costs by wind power are plotted, against the mean annual wind speed at the site, for initial costs of £100, £120, £150 and £200 per kilowatt. Annual charges are taken as 12 per cent. It appears from these curves that, to compete with minimum diesel fuel costs, the mean wind speed at the site of the wind-driven generator should be 30 m.p.h. or over when the initial cost of the plant is £200 per kilowatt, but can be as low as 17 m.p.h. for an initial cost of £100 per

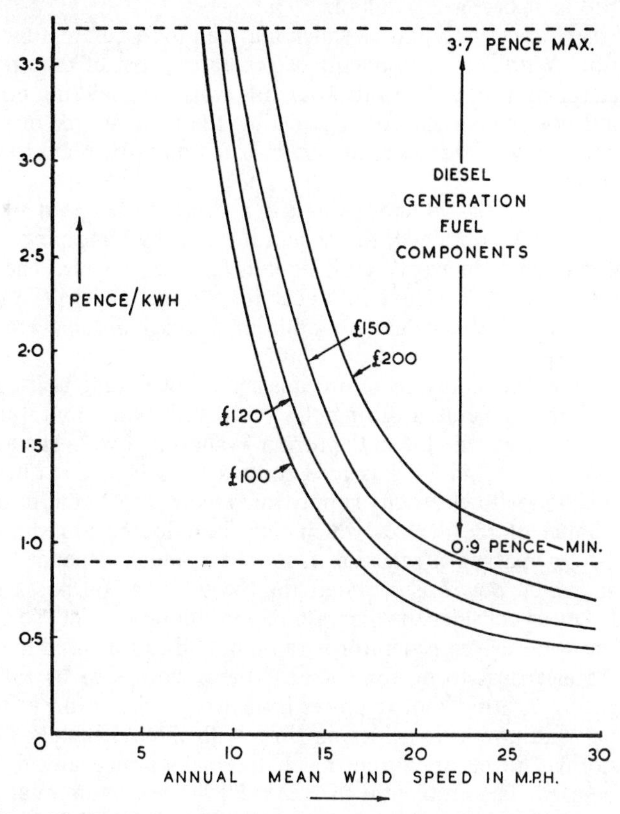

FIG. 99. *Variation of generating cost with annual mean wind speed and with capital cost*

kilowatt. At the other end of the generating cost scale, mean wind speeds of only 9–12 m.p.h. are needed.

Storage devices

It is clear from the preceding paragraphs that where a local electricity supply network is established in an area where the incremental (or fuel component) cost of energy generation does not greatly exceed 1 penny per kilowatt-hour, and an annual average wind speed of 15 m.p.h., or more, can be obtained as a well-chosen site, the most obvious function of wind power is that of a fuel saver. The main load on the system is then carried by the generating plant energizing the network continuously, with the wind-generated energy being absorbed as it becomes available.

But there are circumstances which may call for a different mode of operation. With fuel components of generating cost of the order of 3 pence (or more) per kilowatt-hour it is worth considering how far the wind power plant can be regarded as the main source of power using other power sources as auxiliary or for stand-by plant to cover calm spells.

Reference to Fig. 99 shows that, if a mean wind speed of over 10 m.p.h. can be obtained, the generating cost by wind power may be low enough to be worth utilizing it as fully as possible. The costs on these curves do not take into account storage, however, and are thus only applicable when no additional costs are incurred for storage purposes.

For the best economy, therefore, it is necessary, first, to utilize every unit of energy generated during the year by the wind and, further, to use it as far as possible in the form of random power. The author (Refs. 21, 22, 23 and 24) has paid some attention to this question, in connection with electricity supply for remote farms and for desert areas. Some of the devices which may be adopted are shown in Fig. 100. In making a selection from these, much depends on the form in which power is required for the different purposes to be served. Let us consider the diagram of Fig. 100 in detail. The power from the wind-driven generator is random and can be used immediately, in electrical form, for some of the purposes to be fulfilled, e.g. lighting, heating, motive power from an electric motor, provided that the wind power coincides, in time, with these demands. Otherwise storage, in electrical, chemical, thermal or mechanical form, must be used. The problem is then to fulfil the maximum number of purposes through storage-type equipment and to plan the utilization —perhaps by automatic 'load-dispatching' controls—to distribute the total available energy between the loads in the best way. The greater the number of conversions of energy from one form to another the worse the economy. Thus the aim should be to use the energy with no more than one conversion from the electrical form to that needed for the load.

If a supply of electricity is to be maintained continuously, a battery, charged from the wind-driven generator, is the obvious requirement. But a battery to supply (say) one fifth of the total energy to loads demanding continuity of supply during periods of two or three days of calm weather, will cost between a quarter and half as much as the wind-power plant itself. Again, the annual charges for the battery being about 20 per cent because of its short life, these charges are of the same order as those for the main plant. This means that the energy supplied via the battery (allocating the battery annual charges to it) will cost some six times as much as that supplied directly in the form of random power; or, considering the total annual

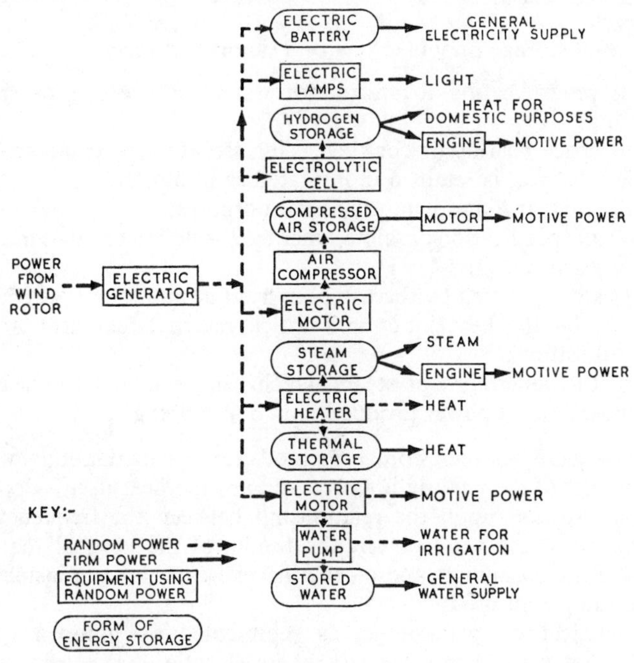

FIG. 100. *Uses of wind power directly and with storage*

energy from the wind power plant, the cost per kilowatt-hour is approximately doubled if even one fifth of the total energy is to be 'firmed' by a battery.

Hydrogen storage is included in the diagram of Fig. 100 because, technically, it is a possible method. Distilled water, the production of which from local (often saline) water would absorb some of the available energy, has to be used in electrolytic cells and the hydrogen generated by electrolysis must then be either compressed and stored in cylinders or stored, at low pressure, in a gas holder. It can be used for heating or cooking or to run an engine. In effect the hydrogen is a 'locally-produced fuel'. Unfortunately, as shown by A. H. Stodhart

(Ref. 25), the number of conversions of energy involved and the associated cost of equipment, particularly of the gas-holder, render the economy of the method doubtful.

Storage by compressed air, using a motor-driven compressor and storage cylinders may be a more economical method to serve the purpose of providing motive power. Here again, there must be several energy conversions but the need for distillation of water, at least, is ruled out.

Thermal storage may take one of a number of forms:

(*a*) water heating to afford hot water for washing or space heating;

(*b*) space heating or cooking, using storage-type equipment in which the heat is retained in insulated solid blocks;

(*c*) soil warming for horticultural purposes;

(*d*) the purification of saline water by distillation or other means;

(*e*) refrigeration;

(*f*) space heating by the use of crystalline substances which are melted by the heat input and which return latent heat at re-crystallization;

(*g*) the generation of steam which can be stored for heating purposes or for power production in a steam engine.

All of these thermal storage methods, except distillation—when the product of the heating is stored and not the heat itself—depend for their success upon the relationship between the frequency of occurrence of the wind power, the timing of the thermal requirements of the consumers and on the thermal capacity and insulation of the equipment used.

Mechanical energy storage, as represented by pumped water (having potential energy because of its elevation), is a very satisfactory method when a water supply is required. The practicable magnitude of such storage is governed by the water requirements of the community to be served, the size of reservoir which can be used and, in some countries with hot climates, by the limited period of time for which it is possible to store water in fresh condition.

For water pumping alone, non-electric windmills, having a slow running, multi-bladed windmill driving a pump, are commonly used. These must be located immediately adjacent to the water to be pumped and may sometimes suffer, therefore, from lack of wind. The alternative of using a wind-driven electric generator located at a more favourable site for wind, at a short distance away from the pumping site, and driving the pump by an electric motor, may often be worth while. This has the advantage of being a multi-purpose

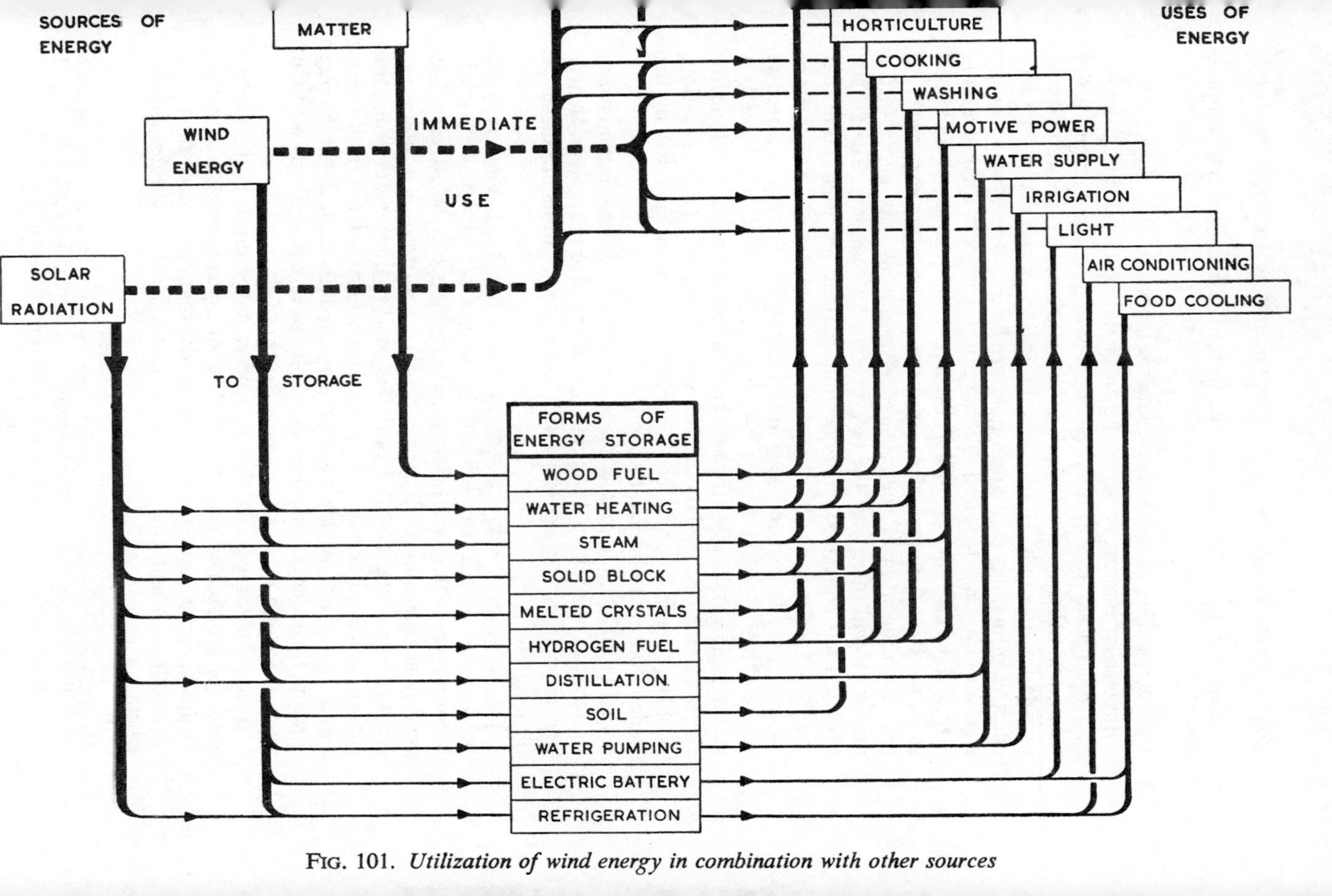

FIG. 101. *Utilization of wind energy in combination with other sources*

machine through its electric output, instead of being solely a pumping plant.

The combination of intermittent power sources

Increasing interest is being shown in the development of underdeveloped areas which might be capable of producing some of the food and other raw materials which the rapidly growing world population demands. Human labour, unaided, is not likely to be very effective in making such areas fully productive; energy supplies for mechanization are required. To import energy from a considerable distance, whether by electrical transmission or in the form of oil to drive a local power plant, may be prohibitively expensive, so that the utilization of such energy resources as may occur in the area itself will often give the best promise of economy. The author (Ref. 22) has considered this question, making the assumption that the three local sources of energy, all of intermittent availability, are wind, solar radiation and vegetable wastes. (The last may be used as fuel so that a steam engine becomes practicable and, indeed, a steam engine and boiler for this application are being developed in Great Britain by the National Research Development Corporation. A solar cooker has also been brought to fruition in India by the Indian National Physical Laboratory.)

In Fig. 101 an attempt is made to show how these three energy sources can be combined to fulfil, directly or through storage as may best serve, the purposes shown in the upper right-hand corner of the diagram.

The diagram is self-explanatory but a few brief notes may be helpful. The dotted lines indicate immediate use of the energy as it becomes available. Wind and solar radiation can be so used but vegetable matter must first pass through a process of burning or, perhaps, fermentation. The full lines passing downwards from the energy sources indicate, when they turn horizontal, the different forms of storage which may be employed in each case. Passing through the storage table, the lines link up storage methods with the purposes which they can be made to serve.

It may be agreed that these purposes cover most of the energy needs of an isolated community but, of course, careful planning, with automatic transference of the loads from one source of supply to another, would be required to make such a scheme operate satisfactorily.

BIBLIOGRAPHY

(1) *Hydro-electric potential in Europe and its gross, technical and economic limits.* Study by the Electric Power Section, Industry Division, United Nations, Economic Commission for Europe (Geneva, May 1953).

(2) *Prospects opened up by technical advances in electric power production.* Study by the Electric Power Section, United Nations, Economic Commission for Europe (Geneva, August 1952).

(3) *Transfers of electric power across european frontiers.* Study by the Electric Power Section, United Nations, Economic Commission for Europe (Geneva, August 1952).

(4) THACKER, M. S. *The role of energy in under-developed areas.* United Nations Educational, Scientific and Cultural Organization, Unesco Discussion Theme 1951, *Energy in the service of man* (Paris, 24th May 1951). UNESCO/NS/78.

(5) SIMON, F. E. *Energy in the future.* United Nations Educational, Scientific and Cultural Organization, Unesco Discussion Theme 1951, *Energy in the service of man* (Paris, 5th February 1951). UNESCO/NS/79.

(6) AILLERET, P. *Energy in its international aspects.* United Nations Educational, Scientific and Cultural Organization, Unesco Discussion Theme 1951, *Energy in the service of man* (Paris, 8th September 1950). UNESCO/NS/77.

(7) EGERTON, A. C. *Civilisation and the use of energy.* United Nations Educational, Scientific and Cultural Organization, Unesco Discussion Theme 1951, *Energy in the service of man* (Paris, 13th February 1951). UNESCO/NS/74.

(8) MCCABE, L. C. *World sources and consumption of energy.* United Nations Educational, Scientific and Cultural Organization, Unesco Discussion Theme 1951, *Energy in the service of man* (Paris, 11th June 1951). UNESCO/NS/75.

(9) EICHELBERG, G. *Utilization of energy.* United Nations, Educational, Scientific and Cultural Organization, Unesco Discussion Theme 1951, *Energy in the service of man* (Paris, 31st August 1950). UNESCO/NS/76.

(10) PUTNAM, P. C. *Power from the wind.* Van Nostrand (1948).

(11) THOMAS, P. H. *Electric power from the wind.* Federal Power Commission (1945).

(12) FATEEV, E. M. and ROZHDENSTVENSKII, I. V. Achievements of Soviet wind power engineering. *Vestnik Mashinostroeniya*, No. 9, pp. 24–27 (1952). (In Russian.)

(13) VEZZANI, R. Un impianto aeroelettrico pilota di media potenza con accumulo idrico di pompaggio. *L'Elettrotecnica*, Vol. XXXVII, No. 9, pp. 398–419, Del 15–25 (Settembre, 1950).

(14) HAMM, H. W. Pumped storage power plants in Europe. *Fiat Final Report* No. 1060, April 1947. (Office of Military Government of Germany (U.S.).)

(15) Final report on the wind turbine. *Research Report*, PB25370, Office of Production, Research and Development, War Production, Washington, D.C. (1946).

(16) LINNER, L. Der Parallelbetrieb eines Synchrongenerators an einen unendlich starken Netz bei Antrieb durch eine Honnef Gross-Windturbine. *E.T.Z.*, Vol. 9 (September 1948).

(17) WALL, T. F. Large wind-driven synchronous generators. *Engineering*, Vol. 155, pp. 421–3 (May 28th); pp. 461–3 (June 11th); pp. 501–3 (June 25th 1943).

(18) ANDRIANOV, V. N. and BYSTRITSKII, D. N. Parallel operation of a wind power station with a powerful grid. *Elektrichestvo*, No. 5, pp. 8–12 (1951). (In Russian.)

(19) SEKTOROV, V. R. The present state of planning and erection of large experimental wind-power stations. *Elektrichestvo*, No. 2, pp. 9–13 (1933). (In Russian.)

(20) ANSCOMBE, L. D. and ELLISON, A. J. Technical aspects of interchange with the grid. *Journal of the Institute of Fuel* (June 1949).

(21) GOLDING, E. W. Harnessing the wind. *Discovery*, pp. 373–378 (December 1953).

(22) GOLDING, E. W. Local energy sources for underdeveloped areas. *Impact*, Vol. V, No. 1, pp. 27–46 (Spring 1954).

(23) GOLDING, E. W. Economic aspects of the utilization and design of wind power plants. World Power Conference, Brazilian Sectional Meeting (1954).

(24) GOLDING, E. W. The utilization of wind power in desert areas. *Proceedings, International Symposium on Desert Research* (Jerusalem, 7th–14th May 1952).

(25) STODHART, A. H. A comparison of the costs of providing heat, light and power by hydrogen and oil. Electrical Research Association, *Technical Report* Ref. C/T111 (September 1953).

(26) SIL, J. M. Windmill power. *The Indian Journal of Meteorology and Geophysics*, Vol. 3, No. 2 (April 1952).

(27) ANDRIANOV, V. N. Stability of a synchronous generator driven by wind power and working on a power system. *Elektrichestvo* (October 1949).*

(28) SECTOROV, V. R. Operating conditions and types of wind power installations for rural districts. *Elektrichestvo* (October 1949).*

(29) ANDRIANOV, V. N. and POKATAIEF, A. J. Regulation of the output of a wind power station. *Elektrichestvo*, No. 6 (1952). (*Elektrotechnik und Maschinenbau*, No. 11 (Vienne, 1953).)*

(30) DUQUENNOIS, H. *Association de groupes anémo-électriques avec les usines hydro-électriques régularisées*. Paper given at the World Power Conference (Brazil, 1954).

(31) KROMS, A. Wind power stations working in connection with existing power stations. *A.S.E. Bull.*, Vol. 45, No. 5, pp. 135–144 (6th March 1954).*

(32) GOLDING, E. W. *The economic utilization of wind energy in arid areas*. Paper given at the UNESCO Symposium on Wind and Solar Energy (New Delhi, 1954).

(33) GOLDING, E. W. and THACKER, M. S. *The utilization of wind, solar radiation and other local energy resources for the development of a community in an arid or semi-arid area*. Paper given at the UNESCO Symposium on Wind and Solar Energy (New Delhi, 1954).

(34) KLOSS, M. The direct drive of synchronous generators by large wind power stations operating in parallel with a system of fixed frequency. *E.T.Z.*, Vol. 63 (1942), Nos. 31–32, pp. 362–367; Nos. 33–34, pp. 388–392.*

(35) CHRISTALLER, H. Utilization of the wind energy. *Elektrizitatswirtschaft*, Vol. 50, No. 11, pp. 320–322 (1951).*

(36) KROMS, A. The co-operation of power stations. *Technik*, Vol. 8, No. 6. pp. 395–406 (1953).*

(37) KROMS, A. Utilization of the excess power of interconnected systems. *Technik*, Vol. 7, No. 8, pp. 435–444; No. 10, pp. 581–586 (1952).*

(38) KROMS, A. Choice of the parameters of hydro-electric power stations. *A.S.E. Bull.*, Vol. 42, No. 18, pp. 698–706 (1951).*

(39) GOLDING, E. W. and STODHART, A. H. An energy survey in the Somaliland Protectorate. Electrical Research Association, *Technical Report*, IB/T14 (1954).

* Translated title.

CHAPTER 18

RESEARCH AND DEVELOPMENT

RESEARCHES on the performance of the old windmills and experiments aiming at their improvement have already been described: we have the work of J. Smeaton, Sir William Cubitt, A. Meikle and others in England, of the Dutch millwrights and, much later, of P. La Cour (Ref. 1) in Denmark. These were, in the main, individual efforts devoted to improving the construction of the mills, increasing their output and rendering them more convenient to operate for corn grinding or pumping. As far as can be ascertained, little research was done to develop a technique of site selection, probably because of the very limited choice of site imposed by the need to establish the mill near to its work. Again, there being no competition from other sources of power, and the materials of construction, as well as labour, being plentiful, intensive studies of wind behaviour were not called for: it was easier to construct the mill very solidly to withstand any wind forces which were likely to occur, with an ample margin of safety, and to accept a relatively low efficiency.

Since those early days research and development work of good quality has certainly been carried out during the last two or three decades by manufacturers of small and medium-sized windmills, both for water pumping and for electricity supply to isolated premises or communities (Ref. 2). Concurrently, also, there have been produced a number of designs—and some prototypes—for much larger aerogenerators and their authors have sometimes accompanied them by wind studies based on such meteorological records as have been available in the countries of their origin (Refs. 3–8).

These efforts, which have called for a very considerable amount of thought on the subject, should not be overlooked; but it seems clear that, if wind power is to be exploited on a significant scale in those parts of the world where it is likely to be economic, a much more concerted attack must be made on the problems involved.

These have already been discussed in some detail but it is worth while summarizing them as a basis for the formulation of a research programme.

Problems

They may be divided into two main categories containing (*a*) problems of construction and operation and (*b*) those concerned with wind behaviour.

(*a*) Construction and operation

The requirements for any wind turbine are that it shall be as cheap as possible in capital cost, robust enough to have a long life under vigorous climatic conditions and—for a large plant—that it shall operate in connection with a supply network automatically and with a low maintenance cost. These requirements call for theoretical and practical studies under the following heads:

(i) *Type of plant*. Clearly it is first necessary to decide upon the type which appears most likely to be worth detailed study. The propeller type, which is the modern counterpart of the old-fashioned sail windmills, certainly shows most promise, especially for large-scale operation. Its main advantage is the small amount of metal required to extract the energy from a considerable swept area in the wind. For small or medium size units, however, it is at least possible that another type having, perhaps, a vertical-axis rotor may be more economic.

Again, assuming a propeller type, what form of drive should be adopted for the generator itself—mechanical, hydraulic or pneumatic? Should the generator be placed aloft on the supporting tower or on the ground? What form of supporting structure is the best: a steel lattice tower, a concrete tower, a mast supported by guys, or some form of tripod structure?

Many suggestions have been put forward in an endeavour to reduce the cost of construction or to increase the output for a given size of rotor. Thus, is it worth while making the turbine orientate to face the wind or will it be so much cheaper to have the turbine facing in a fixed direction that the inevitable loss of output when the wind direction changes is more than counterbalanced by the reduced cost of the plant? Can some arrangement of ducting be used to increase the wind speed through the rotor and so reduce its diameter and cost?

(ii) *Size of plant*. On the assumption that the output from any practicable size of wind-driven plant can be absorbed, as it certainly could be when the plant is connected to a large network, the most economic size of unit must be determined. There is sufficient evidence that the larger the plant the cheaper it is likely to prove per kilowatt of capacity, although there must be some upper limit to this. It is probable, however, that in the present state of knowledge the limit to the size of wind turbine will be fixed by technical considerations rather than by economic.

(iii) *Design*. By costing studies, without entering into detailed design, a decision on both the type and size of plant can doubtless be reached. But this leaves unanswered many difficult technical questions concerning the effects of non-uniform distribution of wind speed over the swept area, of high-speed gusts, of rapid changes in wind direction and the form of control gear to relieve sudden heavy stresses and to maintain smooth operation when supplying load to a network of constant frequency.

Flexible blades (in the propeller type machine) with arrangements for very rapid, and closely-controlled, pitch changing are called for, but these are bound to introduce vibrational problems which, it is generally agreed among designers who have considered wind turbine construction, may be the most difficult part of the whole job. What are the possibilities of using fixed pitch blades with bracing (as used by Juul in Denmark (see p. 214)) and up to what rotor size could this form of construction be adopted economically?

Automatic operation of large wind turbines to allow them to cut in when the wind speed is high enough for them to supply load, and to shut down when the wind speed rises above a safe value, again presents problems, as does the question of protection on the electrical side to prevent unstable conditions or damage to plant in the event of faults on the system.

The materials to be used in the construction of wind-driven plant need careful consideration, particularly those for the blades and other rotating parts aloft, for the supporting structure and the protective coverings. It is, of course, essential that the plant shall be installed in well-exposed positions in whatever part of the world it may be built and this will imply, in different instances, the need to withstand the effects of snow and ice, of salt-laden winds and spray, of hot winds and sand storms and possibly also of chemical fumes and other atmospheric contaminations.

Erection and transport costs forming, as they do, an appreciable part of the total, the designer must adopt constructional methods which minimize these costs both at initial erection and for subsequent maintenance, including replacement of parts.

Clearly, the smaller the number of modifications in the main design features to cater for differing climatic and economic conditions in the various countries of their utilization, the greater the possibilities for series production with consequent reduction of manufacturing costs. For a given capacity (in kilowatts) the two main variables are the diameter of the wind rotor and the rated wind speed, and the aim would seem to be to cover all requirements in each category of utilization—large, medium or small-scale—by using not more than two or three values of each of these variables. Taking an overall power coefficient of 0·40 for capacities from 3000 kW down to

100 kW, 0·35 for 50 kW and 0·30 for 10 kW, the diameters for rated wind speeds of 35, 30 and 25 m.p.h. are as given in Table XXXVIII.

It can be seen that for a diameter around 200 ft the capacity can be made either 3000 kW, 2000 kW or 1000 kW according as 35 m.p.h., 30 m.p.h. or 25 m.p.h. is chosen for the rated wind speed.

Within close enough limits,

$$\frac{35^3}{28^3} = \frac{34^3}{27^3} = \frac{33^3}{26^3} = \frac{32^3}{25^3} = \frac{2}{1}$$

TABLE XXXVIII

Capacity (kW)	Diameter in feet for different rated wind speeds		
	35 m.p.h.	*30 m.p.h.*	*25 m.p.h.*
3000	205	—	—
2000	167	210	—
1000	118	149	195
500	83·5	106	138
100	37·5	47	62
50	28·5	35·5	47
10	13·7	17·2	22·5

so that a 2 to 1 ratio of capacities can be obtained by using the same diameter but different rated wind speeds such as one of the pairs 35/28, 34/27 etc.

(*b*) **Wind behaviour**

Investigations under this heading can be classified in accordance with their bearing upon (i) regional winds, (ii) the effects of local topography or (iii) wind behaviour at a chosen site.

(i) *Regional winds.* In most countries meteorological services have made wind measurements over a sufficient number of years for annual mean wind speeds occurring in different parts of the country to be known. Provided that the measuring methods used at the observation stations are standardized, isovent maps (see p. 68) can be drawn. These serve as a useful basis for more specialized studies of wind speeds. Meteorological stations have to serve several different purposes; wind measurements form only one part of their work. Their site is very seldom chosen for its exceptional windiness and, often, for convenience in making continuous observations the stations are established at such places as agricultural research establishments, and coastguard stations. It frequently happens, therefore, that the location of the anemometer is by no means ideal even for measurements of the wind in the locality of the station, e.g., it may be mounted at a low height above a roof top or may be partially screened by trees or buildings. Thus, some caution should be used in applying the results of measurements from any particular station unless it is known that the anemometer is well sited.

The greatest value of such measurements is that they are available for a long period and thus indicate the variations in wind speed which may be expected from year to year. They also give information on the duration of calm spells, which is important if energy storage is to be considered, and on the time of the day, in different seasons, when winds regularly occur. The diversity of occurrence of wind in widely separated districts of a country is also important if there is a prospect of aerogenerators being installed to feed their output into an extensive network.

(ii) *Effects of local topography.* The behaviour of wind flowing over flat country, and the increase in its speed with altitude, are known well enough, but comparatively little precise information exists on wind flow over hills, of the effects of different contours and of the screening effect of high ground in the vicinity.

The relative merits of ridges and isolated hills as wind-power sites, their optimum shapes and the possible advantages of gaps between hills, or near their summits, are matters needing study. Laboratory measurements on scale models of hilly country, using the technique suggested in Ch. 7, may throw light on them but in the final stages field measurements will be necessary.

(iii) *Wind behaviour at chosen sites.* The investigations mentioned in the preceding paragraph are concerned with the technique of site selection. Having chosen a site, knowledge of the behaviour of the wind there is valuable.

Measurements of wind speed and direction which may be needed are:

(*a*) hourly mean wind speeds at different heights;
(*b*) hourly wind directions;
(*c*) gust measurements to determine the magnitude, and rate of change of magnitude and direction, of gusts.

Such measurements might be in two directions at right angles in the horizontal plane and in the vertical plane.

It may be that not all of these measurements will be required at each chosen site: indeed it would be costly to undertake them frequently and it is probable that hourly mean wind speeds and directions at heights up to that of the top of the circle which would be swept by the wind-turbine rotor will generally suffice. Gust data may not differ appreciably from one hill to another in the same region but, of course, this remains to be proved.

Prototypes and their testing

In Ch. 12 and 13 various designs for wind turbines which have been put forward from time to time were discussed. Although the number of such designs is legion, and most of these would doubtless result in a workable plant, they can all be placed in quite a small number of categories several of which can quickly be ruled out as unsuitable for large-scale plant and as showing no advantages over the commonest—the propeller type—even for medium or small-scale use.

Although there may be room for further design studies it appears at present unlikely that any other type of plant will prove generally superior to the propeller type. But this still leaves considerable scope for inventiveness in the design of the various components so that a number of different machines may emerge, through several stages, to their final form.

The number of stages required will depend on the target size; the largest may call for two or three while one may suffice for the smallest. Opinions differ on the question of prototypes. Some designers contend that the problems to be solved with large plant, of several thousand kilowatts capacity, are not met in prototypes which are scaled down to perhaps one tenth, or some even smaller fraction, of the eventual size, and that it is therefore pointless to construct them. In brief, build full scale or not at all. This view appears to have been held in the past in Germany and, to some extent, in the United States.

While undoubtedly this is a bold attitude to adopt, it is one which demands large financial resources for its support and which calls for collaboration by a considerable number of scientists and engineers expert in their particular field. Even so, it is a matter of pushing forward right up to, or even beyond, the frontiers of existing knowledge of aeronautical engineering. It is probably safe to say that no such team of designers could carry through a project of this kind without a large amount of experimentation to support calculations and it becomes a question of choosing the point at which these experiments should start.

With sufficient support from public funds, either Governmental or from a large public utility authority, such a method could be followed, but it must be recognized that a large sum of money would be involved, accompanied by the risk of no very tangible result for some years.

The alternative is to develop step-by-step through smaller prototypes which, in themselves, may have applications in medium-scale use. This more cautious method limits expenditure in development at the outset leaving the major expenditure to be incurred after much experience, both in construction and performance testing, has been gained. It has the advantage that it can be followed by manufacturing firms with co-operation from electricity authorities, thus making possible the trial of more than one design of plant and providing visible results in a reasonably short time. Unlike some other new sources of energy, wind power lends itself to this stepped form of development. It is not essential to expend very large capital on an initial venture which must always have an attendant risk. Again, it is improbable that there is but one answer to the problem of economic development of wind power plant, and competition between several independent, or semi-independent, groups is healthy.

However the question resolves itself, it is certain that a number of prototypes of different designs will be built and tested. Ideally, though it is perhaps impracticable, they should be tested under identical conditions for a sufficient length of time to discover, not only their performance in producing power at given wind speeds, but in giving satisfactory service in practical operation with little maintenance. With machines of a few kilowatts capacity it is possible to envisage simultaneous testing on the same site when, after the inevitable 'teething troubles' have been overcome, a useful comparison between them can be made.

For larger plants the comparison will have to be through tests of their performance in the different wind régimes occurring at their sites and it is clear that, to avoid the results being misleading, as nearly as possible identical methods of measurement, both of wind and of electrical output, must be used.

There is some evidence (see Ch. 9) that, at least for a few seconds at a time, wind speeds steady to within 2 or 3 m.p.h. exist over an area corresponding to the swept circle of the wind turbine rotor. This should make it possible, by waiting for these periods of steady wind, to obtain a curve of power output at different wind speeds—by a 'natural wind tunnel' test.

In addition, a criterion of successful operation must always be the total annual output of energy in a known wind régime. Care will be needed in assessing the economy of operation from results of this kind. Sites may not be alike in gustiness so that one plant may be subjected to a more severe test than another, even though this is not apparent from measurements of hourly wind speeds. Again, the electrical operating conditions on different parts of the network will vary as well as the setting of the output controls of the machines themselves. Thus it may be difficult to determine the relative economy expressed in terms of annual output per unit of capital cost, especially as the latter will not be known with any precision as applying to series production. Nevertheless judgments could be made of the adequacy of the design to meet the site conditions and of its potential economy if development were pursued to the ultimate stage.

The probability is that such tests will be most valuable in suggesting ways of improving the construction and of reducing its costs before too much expenditure in effort and finance has been incurred.

BIBLIOGRAPHY

(1) La Cour, P. *Die Windkraft und ihre Anwendung zum Antrieb von Elektrizitäts-Werken.* M. Heinsius Nachfolger (Leipzig, 1905).

(2) Stodhart, A. H. A comparison of the costs of providing heat, light and power by hydrogen and oil. Electrical Research Association, *Technical Report*, Ref. C/T111 (September 1953).

(3) Sektorov, V. R. The present state of planning and erection of large experimental wind-power stations. *Elektrichestvo*, No. 2, pp. 9–13 (1933). (In Russian.)

(4) Hamm, H. W. German wind-turbine projects planned during the Hitler era. *Fiat Final Report* No. 1111 (May 1947).

(5) Witte, H. *Windkraftwerke.* Rudolf A. Lang (Possneck, 1950).

(6) Heys, J. W. van. *Wind und Windkraftanlagen.* Georg Siemens (Berlin, 1947).

(7) Vezzani, R. Un impianto aeroelettrico pilota di media potenza con Accumulo idrico di pompaggio. *L'elettrotecnica,* Vol. XXXVII, No. 9, pp. 398–419 Del 15–25 (Settembre 1950).

(8) Juul, J. Application of wind power to rational generation of electricity. *Elektroteknikeren*, 43, 137–48 (7th August 1947). (In Danish.)

(9) Golding, E. W. Electricity generation by wind power. *Research* (April 1953).

(10) Golding, E. W. Large-scale generation of electricity by wind-power—preliminary report. Electrical Research Association, *Technical Report,* C/T101 (1949).

(11) Organization for European Economic Co-operation. Committee for Productivity and Applied Research, Working Party No. 2 (*Wind Power*) (1954).

CHAPTER 19

INTERNATIONAL CO-OPERATION

IN earlier chapters the efforts which have been made in the past to develop wind power utilization in various countries have been mentioned. These efforts have led to the construction of wind turbines of medium or large-size in the United States, Russia, Germany and Denmark and to designs for much larger plants in these and other countries. But until recently there had been no concerted attempt at international co-operation in wind power research and development.

The subject had been discussed at international meetings such as those of the World Power Conference and, since the World War II, at several semi-private conferences of a national or international character particularly concerned with post-war development of energy resources. In 1950 the Organization for European Economic Co-operation (O.E.E.C.) established a wind power group to promote co-operation in research between European countries interested in the subject. Before discussing the work of this group it might be well to review briefly the reasons why some degree of co-operation appears to be desirable.

The need for co-operation.

The full development of wind power comprises three main fields of endeavour: (*a*) the design and construction of the power-producing plant; (*b*) study of the behaviour of the wind and of the influence of that behaviour upon power potentialities in different localities and (*c*) the technique of utilizing the available energy in the most economic way under different geographical and social conditions. The three are, of course, interdependent and information gained from investigations or experience in any of them must be of interest to those mainly occupied in the others. Nevertheless each demands specialized knowledge and, in general, separate groups of investigators who can collaborate to ensure that each group works on lines which are most likely to be effective. Some co-ordination is obviously necessary and it is a question how this can best be achieved.

In the more industrialized countries, with large and adequately financed electricity supply and research organizations, such co-ordination is a relatively simple matter but, even so, consideration must be given to the fact that the three necessary parts of the development will not usually be under the same controls. There will be a division between the user of the plant and its manufacturer, although it cannot be a very sharp division since the two must work closely together if successful development is to take place. The user needs plant which can be supplied at a low enough constructional cost to be economic in his particular circumstances of climate and of utilization, while the manufacturer must be informed of those circumstances and must be sufficiently assured of demand to enable him to plan for production in quantity so as to reduce production costs.

(*a*) *Design and construction*. The main responsibility for the first of these three components of development—the design and construction of plant—must rest with the manufacturing firms who undertake it. For the production of small and medium-sized machines the incentive of an adequately large market should be sufficient since the capital required for development will not be very large. But the finance needed for the construction of aerogenerator units of 1000 kW or more is of a different order: even in the post-development stage the cost per unit may lie between £50,000 and £100,000 and the first full-scale prototype will cost considerably more. Clearly, therefore, such a project cannot be undertaken by a manufacturer without either financial support from public funds or, at least, some guarantee that development costs can be recovered from further orders by electricity supply authorities. It is possible that a major experiment, involving the building of a large aerogenerator, could be undertaken on an international basis with the participating countries sharing the cost through an agreement between Governments or between appropriate national organizations. But there are obvious difficulties, both political and administrative, in such a plan and, since the costs involved are not too large to be borne nationally, it may be best avoided. There is little doubt that, with sufficient conviction concerning the economic potentialities of large-scale wind power generation the finance needed for its development can be provided. Again, it would not be realistic to suppose that all countries who may be interested in the use of such large sets would be equally interested in their manufacture for which, undoubtedly, very considerable engineering and aeronautical knowledge and facilities must be available. The scope for international co-operation in this sphere, at least in the initial stages of development, thus appears to be limited, although the need for it might arise later.

(*b*) *Studies in wind behaviour*. Since it must be acknowledged that the successful application of this form of power depends largely upon accurate information on wind behaviour, both for the design and siting of the plant, it follows that such information should be obtained for all interested countries. Further, there is no apparent reason for keeping it secret but, on the other hand, there are good reasons for making it generally known. Wind surveys to determine mean wind speeds at potential power sites in different parts of a country, as well as measurements of the detailed structure of the wind over terrains of varied character, should be made nationally. But the methods used, and the form in which the results are expressed, should be internationally accepted if the investigations are to serve their full purpose.

There is, of course, wide variety in wind-measuring equipment and perhaps it would be presumptuous to suggest that any one form should be standardized. But, at least, intercomparison of instruments is necessary and there should be sufficient similarity between their methods of use to ensure that the results expressed, for example, as annual mean wind speeds, are strictly comparable and apply to the same height above ground in open situations. Otherwise the position becomes chaotic. Imagine, for instance, attempting to relate mean wind speeds measured in different parts of the world, in one case by a cup-contact anemometer erected on a 10 ft pole on a smooth hill-top, in another by a cup, generator anemometer at perhaps 10 m above broken flat country and read at the end of each six hours or less frequently, and in a third by a Dines anemometer the head of which is a few feet above a roof top surrounded by trees at no great distance. Effective comparison becomes a hopeless task: yet the instances chosen are not exaggerations.

Without over-emphasizing the point, it is obvious that there is scope here for international agreement and co-operation both in deciding upon the measuring methods and in defining the results. At the beginning of this new work—or, at least, this study of a new aspect of an age-old natural source of energy—definitions of many quantities to be dealt with are important to avoid subsequent confusion.

(*c*) *Technique of utilization.* There are two parts to this subject: (i) large-scale utilization and (ii) medium and small-scale utilization.

As for the first, strictly speaking, there is no problem connected with the use of the energy generated because this will be fed into the network to be used in the same way as energy supplied to the network from other sources. The problems here are rather those of operation of a fairly large generator which will cut in or cut out without warning when the wind speed is variable about the lowest operating limit. This, and the possibility of such generators being located at the far ends of transmission lines or spurs, may lead to electrical disturbances on the system although it is not generally anticipated that these will be serious.

The operation of a fairly large volume of wind power, together with other forms of generating plant, may call for some thought to ensure the maximum operating efficiency, especially if hydropower is involved. Occasional international meetings for the exchange of experiences would thus be useful.

With medium and small-scale wind-driven plant, the problems are more definitely those of using the random-occurring energy fully, although if they are used in conjunction with small networks, with an alternative source of supply, there might be problems in the control of the power sources to maintain constant voltage or frequency.

Different climatic conditions and modes of life in various countries where these smaller sets may be used afford the opportunity of obtaining a mass of information on utilization which should be made freely available. Mutual advantage would result from comparison of methods and the discussion of possible difficulties.

The organization of international co-operation

As already mentioned, an O.E.E.C. group was established early in 1950 to encourage co-operative research and development in wind power.

The group has held several meetings, in Paris and in London, which have been attended by representatives of a number of European countries that have wind power potentialities. National wind-power research centres have been designated and arrangements have been made for the collection and exchange of information and for its decimal classification to facilitate filing. Wind-measuring methods and equipment have been discussed and, to some extent, standardized. There have also been some exchanges of anemometers to obtain an intercomparison of the various types in use.

Perhaps the most important outcome of the group's establishment has been the free exchange of information on the national activities in the subject and of ideas for future developments (Ref. 1). Since, however, many countries outside Europe have become seriously interested in wind power during the last year or two it appears desirable to widen the membership of the group, for which purpose some new form of organization may be needed and is, in fact, being discussed.

UNESCO

UNESCO, as part of the plan for assistance to underdeveloped countries, under its Arid Zone Committee has displayed an interest in the subject and has sponsored some preliminary work on it.

A UNESCO Symposium on Wind and Solar Energy was held in New Delhi in October, 1954 (Refs. 2, 3, 10 to 15).

World Power Conference

As part of the activities of its section on less conventional sources of energy, World Power Conference has had several papers on wind power presented to it. At the Sectional Meeting in Brazil in July, 1954, one complete session of the Conference was on different aspects of the subject of wind power (Refs. 4 to 9).

BIBLIOGRAPHY

(1) Organization for European Economic Co-operation. Committee for Productivity and Applied Research, Working Party No. 2 (*Wind Power*) (1954).

(2) Lacroix, G. *L'utilisation de l'énergie éolienne: les machines éoliennes.* Organisation des Nations Unies pour l'Education, la Science et la Culture (Paris, le 21 Août 1953). UNESCO/NS/AZ/143.

(3) Golding, E. W. *The economic and practical aspects of utilizing wind energy in arid areas.* United Nations Educational, Scientific and Cultural Organization (Paris, 10th August 1953). UNESCO/NS/AZ/139.

(4) Golding, E. W. *Economic aspects of the utilization and design of wind power plants.* Paper given at the World Power Conference (Brazil, 1954).

(5) Andrianov, V. N. and Sazonov, N. A. *Utilization of wind energy for the electrification of agriculture in the U.S.S.R.* Paper given at the World Power Conference (Brazil, 1954).

(6) Duquennois, H. *Association de groupes anémo-électriques avec les usines hydro-électriques régularisées.* Paper given at the World Power Conference (Brazil, 1954).

(7) Oniga, T. *Características brasileiras para o aproveitamento da energia eólica.* Paper given at the World Power Conference (Brazil, 1954).

(8) Hütter, U. *The use of wind energy for generating electric current in Western Germany.* Paper given at the World Power Conference (Brazil, 1954).

(9) Vezzani, R. *Organisation systématique des observations concernant la disponibilité et l'utilisation de l'énergie éolienne pour la production de l'électricité—centrales aéro-électriques.* Paper given at the World Power Conference (Brazil, 1954).

(10) Hütter, U. *Planning and balancing of energy of small-output wind power plant.* Symposium on Wind and Solar Energy (New Delhi, 1954). UNESCO/NS/AZ/191/Ann.14.

(11) Golding, E. W. *The economic utilization of wind energy in arid areas.* Symposium on Wind and Solar Energy (New Delhi, 1954). UNESCO/NS/AZ/191/Ann.3.

(12) Golding, E. W. and Thacker, M. S. *The utilization of wind, solar radiation and other local energy resources for the development of a community in an arid or semi-arid area.* Symposium on Wind and Solar Energy (New Delhi, 1954). UNESCO/NS/AZ/191/Ann.7.

(13) Dresden, D. *Wind energy: general review paper.* Symposium on Wind and Solar Energy (New Delhi, 1954). UNESCO/NS/AZ/191/Ann.1.

(14) Frenkiel, J. *Wind power research in Israel.* Symposium on Wind and Solar Energy (New Delhi, 1954). UNESCO/NS/AZ/191/Ann.22.

(15) Juul, J. *Wind machines.* Symposium on Wind and Solar Energy (New Delhi, 1954). UNESCO/NS/AZ/191/Ann.11.

(16) World Meteorological Organization. Energy from the Wind. Assessment of suitable winds and sites. *Technical Note No.* 4. WMO—No. 32. TP.10 (1954).

SELECTED BIBLIOGRAPHY OF SURFACE WIND DATA

GENERAL

LONDON, AIR MINISTRY, METEOROLOGICAL OFFICE, Handbooks of weather over the oceans and coastal regions:

Weather on the west coast of tropical Africa from 20° N. to 20° S. including the Atlantic Ocean to 25° W. (M.O. 492, 1949.)

Weather in the China seas and in the western part of the North Pacific Ocean (M.O. 404).

Weather in home waters and the north-eastern Atlantic (M.O. 446).

Weather in the Indian Ocean to latitude 30° S. and longitude 95° E., including the Red Sea and Persian Gulf (M.O. 451).

Weather in the Mediterranean (M.O. 391).

KÖPPEN and GEIGER. *Handbuch der Klimatologie:*

Band II, Teil G, Klimakunde von Südamerika, von K. Knoch.

Band IV, Teil H, Klimakunde von Mittelamerika, von K. Sapper.

Band II, Teil I, Westindien, by R. de C. Ward and C. F. Brooks.

Band II, Teil J, The climates of North America. Lief i. Mexico, U.E., Alaska, by R. de C. Ward and C. F. Brooks.

Band II, Teil K, Klima des Kanadischen Archipels und Grönlands, by H. U. Sverdrup, J. Petersen and F. Loewe.

Band III, Teil L, Klima von Nordwesteuropa und der Inseln von Island . . ., von B. J. Birkeland und N. J. Foya.

Band III, Teil M, Klimakunde von Mittel—and Sudeuropa, von E. Alt (Berlin, 1932).

Band III, Teil N, Klimakunde von Russland in Europa und Asien. 2 Hälfte Tabellen, von W. Köppen (1939).

Band IV, Teil R, Klimakunde von Hinterindien und Insulindien von C. Braak (1931).

Band IV, Teil S, Australien und Neuseeland, by Griffith Taylor and E. Kidson (Berlin, 1932).

Band V, Teil X, The climate of Rhodesia, Nyasaland and Mozambique Colony, by C. L. Robertson and N. P. Sellick.

HANN, J. *Handbuch der Klimatologie*, Band 3 (Stuttgart, 1911).

HANN, J. and SURING, R. *Lehrbuch der Meteorologie*. 4th edn., p. 448 (Leipzig, 1926).

BUCHAN, A. Report on atmospheric circulation. *Report on Sci. Results of voyage of H.M.S.* Challenger, Vol. 2, Part 5 (London, 1889).

WEGENER, K. Deutschen Grönland-Expedition A. Wegener 1929 und 1930–1931 Wissenschaftliche Ergebnisse. Bd. 4. *Meteorologie*, Z. Halbband. *Die Ergebnisse*, by R. Holzapfel, W. Kopp and K. Wegener (Leipzig, 1939).

Pilots, published by London, Admiralty, Hydrographic Department. (There is a large number of these pilots covering most of the world.)

EUROPE

General

London, Air Ministry, Meteorological Office. Meteorological report on the English Channel and northern France, 2nd edn., *Aviation Met. Rep.*, No. 9 (London, 1943). (M.O.M. 365/a.)

London, Air Ministry, Meteorological Office. Meteorological report on S. Scandinavia and Denmark. *Aviation Met. Rep.*, No. 8 (London, 1940). (M.O.M. 365/8.)

London, Air Ministry, Meteorological Office. Meteorological report on Arctic Scandinavia. *Aviation Met. Rep.*, No. 22 (London, 1944). (M.O.M. 365/22.)

London, Air Ministry, Meteorological Office. *Meteorological report on Scandinavia, Finland and Denmark.* (London, 1947).

London, Air Ministry, Meteorological Office. Meteorological report on Poland and East Prussia. *Aviation Met. Rep.*, No. 6 (London, 1939).

Thiemann, K. Klimatologische Unterlagen zur Feststellung der Anwendbarkeit kunstlicher Vernebelung in europaischen Kustengebieten. *Marineobs. Ber.*, No. 4 (Hamburg, 1944).

Austria

Dimitz, L. Die regionale verteilung und der Jährliche Gang der Windgeschwindigkeit im Österreich. *Wien. Jb. ZentAnst. Met. Geodyn.*, 86, 1949, *Anh.* 4, p. D22 (1938–1948).

London, Air Ministry, Meteorological Office. Meteorological report on Bohemia, Moravia and Austria (excluding Tyrol). *Aviation Met. Rep.*, No. 11 (London, 1941). (M.O. Misc. 365/11.)

London, Air Ministry, Meteorological Office. Meteorological report on the middle Danube area. *Aviation Met. Rep.*, No. 23, F°., pp. 53 (London, 1944). (M.O.M. 365/23.)

Belgium

Peppler, W. Aerologische und Hydrographische Beobachtungen der D. Marine-Stationen während der Kriegszeit 1914–1918. See und landbrise an der flandrischen Kuste. *Aus. d. Arch. dtsch. Seew.*, 47 (Hamburg, 1922).

Lancaster, A. La force du vent en Belgique. *Ann. meteor.*, p. 221 (1903).

Paris, Office National Météorologique. *Étude climatologique sommaire sur la Belgique* (Paris, 1944).

Godart, O. Force moyenne du vent. Bruxelles, Régie Voies Aérien., *Serv. Mét. B. Mens.*, pp. II–XX (Mars/Mai, 1951).

British Isles

Gold, E. Wind in Britain. The Dines anemometer and some notable records during the last 40 years. *Q.J. R. Met. Soc.*, 62, pp. 167–206.

London, Air Ministry, Meteorological Office:

Monthly percentage frequencies of surface and upper winds over the British Isles between the years 1920 and 1934. (M.O. 433 (1939).)

Table of wind direction and force over the British Isles. (M.O. Misc. 370 (1939 and 1943).)

Meteorological report on southern England. *Aviation Met. Rep.*, No. 15 (1941). (M.O.M. 365/15.)

Meteorological report on Scotland. *Aviation Met. Rep.*, No. 13 (1941). (M.O.M. 365/13.)

Meteorological report on Northern Ireland. *Aviation Met. Rep.*, No. 16 (1942).

Bulgaria

KASSNER, C. Über die Winde an der bulgarischen Schwarzmeerkuste. *Ann. Hydr.* 59, p. 403 (1931).

LONDON, AIR MINISTRY, METEOROLOGICAL OFFICE. Meteorological report on Bulgaria. *Aviation Met. Rep.*, No. 20 (1943). (M.O.M. 365/20.)

Danzig

STABEN, J. Beiträge zum Klima von Danzig. *Veröff St. Obs.* (Danzig, 1939).

STABEN, J. Zum Klima von Danzig-Neufahrwasser. *Schr. naturf. Ges., Danzig.*, 17, p. 273 (1926).

Denmark

COPENHAGEN, DET. DANSKE METEOROLOGISKE INSTITUT. *Danmarks-Klima, Tabeller og kort* (1933).

Finland

JOHANSSON, O. V. Suomen ilmasto. The climate of Finland. *Mitt. Met. Inst. Univ.*, No. 33 (Helsingfors, 1936), p. 49 and G. Handb. Finnlands, p. 203 (Helsinki, 1936). (Translation available.)

KUUSKOSKI, U. Die Windverhältnisse in südöstlichen Finnland. *Mitt. Met. Inst. Univ.*, 34, pp. 1–186 (Helsingfors, 1936). (German summary, pp. 193–199.)

France

BIGOURDAN, G. *Le climat de la France. Témperature, pression, vents* (Paris, 1916).

ROUCH, J. *Notice météorologique sur les côtes de France et d'Algérie* (Paris, 1919).

ANGOT, A. Études sur le climat de la France. Régime des vents. *Ann. Bur. Cent. Météor.* (Paris, 1907). (1881–1900.)

Germany

LONDON, AIR MINISTRY, METEOROLOGICAL OFFICE. Meteorological report on northern Germany. *Aviation Met. Rep.*, No. 7 (1939). (M.O.M. 365/7 (1943).)

ASSMANN, R. Die Winde in Deutschland. *Im Auftrage der Motorluftschiff-Studiengesellschaft in Berlin* (Braunschweig, 1910). (Seasonal.)

HANKOW, G. SICHT. Windrichtung und Windstärke im Nordwest—und West-deutschland. *Erfahrb. D. Flugw.*, 7, No. 10, p. 65 (Berlin, 1932).

BERLIN, REICHSAMT FÜR WETTERDIENST. Klimakunde des Deutschen Reiches. Bd. II, *Tabellen.* Deitrich Reimer (Berlin, 1939).

ARNOUX, J. *Éléments de climatologie des Pays Rhénans du Sud.* Palatinat Rhénan-Bade-Sud Wurtemberg. Published by Mission Météorologie Française en Allemagne et en Autriche (1949). (Many tables, 1881–1930.)

FRANKENBERGER, E. and RUDLOFF, W. *Windmessungen an den Quickborner Funkmasten*, 1947–48 (Hamburg, M.A.N.W.D., 1949). (Monthly means with velocity many stations.)

Holland

VISSER, S. W. Het klimaat van nederland. *Hemel dampkr., Den Haag, Suppl.* No. 2, p. 27 (1949).

LABRIJN, A. The climate of the Netherlands during the last two and a half centuries. *De Bilt, Meded. Verh.*, No. 49 (1945). (Monthly, annual and seasonal means . . . wind 1700–1944.)

LABRIJN, A. Het klimaat van Nederland. Temperataur Neerslag en wind. *De Bilt. Meded. Verh.*, Ser. A, No. 53 (1948).

Hungary

RONA, A. Das Klima von Ungarn. *Met. Zs.* 28, pp. 16–28, 53–66 (1911).
RÉTHLY, A. Das Klima von Ungarn. Budapest Klein. *Veröff. Reichanst. Met. Erdmag.* (*Neue Reihe*), No. 3 (1937).
KAKAS, J. Wind direction frequencies on Hungarian arifields. *Idöj.*, 51, pp. 58–68 (Budapest, 1947). (Hungarian. Abs. English, p. 101.)

Iceland

LONDON, ADMIRALTY, NAVAL METEOROLOGICAL BRANCH. *Wind in Icelandic fjords.* Memo. 108/42 (London, 1942), also by G. H. T. Kimble, *et al.*, in *B. Amer. Met. S.* 27, p. 216 (1946).

Italy

EREDIA, F. *I venti in Italia* (Rome, 1909).
MARINA, L. Nota climatiche per le principali citta costieri dell' Adriatico. Boll. bimens. . . .
CRESTANI, G. Climatologia (*trattato Italiano de Igeiene, Monografia* No. 17). (Torino, 1931.)
EREDIA, F. Sulla variazione della frequenza dei venti. *Riv. Met. Aeron.* 4, No. 4, p. 43 (Rome, 1940).
LONDON, AIR MINISTRY, METEOROLOGICAL OFFICE. Meteorological report on the Alps. *Aviation Met. Rep.*, No. 21, M.O.M. 365/21 (1944).

Lithuania

PAKSTAS, K. *Le climat de la Lithuanie* (Klaipeda, 1926).

Luxembourg

LEHR, E. Un siècle d'observations météorologiques appliquées à l'étude du climat Luxembourgeois. *Serv. Mét. Hydr.* (Luxembourg, 1950).

Norway

SPINNANGR, F. Om terrengets virkning på vindene i Vest-Norge og fiskenbankene utafor (on the influence of the terrain on winds of W. Norway and fishing banks off coast). *Norsk Tidsskr. Sv., Horten*, 57, p. 434 (1939).

Poland

BARTNICKI, L. Les courants atmosphériques en Pologne. *Étud. géophys., Varsovie*, Fasc. 3, p. 3 (1930).
DIEKERMANN, C. Die Windverhältnisse an der vorpommerschen Küste mit besonderer Berücksicktigung der Anemometer-Beobachtungen von Wustrow 1887 bis 1910. *Ann. Hydrogr.*, 42, p. 581 (Berlin, 1914).
GIERE, W. Die Windverhältnisse an den ostbaltischen Küsten. *Arb. NaturfVer.*, Heft 20 (Riga, 1933).
KOSCHMIEDER, H. Danziger Seewindstudien I. *ForschArb. Met. Inst. Danzig*, Heft 8 (Leipzig, 1936).

Portugal

LONDON, AIR MINISTRY, METEOROLOGICAL OFFICE. Meteorological report on Spain and Portugal. *Aviation Met. Rep.*, No. 12 (1941).
CARVALHO, A. F. *de Clima de Coimbra.* Observatória Meteorológico da Universidade de Coimbra (Lisbon, 1922).

FERREIRA, H. A. *O clima de Portugal.* Valores médios dos elementos climáticos no periodo, Fasc. I (1901–1930). (Lisbon, *Obs. Inf. D. Luiz*, (1942), Fasc. II (1942), Fasc. III (1943), Fasc. IV (1945), Fasc. VI (1950).

PEREIRA, A. B. *Clima de Coimbra.* Normais e valores dos elementos principais. *Inst. Geofis.* (Coimbra, 1942). (1886–1940.)

Roumania

BUCHAREST, INSTITUTUL METEOROLOGIC CENTRAL. *Date climatologice*, Vol. I, No. I; I. *Moyennes déduites de la période 1896–1915 de 32 stations de IIe ordre;* III. *Moyennes horaires des élements météorologiques de Buca rest.* Recueillies par Const. A. Dissesco (Bucarest, 1931).

Russia

LENINGRAD, GEOPHYSIKALISCHES ZENTRAL OBSERVATORIUM. *Klima der Unia der Socialistischen Sowjest-Republiken.* Teil II, Luftruck und Wind in der U.S.S.R., von A. Kaminsky. Text and atlas (Leningrad, 1932).

Sweden

ÅNGSTRÖM, A. Sveriges klimat. *Generalstat Litogr. Anst.* (Stockholm, 1946).

STOCKHOLM, SWEDISH MET. AND HYDROLOGICAL INSTITUTE. *Contributions to aeronautical climatology of Sweden*, Pt. I (Stockholm, 1949).

OSTMANN, C. S. Om storman vid. Svealands och Götalands Kuster. *Medd. met.-hydr. Anst., Stockholm*, 1, No. 4 (1922).

ÅNGSTRÖM, A. Some characteristics of the climate of Stockholm. *Geogr. Ann., Stockholm*, 14, p. 165 (1932).

WALLÉN, A. *The climate of Sweden.* Statens Meteorologist-Hydrografiska Anstalt (Stockholm, 1930).

Yugoslavia

VUJEVIČ, PAUL. Climate (of Yugoslavia). (Extrait du *Royaume de Yougoslavie, Aperçu Géog. et Ethnogr*) (Belgrade, 1930).

SPLIT, HYDROGRAFSKI INSTITUT, METEOROLOSKI OTSJEK. *Klimatoloski Konsultacija.* (Climatological summary.) Year 1, No. 1–6 (Spranj) (July 1946–December 1946). Became: *Vremenske prilike* (Monthly bulletin) Year 2, No. 1 (January 1946).

MILOSAVLJEVICA, M. Fizičke osobine vetrova u. Beogradu. (Les qualités physiques des vents à Belgrade.) (*Univ., Prir-Mat. Fak.*) (Beograd, 1950). (French Abs. pp. 65–67.)

AFRICA

PARIS, OFFICE NATIONAL MÉTÉOROLOGIQUE. Études climatologiques des provinces françaises. Pt. 20, *Corse-Afrique du Nord* (*Corse, Maroc, Algérie, Tunisie*). (Paris, 1943.)

BOUGNOL, M. Note sur le régime du vent à Port Etienne. *Publ. Com. hist. sci. A.O.F.*, Sér. B, No. 3, p. 33 (Paris, 1937).

BROOKS, C. E. P. and DURST, C. S. The winds of Berbera. *Prof. Notes, Met. Off.*, 5, No. 65 (London, 1934).

BROOKS, C. E. P. and DURST, C. S. The circulation of air by day and night during the S.W. monsoon near Berbera, Somaliland. *Quart. J.R. Met. Soc.*, 61, p. 167 (London, 1935).

EREDIA, F. Le osservazioni anemologiche eseguite nella Libia a nell'Impero Italiano . . . *Met. prat.*, p. 93 (Perugia. 18, 1937).

PERRET, R. Le climat du Sahara. *A. Geog.*, 44, p. 162 (Paris, 1935).

FANTOLI, A. *Dati medi di alcune localita caratteristiche dell'A.O.I. Africa Italiana*, 2, No. 11 (1939).

PARIS, OFFICE NATIONAL MÉTÉOROLOGIQUE. *Aperçu climatologique sur le parcours projecté du Transsaharien* (Paris, 1941).

Meteorological means for the period 1926–1935. M.O. (London, 1939).

LONDON, ADMIRALTY, HYDROGRAPHIC DEPARTMENT, NAVAL METEOROLOGICAL DEPARTMENT. Summary of weather conditions at certain ports in W. and E. Africa: The Gulf of Guinea including the Gold Coast and W. Nigeria. *Suppl. No. 2 to Memo 102/40* (London, 1941).

CAIRO, PHYSICAL DEPARTMENT. *Climatological normals for Egypt and the Sudan, Candia, Cyprus and Abyssinia* (1938).

CAIRO, MINISTRY OF WAR AND MARINE, METEOROLOGICAL DEPARTMENT. *Climatological normals for Egypt* (Cairo, 1950).

PRETORIA, WEATHER BUREAU. *Surface winds of South Africa* (Pretoria, 1949).

UNION OF SOUTH AFRICA, UNION OFFICE OF CENSUS AND STATISTICS. *Official Yearbook of the Union and of Basutoland, Bechuanaland Protectorate and Swaziland.* Nos. 6, 7, 9, 10, 11, 12.

DAKAR, SERVICE MÉTÉOROLOGIQUE DE L'AFRIQUE. Occidentale Française. *Memto No. 7B, Fréquences* (Rufisque, 1942).

Azores, Madeira and Canary Islands

TANNEHILL, I. R. *Hurricanes: their nature and history* (Princeton and London, 1938).

TANNEHILL, I. R. Zum Klima der Azoren und der Insel Madeira. 3. Hydrometeore und Wind. *Zöst. Ges. Met., Wien*, 6, p. 408 (1871).

FICKER, H. Richtung von Wind und Wolken auf Teneriffa. *S.B. Akad. Wiss. Wien. Abt.* IIa, 135, p. 307 (1926).

HERGESELL, H. Uber lokale Windströmungen in der Nähe der kanarischen Inseln. *Beitr. Phys. frei. Atmos.*, p. 51 (Leipzig, 2, 1906).

AMERICA

North America

Canada

TORONTO, MET. DIVISION. *Climatic summaries for selected meteorological stations in Canada, Newfoundland and Labrador.* Vol. II, *Humidity, wind speed and direction*, by C. C. Boughner and M. K. Thomas (Toronto, 1948).

United States

TROTTER, S. L. Local peculiarities of wind velocity and movement, Atlantic seaboard—Eastport, Me., to Jacksonville, Fla. *U.S. Weath. Rev.*, 48, p. 634 (1920).

MILLER, ERIC R. Monthly charts of frequency—resultant winds in the United States. *M.W. Rev.*, 55, pp. 308–309 (1927).

WASHINGTON, WEATHER BUREAU. *Normal surface wind data for the United States* (Washington, 1942).

VISHER, S. S. Winds of the United States. *Sci. Mon.*, pp. 105–112 (Washington, 57, 1943).

Greenland

PETERSEN, H. The climate of Greenland, Vol. I, p. 257 (Greenland, 1927).

West Indies

REED, W. W. Climatological data for the West India islands. *Monthly Weather Rev.*, 54, p. 133 (1926).

WASHINGTON, HYDROGRAPHIC OFFICE. *Weather summary, West Indies.* For use with naval air pilots H.O. publ. No. 365 and 366, H.O. No. 530 (Washington, 1947).

South America

General

LONDON, AIR MINISTRY, METEOROLOGICAL OFFICE. Aviation meteorology of South America. *Met. Rep.* 1, No. 1 (London, 1948). (M.O. 476a.)

Brazil

RIO DE JANEIRO, SERVIÇO DE METEOROLOGIA. *Normais climatológicas* (Rio de Janeiro, 1941).

Chile

NAVARRETA, J. B. Estudio meteorológico de Chile. Ruta, Arica-Santiago-Magallanes. Santiago de Chile 1930. *Repr. in Proc. 8th Amer. Sci. Congr.*, p. 381, see also p. 377 (Washington, 1940).

FERNANDEZ, D. J. Generalidades sobre el clima de Chile. *Montevideo, Rev. Met.*, 2, p. 7 (1943).

Uruguay

FONTANA, H. V. P. Estudio de correlacion de elementos meteoro-oceanográficos en los puertos de Montevideo y Punta del Este. Año 1938. *Montevideo, Rev. Met.*, 1, No. 2, p. 72 (1942). (Direction and Velocity.)

BERGEIRO, J. M. *Clima de Uruguay* (Montevideo, 1945).

ASIA

General

DEPPERMANN, C. E. *The mean transport of air in the Indian and South Pacific Oceans, Manila, Philippine, Weather Bureau* (1935).

SIMPSON, G. C. The south-west monsoon. *Quart. J.R. Met. Soc.*, 47, p. 151 (London, 1921).

CASTENS, G. Wind und Wetter. *Handbuch für das Rote Meer und Golf von Aden.* 2 Aufl., p. 1 (Berlin, 1926).

LONDON, AIR MINISTRY, METEOROLOGICAL OFFICE. *Monthly normals of percentage frequency of surface and upper winds over Malta, Egypt, Palestine, Trans-Jordan and Iraq, mainly between the years 1921 and 1932* (London, 1937).

POONA, INDIA METEOROLOGICAL DEPARTMENT. *Climatological charts of the Indian monsoon area* (Poona, 1945).

PARIS, MIN. DE LA FRANCE D'OUTRE-MER, SERVICE DES STATISTIQUES, MINISTERE DES FINANCES ET DES AFFAIRES ECONOMIQUES, ET INSTITUT NATIONAL DE LA STATISTIQUE ET DES ETUDES ECONOMIQUES: *Annuaire statistique de l'Union Française Outre-Mer, 1939–46.* Chapter A Climatologie (Paris, 1949).

Borneo

MELBOURNE, R.A.A.F., METEOROLOGICAL SERVICES. *Report on general climatic and meteorological conditions and on the military implications of climatic conditions on the following localities in Borneo:* (1) *Brunei Bay*, (2) *Kudet-Jesselton*, (3) *Sendakan*, (4) *Taraken*, (5) *Balikpapen*, (6) *Bandjirmasin* (Melbourne, 1945).

Ceylon

COLOMBO OBSERVATORY. *The weather along the west coast of Ceylon* (1937).

O'DWYER, D. The climates of Ceylon. Nedlands, Univ. W. Australia, *G. Lab.*, *Res. Rep.* No. 14 (1950).

Hong Kong

HONG KONG, ROYAL OBSERVATORY. *The climate of Hong Kong*, by T. F. Claxton (*appendixes to Hong Kong Royal Observatory*, *Met. Obs.*, 1915 and 1931) (1884–1929).

HONG KONG ROYAL OBSERVATORY. *The winds of Hong Kong* (appendix to Hong Kong Royal Observatory 1921) by T. F. Claxton (Hong Kong, 1921).

HEYWOOD, G. S. P. *Hong Kong typhoons*. Hong Kong Royal Observatory, *Tech. Mem.* No. 3 (1950).

India

CALCUTTA, *Indian Meteorological Memoirs*. On winds at ground level and above, at nine stations in India, by J. H. Field. *I.M.M.* 22, Pt. 4, p. 505 (1920).

POONA, INDIA METEOROLOGICAL DEPARTMENT. Tables of monthly average frequencies of surface and upper winds up to 3 km. in India. Pts. A, B, C and D. *Sci. Notes*, Vol. II, No. 17 (Poona, 1930).

Indochina

BRUZON, E., *et al.* Typhons d'Indochine. *A. Phys. Globe France d'outre-mer*, 1, p. 119 (Paris, 1934).

BRUZON, E., CARTON P. and ROMER, A. *Le climat de L'Indochine et les typhons de la Mer Chine*. 3rd edn., Tome I. Le climate de l'Indochine. Par P. Carton Hanoi, 1940).

BRUZON, E., CARTON, P. and ROMER, A. Le climate de l'Indochine. Aperçu générale et régime des vents. *Serv. Met.* (Saigon, 1950).

Iraq

BAGHDAD, METEOROLOGICAL SERVICE. *Climatological means for Iraq*. Publ. 7 (Baghdad, 1942).

BAGHDAD, METEOROLOGICAL SERVICE. *Climatological means for Iraq*. Publ. 9 (Baghdad, 1950).

Japan

TERADA, TORAHIKO and TATUO KOBAYASI. On the diurnal variations of winds in different coastal stations of Japan. *Rep. Aero. Research Inst.* (Tokio, 1, 1, 1923).

WASHINGTON, HYDROGRAPHIC OFFICE. *Meteorological data for Japan*. Prepared by the Weather Bureau (Washington, 1943).

SCHWIND, M. Die Gestaltung Karafutos zum Japanischen Raum. *Peterm. Mitt.*, *Ergh.* 239 (1942). (Chap. III, *Das Klima*—translation available.)

Madagascar, Comores and Réunion

RAVET, J. Statistiques du vent au sol à Madagascar aux Comores et à la Réunion. Tananarive, *Publ. Serv. Mét.*, Madagascar, No. 17, 1949 (1938–47).

Malaya

WATTS, W. L. The climate of the Malaya Peninsula. *Calif. Inst. Tech. Met. Dept.* (Pasadena, 1942).

Netherlands Indies

BRAAK, C. The climate of the Netherlands Indies. *K. Magn. Meteor. Obs.* Vol. I, 8 pts., Vol. II, 3 pts. Verh. No. 8 (Batavia, 1921–29).

MELBOURNE R.A.A.F. METEOROLOGICAL SERVICES. *Maximum surface wind speeds in the W. Pacific and N.E.I.* (Melbourne, 1945).

BATAVIA, ROYAL MAGNETIC AND METEOROLOGICAL OBSERVATORY. *Observations at secondary stations in the Netherlands Indies*, Vol. XIX (periods ending 1938) (Batavia, 1940).

WASHINGTON, WEATHER BUREAU. Climate and weather of Southeastern Asia. Pt. II. Father India and the Netherlands East Indies. *U.S.A.A.F., Wthr. Inf. Branch Publ.* 5, No. 3 (Washington, 1943).

Pakistan

KRISHNA RAO, P. R. Squalls at Karachi. *India Met. Dept. Scientific Notes*, Vol. VII, No. 75 (Poona, 1938).

Persia

LONDON, AIR MINISTRY, METEOROLOGICAL OFFICE. Meteorological report on W. Persia. *Aviation Met. Rep.*, 33 (London, 1947).

Philippines

SELGA, MIGUEL. Frecuencia relativa de vientos duros en Manila. *Rev. Soc. Astron. Espana, America*, 14, p. 24 (Barcelona, 1924).

SELGA, M. *The velocity of the wind at Manila, Baguio, Iloilo and Celeu* (Manila, 1931).

SELGA, M. List of remarkable typhoons in Philippines, 1919–30. *Manila, Met. B.*, p. 157 (1931).

SELGA, M. List of ordinary typhoons or depressions in the Philippines, 1919–30. *Manila, Met. B.*, p. 189 (1931). (Met. Notes 5 and 6 respectively.)

SELGA, M. Wind roses of ideal marine stations in and near the Philippines. *Manila Obs. Publ.* 3, No. 2 (1931).

Siam

BUNNAG, C. V. Climates of Siam. *Met. Dept. R. Siamese Navy, Sc. Articles* No. 4 (Bangkok, 1947).

Turkey

WEICKMANN, L. *Zum Klima der Turkei.* Ergebnisse dreifähriger Beobachtungen 1915–1918. Luftdruck und Winde im östlichen Mittelmeergebeit (Munchen. S.a.).

Australasia

MELBOURNE, BUREAU OF METEOROLOGY. *Seasonal and diurnal wind roses for selected Australian stations* (Melbourne, 1928).

GLOSSARY OF TERMS USED IN THE STUDY OF WIND POWER

French	***English***	***German***
Généralités	*General*	*Allgemeines*
soufflerie *ou* tunnel aérodynamique	wind tunnel	Windkanal
station de jaugeage	measuring station	Mess-station
bande d'enregistrement	recorder chart *or* recorder tape	Registrier-streifen *oder* Registrierband
sillage	wake	nachstrom
frais d'entretien	maintenance costs	Unterhalts-kosten
face au vent	facing the wind	gegen den Wind
filet d'air	air stream, filament	Luftstrom Luftschichtung
chute de pression	pressure drop	Druckabfall
énergie éolienne *ou* énergie du vent	wind power *or* wind energy	Windkraft *oder* Windenergie
par la tranche	edgewise	im Schneidenrichtung
facteur de puissance	power factor	Wirkungsgrad
caréner	to stream-line	durchfurchen
débit d'air	air flow	Luftmenge in der Zeiteinheit
surface balayée *ou* surface offerte au vent *ou* surface soumise au vent	swept area	bestrichene (kreis-) fläche
Metéorologie ou Aérologie	*Meteorology*	*Meteorologie*
rose des vents	wind rose	Windrose
structure du vent	wind structure	Windstruktur
vitesse moyenne du vent	average wind speed $V_M = \frac{\int_0^T v\,dt}{T}$	mittlere Windgeschwindigkeit (or Windmittel)
vent dominant	prevailing wind	vorherrschende windrichtung
vitesse quadratique	root mean square wind speed $\sqrt{\frac{\int_0^T v^2\,dt}{T}}$	quadratisches geschwindigkeitsmittel

French	*English*	*German*
vitesse cubique	cubic root mean wind speed $\sqrt[3]{\frac{\int_0^T v^3\,dt}{T}}$	kubisches Geschwindigkeitsmittel
rafale *ou* coup de vent	gust	Bö
en amont, *ou* à l'amont, *ou* au vent	up wind *or* windward	luv
en aval, *ou* à l'aval *ou* sous le vent	down wind *or* leeward	lee
le comportement des vents	wind behaviour	Windverhalten
courbe des vitesses classées	velocity duration curve	Geschwindigkeits-dauerlinie
courbe des fréquences classées	velocity frequency curve	Häufigkeitskurve der Windgesch windigkeit
courbe des puissances classées	power duration curve	Leistungsdauerlinie
coefficient d'irrégularité	energy pattern factor $\frac{\int_0^T v^3\,dt}{T \,.\, V_M{}^3}$	Energiemittel faktor
anémomètre à coupe *ou* à coquille	cup anemometer	Schalenkreuz-anemometer
tourbillon	eddy	Wirbel
girouette	wind-vane, weathercock	Windfahne
remonter le vent	to move against the wind	sich gegen den Wind bewegen
Machines	*Machines*	*Maschinen*
énergie annuellement produite en kWh per kW installé	specific output (kWh/year/kW)	Benützungsdauer (Spezifische Leistung)
puissance installée	installed capacity	Installierte Leistung
énergie annuellement produite par unité de surface balayée	annual output per unit swept area	Spezifische Jahresleistung (bezogen auf die Einheit der bestrichenen Kreisfläche)
rendement global	overall efficiency	Gesamtwirkungsgrad
couche limite	boundary layer	Grenzschicht
queue en bois	wooden tailpole	Hölzerner Windfahnenträger
rose des vents	fantail	Seitenrad
gouvernail	tailvane	Windfahne

French	*English*	*German*
couple moteur	torque *or* turning moment	Drehmoment
changement de pas *ou* variation d'incidence	pitch changing	Verstellung des Blattwinkels
paramètre de fonctionnement	tip-speed ratio $\frac{\text{blade tip speed}}{\text{free air wind speed}}$	Schnellaufzahl
dispositif d'effacement	furling device	Orkanstop
finesse	the lift-drag ratio	Gleitzahl
portance	lift	Auftrieb
traînée	drag	Widerstand
couple au démarrage *ou* couple de démarrage	starting torque	Anlaufmoment
pale à incidence variable *ou* pale à pas réglable	variable pitch blade	Verstellflügel
hélice aérienne	airscrew	Luftschraube (Windrad)
pylône en treillis	lattice tower	Gittermast
volet à freinage	air brake, braking vane	Luftbremse (Bremsklappe)
hauban	stay *or* guy wire	Abspannseil
tirant	tie-rod	Abspannstange
bord de fuite	trailing edge	Endkante
bord d'attaque	leading edge	Profil nase, Eintrittskante
aile en toile	cloth sail	Tuchsegel
roue éolienne	wind wheel	Windrad
désorienter le rotor dans le vent	to turn the wheel out of the wind	Das Lad aus dem Wind drehen
angle de torsion	angle of twist	Verdrehungswinkel
s'emballer	to race	durchgehen
aile	wing *or* sail	Flügel, Flügelblatt
pale	blade	Blatt
moulin à vent	windmill (grinding)	Windmühle
éolienne	water pumping windmill	Windpumpanlage
aéromoteur	wind-driven generator	Elektro-Windkraftanlage
turbine à action	impulse turbine	Aktionsturbine

French	***English***	***German***
turbine à air *ou* turbine atmosphérique	air turbine (*i.e. a machine with a multi-bladed rotor*)	Windturbine
panémone (*In French this term applies particularly to machines whose rotor elements run in the opposite direction to the wind during one half revolution, e.g. a cup anemometer*)	wind-driven machine with a vertical axis	Windrad mit vertikaler achse
engrenage multiplicateur	step-up gear	(Übersetzungs) getriebe
palier	foot-step bearing	Fusslager
pignon d'angle	bevel wheel	Kegelrad
désorienter par rapport au vent	to turn out of wind	aus dem Wind drehen
volet à charnière	hinged shutter	jalousie
effort de flexion	bending stress	Biegespannung
emplanture	blade root	Blattwurzel
vitesse de démarrage, vitesse du vent provoquant le démarrage	starting velocity	Anlauf geschwindigkeit
mise en drapeau	feathering	auf Fahne *oder* auf Fahne drehen
vitesse du vent pour laquelle un aérogénérateur commence à débiter	cut-in velocity	Einschaltgeschwindigkeit
vitesse d'arrêt	furling velocity	Windgeschwindigkeit bei der die Anlage stillgelegt wird
vitesse minimum du vent pour laquelle un aéromoteur donne sa puissance nominale	rated wind speed	Auslege-geschwindigkeit
puissance nominale	rated power	Nennleistung

APPENDIX

RECENT DEVELOPMENTS IN WIND-POWER RESEARCH

Introduction

The Generation of Electricity by Wind Power was originally published in 1955, and the contents represent the state of knowledge concerning wind power generation at that time. When the publishers, in response to a growing demand, decided to reprint Golding's classic text, it was thought appropriate to add an appendix to indicate to the contemporary reader those parts of the text which have now become dated as a result of the advance of knowledge in the last twenty years. As far as possible these matters will be treated in the same order that they occur in Golding's original text.

Wind characteristics, wind data and energy estimation

These matters are treated by Golding in Chapters 3 to 6 inclusive and in Chapter 10. During the last twenty years there has been a growth of interest in wind data and consequently a considerable expansion in the network of meteorological stations collecting such data. This interest has arisen not because of wind power applications, but because of an interest in wind loading on buildings and structures and because of a realisation of the economic importance of meteorological data to a number of industries including construction, agriculture and horticulture.

In 1955 the number of anemograph stations (those where wind speed and direction are continuously recorded on a paper chart running at 1 in/hr) in the U.K. meteorological network was 56. The comparable figure for 1975 is 112. During this period the Dines pressure tube anemometer, with its limitations of accuracy and response, has been wholly replaced by the electric cup generator anemometer.

Moreover, the same period has seen the development of the electronic digital computer and its routine use by the meteorological services of many countries for processing data. For many anemograph stations several years' data is available in computable form, as punched cards or on magnetic tape. Consequently, it is now a comparatively simple matter to produce and plot automatically velocity/duration or velocity/frequency curves, and to produce statistics relating to the duration of calms, which are necessary for the design of energy reserve storage systems. None of these ideas are new – they were all discussed by Golding. The difference is that what used to require many hours of laborious analysis and hand computation, can now be accomplished in a few minutes. Where such data is available the designer of wind power plant can now forecast the performance of his design at a given site much more accurately. For the same reasons, concepts such as 'energy pattern factors' now receive less emphasis. They were primarily designed to minimise hand computations by providing an empirical comparison between one site and another on the basis of simpler wind data such as the annual mean wind speed.

In the absence of computerised data, the relationship between specific output and annual mean wind speed (Fig. 57 p. 156) is still the most useful method for estimating the wind power potential of a site. Incidentally, the techniques advocated by Golding on an empirical basis, have recently been shown to have a sound theoretical basis, since wind velocity/frequency curves can approximately be fitted by a Weibull statistical distribution.

The development of extreme value statistics and their application to wind data has largely solved another problem – that of forecasting the worst wind conditions likely to be encountered by a machine during its design life. By these techniques, a probability can be associated with a given value of wind load. It should be emphasised that these techniques can be applied in two situations.

Firstly, as an aid to forecasting the loads which a machine will have to survive in a non-running condition, that is when it has already been furled because the wind speed is above the maximum value in which the machine has been designed to work. In such conditions it is clearly necessary that the machine be able to survive as a static structure.

Secondly, extreme value statistics can be used to forecast the probability of occurrence of a given (high) value of hourly mean wind speed. According to modern ideas concerning wind structure, knowledge of the hourly mean is a necessary stage in the prediction of the gust structure and hence the stresses and control problems which arise as a result of gusts.

Wind flow over level terrain and hills

The increase of wind speed with height over reasonably level terrain of fairly uniform roughness has been studied extensively since 1955 and is now well understood. In order to discuss the problem quantitatively, it is necessary to consider the mean speed, and to specify the time interval over which the wind speed is averaged when this mean speed is determined, since means taken over different time intervals show differing rates of increase with height. The variations in wind speed evident in a typical record arise from the 'gusty' nature of wind which contains disturbances of a wide range of time scales from fractions of a second through several minutes to some hours. From modern work on wind structure, formulae are available which are most useful if applied to means taken over an interval between about 15 minutes and 2 hours. This should not be taken to imply that the fluctuations of shorter time scales than 15 minutes are unimportant, but merely that there are more appropriate ways of describing the variation of their properties with height.

For describing the variation of mean wind speed with height, with the meaning interval as just described, the Hellman Formula (p.78) is no longer used. Instead two other formulations are common:–

(i) The power law $V \propto z^{\alpha}$

(ii) The Logarithmic law $V \propto \ln(z/z_0)$

In these laws, z is the height above ground level, and the index a and the roughness length, z_0, vary according to the nature of the terrain. Typically the increase of mean speed with height over grassland could be described by a power law with $a = 0.12$ or a log law with $z_0 = 0.003$m, whilst the mean wind profile over rougher terrain, such as suburban housing, would correspond to values of $a = 0.25$ or $z_0 = 0.25$m. The power law has traditionally been used by structure engineers and provides the best fit to data over the height range 10–200m above ground level, while the log gives the best results if applied over the height range 0–40m. Further details are to be found in the literature.

The above formulae describe the variation of mean wind speed with height over level country. The problem of predicting the wind flow over hills and ridges has seen marked advances in the last ten years. These have been in laboratory and computational techniques – good measurements of profiles above real hills are almost as scarce as they were when Golding was writing.

In the laboratory, it is now possible to use wind tunnels to simulate with acceptable accuracy the flow in the bottom few hundred metres of the earth's atmosphere. These techniques use special boundary layer wind tunnels with very long working sections. The floor of the working section is covered with suitable roughness elements and a thick boundary layer is allowed to grow naturally as the flow passes down the tunnel, the models under test being located at the far end of the working section. In such a tunnel models of hills can be studied at scales of typically 1/500 to 1/2000 and measurements made of the profiles. Because of the scarcity of good 'full-scale' data it is not certain how accurate the results are, but they represent considerable improvement on the results obtained in smooth flow aeronautical-type wind tunnels which were the only ones available in 1955.

The analogue studies in the electrolytic tank discussed by Golding (p.96) were a disappointment. It proved to be a cumbersome technique with quite severe problems of measurement to a satisfactory accuracy. In any case, even if the technique had been successful, the results would have been of little value, since it gives a potential flow approximation, the equivalent to omitting viscosity from the flow equations. Since most designers of wind power plant are concerned with the wind flow fairly near to the surface, where the errors due to omission of the viscous terms are largest, the limitations of the method are obvious.

It is in computational techniques that the other major advances towards obtaining a solution of the problem of wind flow over a hill have been made. The development in the last five years of very large and fast electronic digital computers has made possible a direct attack on the problem by numerical solution of an approximate form of the Navier-Stokes equations – the mathematical equations governing the flow problem. Such an approach would have been quite impractical even with the digital computers available in the 1960's. Several quite effective methods are available for solving the problem of two-dimensional flow, that is over a ridge as opposed to an isolated hill, and it is possible to estimate correction factors to allow for three-dimensional effects. A typical two-dimensional solution takes about 10 minutes on a medium power modern computer such as an ICL 1903T.

Measurement of wind velocity

As already mentioned the Dines anemograph is now obsolete and has been phased out of the U.K. observational network. For long term energy survey measurements, a cup type of instrument would now always be used. The cup anemometer is now the standard instrument used in meteorology throughout the world.

In locations where no form of power supply is available the counter type of cup anemometer is still the best solution. A version is now available with the counter dial inclined towards the ground, so that it can more easily be read from ground level at the base of the pole on which it is mounted.

Wherever possible, records of hourly mean wind speeds should be collected on a continuous hour-by-hour basis so that velocity-duration or frequency curves can be computed. It is obviously more effective if these records can be collected in a form suitable for computer processing. There are now commercially available a number of systems consisting of cup anemometers with low drag photo-electric readout onto punched paper tape. The development of low current drain transistorised electronics allows some of these systems to run unattended from batteries for considerable periods. Consequently, a wind survey is now much simpler. As a result of these advances, the various special purpose instruments designed to record wind energy directly, which were discussed by Golding, are now obsolete. Strangely, the same degree of technical advance does not seem to have occurred with wind vanes, which still usually require a mains supply, although some automatic readout versions are now available.

For the determination of the detailed structure of gusts on short time scales down to fractions of a second, there are now a number of instruments available all of which represent considerable improvements on the technology available to Golding.

The Electrical Research Association gust anemometer underwent considerable development, and the final instrument which emerged used a perforated sphere (table-tennis ball) and an entirely different read-out system from that described by Golding (p. 148). For its performance, it is a relatively cheap instrument, and quite large arrays (up to 42 at a time) have been used for research into wind structure related to structural loading problems. The later ERA instruments measured only the horizontal components of wind speed.

Another perforated sphere anemometer was developed at the Central Electricity Research Laboratories using a larger sphere (plastic ball-cock) as a sensor, which also contained the read-out system. This instrument measures all three components, but is about ten times as heavy and five times as expensive as the ERA instrument, both of which are obvious disadvantages.

The propellor type of anemometer has been developed and there is now available a version with three propellors orthogonally mounted. Ingenious propellor design has secured a sufficiently accurate cosine-law response in yaw to the wind velocity, so that all three components of the wind velocity can be measured. Some care is, however, needed to avoid mutual interference effects between the propellors.

Finally, the most recent development of all is the acoustic anemometer, which works on the principle of measuring the difference in transit time of a sound/pulse over a fixed distance, in and against the instantaneous wind direction. One, two and three-component versions are commercially available and the response time is superior to that of any other anemometer except hot wire instruments, which are not really suitable for sustained outdoor use.

Wind structure, gusts

Perhaps the part of Golding's work which has 'dated' most in the last twenty years is the section concerned with the treatment of gusts (Chapter 9). Since 1955, a great deal of effort has been devoted to both experimental measurement and theoretical studies of the structure of gusts (turbulence) in natural wind, motivated by the need to solve wind loading problems.

As Golding correctly recognised, the major problem is to describe the variations in wind speed both in space, that is over the swept area of a wind turbine, and in time, that is the fluctuations which occur over a few seconds. As has already been noted, the mean wind speed (taken over an averaging time of between 15 minutes and 2 hours) is a well-defined function of height, which may be specified for terrain of a given roughness. The major conceptual advance has been in the treatment of the gust component, that is the difference between the instantaneous value of the wind speed at a given point, and the mean value at the same point. It has been recognised that this gust component varies both in space and time in a random manner, and consequently it is pointless to try to describe these variations deterministically by a mathematical formula expressing the gust speed as some function of space and time. Instead, the random fluctuations have to be analysed by statistical methods. Methods of random signal analysis, originally developed by communications and radar engineers during and after the Second World War, have been applied to the analysis of the structure of atmospheric turbulence.

The variations in space and time of the gust component are described by two methods. The first method is in terms of space-time correlations. In essence, these provide a quantitative answer to the question 'if we measure the gust component at point P at time t_1, how much does this tell us about what the value of the component will be at some other point P' and some other time t_2'. The second method analyses the fluctuations in terms of their frequency content – the analogy here being the way that a prism splits up white light into its constituent colours or wavelengths. These two methods of analysis are not, of course, independent, and mathematical relationships are available which relate the two descriptions of gust structure which result.

For wind turbines, the new methods and the data on the structure of atmospheric turbulence in principle allow two problems to be solved which were not amenable to analysis in 1955:

(i) It is now possible to predict the structural loads in the blades, tower and other components of a wind turbine, either when the machine is generating in wind speeds within its normal range of operation, or under survival conditions in very strong winds when the machine is furled.

(ii) It is now possible to relate the fluctuations in the power output of the machine due to gusting, to the natural velocity fluctuations in the airstream incident upon the wind turbine.

To avoid misunderstanding, it must be remarked that at the time of writing, no solutions to these problems based on the new methods have appeared in readily available published literature.

Testing of wind machines

The general approach described by Golding remains valid, and to test a wind power plant in 1976 would involve measurements of the same variables enumerated by Golding. The difference clearly would be that owing to advances in instrumentation, recording and mini-computer technology, the actual methods used would now be rather different.

Perhaps the only major difference in approach is that there would now be less emphasis on measurements of gusts in the incident wind over the swept area of the machine as part of the test procedure. Because there is now a better understanding of the structure of natural wind, elaborate measurements of this sort would be unnecessary, and indeed pointless. Instead, it would be reasonable to relate power output to a single wind measurement, for example, at the hub height of a conventional horizontal axis propellor machine. For the same sort of reason, measurement of power coefficients would now be done by averaging both power output and wind speeds over periods of at least 15 minutes, rather than by attempts to relate instantaneous 'spot' values of one parameter to the other.

Design and Economics of wind machines

The general principles for the design of wind machines and for assessing whether wind power constitutes an economically sensible solution in a given application, remain as Golding described them. Clearly, the actual numerical values quoted both for the costs of wind power plant, and for the competing costs of power generation by other means, have been overtaken in the twenty years which have elapsed by the general cost inflation suffered by all advanced industrialised nations.

Any contemporary wind-power project, if it is not to be merely an environmental toy should be justified by a careful performance/cost analysis of the type described by Golding. The key questions remain the same:

(i) The cost of wind power generation as opposed to those of alternative methods.

(ii) The relation between the wind power generating potential available at a site and the capital cost of the machine required to extract that power.

(iii) For a given site wind regime, the balance between the efficiency and reliability of the wind turbine and its capital cost.

These have been the major factors in deciding the progress of wind power generation since 1955, which is discussed in the next section.

Wind power since 1955

Golding's text makes only passing reference to the wind power projects which were in progress in the period following the Second World War. Many projects were, of course, still in progress at the time that he was writing and it was not clear how the subject was going to progress.

Small machines of a capacity from a few hundred watts up to about 2kW have been and still are commercially available and in use. The main applications are for battery charging to supply radio telecommunications or emergency lighting at remote sites, or to drive small pumps on farms.

As far as the larger sizes of machines are concerned, the work on wind power generation since World War II divides into two distinct periods.

The first of these was in the immediate post-war period and extended up to the early 1960's and included all the work mentioned by Golding. This work was almost exclusively European, motivated by the desperate fuel shortages in the post-war period in most industrialised countries in Western Europe. In North America, which did not suffer from these problems to the same extent, there was little interest following the conclusion of the work initiated during the war under the auspices of the Federal Power Commission, and the failure of the Smith-Putman machine.

Tables giving details of those machines which were actually built (as opposed to paper designs) but which are not included in Golding's Table XXXII (p. 220) are given at the end of this Appendix. Considerations of space preclude comment on the individual projects, but some general patterns emerge. Some projects were technically unsuccessful because design faults led to failures of blade systems and other components, and the machines were wrecked. The machines which were technically successful, were, not surprisingly, those which were constructed to higher standards, often using techniques such as hydraulically operated pitch controls, borrowed from aircraft propeller practice. Inevitably, these machines were commercially unsuccessful, because the higher capital costs rendered the electricity they produced uncompetitive compared with energy generated in modern post-war oil-fired power stations. Attempts to cheapen the capital costs to render the machines more competitive inevitably led back to technical failures. In the middle 1960's with the prospect of relatively cheap power from thermal stations fired by Middle East oil, and in the face of optimistic forecasts for the future capacity and economics of nuclear power generation, interest in wind power generation in Western Europe waned, and work on all large projects virtually ceased. There continued to be an interest in wind power in Third World countries many of which lacked fossil fuel resources, and, because the distribution of their populations and the cost, found investment in sophisticated European type grid systems impracticable.

The second wave of interest in wind power started as part of the general upsurge in interest in environmental matters, which led to the United Nations conference in Stockholm in 1972. The basis of interest here was in wind power as a pollution-free source of energy and also as a means by which technologically advanced countries could assist the development of the Third World. This was closely followed by the October 1973 Middle East War and the subsequent increases in oil prices, which led many countries to search for alternative sources of energy, particularly those like wind power, which would not depend upon fossil fuel supplies from a foreign source. The growth of interest can almost be called explosive. It is too early to say where this generation of work will lead. Many different ideas are being pursued in parallel, and only after a pattern emerges will it be possible to judge which, if any, of these ideas are likely to lead to large-scale application of wind power.

Apart from small conventional propellor machines the main lines of development are:

(i) Small machines where aerodynamic efficiency has been sacrificed in the interests of simple and rugged construction. These machines are primarily intended for use in remote sites and developing countries, the intention being that they can be built and maintained with locally available materials and personnel.

(ii) A fresh attempt to design and construct machines of large capacity driven by an efficient propellor rotor, rotating about a horizontal axis. The hope here is that improved materials such as glass-fibre reinforced plastics and similar composites, and light metal alloys which have all become generally available since the first post-war generation of machines were constructed, will now enable machines to be constructed which are competitive in terms of capital cost, without sacrifice of efficiency or, more importantly, reliability. At the time of writing the most advanced project in this category is the 100kW machine designed and constructed in the United States under the auspices of the National Science Foundation and the National Aeronautical and Space Administration.

(iii) A new attempt to produce a design of vertical axis machine using the basic principles originally patented by Darrieus. Vertical axis rotors have some important cost advantages in that they are omnidirectional and hence do not require yawing gear, and cannot suffer from gyroscopically induced stresses caused by changes in wind direction at high wind speeds and hence high speeds of rotation. There is also a further potential cost saving in the support structure which does not have to carry the load of the generator, which can conveniently be located at or near to ground level. Previous vertical axis machines working on the drag principle such as the panemones discussed by Golding (p. 195) have suffered from much lower aerodynamic efficiency than a good propellor design, and consequently this has rendered them less economically attractive in terms of capital cost per unit of energy generated. However, recent work at the National Research Council of Canada has evolved a design of vertical axis machine with an efficiency comparable to a good conventional propellor design, which appears to offer potentially low capital costs per unit of energy generated. Two machines of this design rated at 2kW and 6kW respectively are being manufactured.

(iv) New methods for the storage of energy generated by wind and other random 'natural' sources. Most of the methods being investigated are not new, and were briefly discussed by Golding (Fig. 100, p. 277), but again the impetus has come from technological developments in the last twenty years. Methods currently being considered include:

(a) The use of higher energy density batteries such as sodium/sulphur.

(b) Storage as hydrogen produced by electrolysis. There have been several significant developments, including electrolysers which produce hydrogen directly at storage pressure, avoiding the cost and difficulties associated with compressing the gas from atmospheric pressure.

Methods for the storage of hydrogen other than as liquid or compressed gas are also available; these include metallic hydrides and 'hydrogen sponge' materials such as LaNi alloy. These latter are remarkable in that they are capable of storing hydrogen as absorbed gas at greater densities than liquid hydrogen.

Associated with these improvements in hydrogen storage is work on fuel cells, so that the stored gas can be directly converted back to electricity, thus avoiding the heavy penalty which has to be paid in terms of overall efficiency, if conversion back to electricity is via a heat engine.

(c) Storage as pumped water. Where suitable reservoirs are naturally available this is one of the best methods (Golding p. 278). Recent work has lead to the development of small reversible pump/turbine units which may make this method more attractive economically.

(d) Storage as heat. Studies have been made of the direct generation of heat by coupling a heat 'churn' direct to the wind turbine. This offers the attraction of easy matching between the power output/rotational speed characteristics of the wind machine and the power absorption/rotational speed characteristic of the 'churn', thus enabling the wind machine to operate at or near its maximum power coefficient. Where a continuous demand for heat is available, this method appears to offer an economical solution.

(e) Storage as compressed air or hydraulic power. Golding (p. 278)

considered compressed air as a means of energy storage and this remains an available option. An alternative which is now under study is to couple the wind turbine directly to a hydraulic pump and reservoir. This then forms an effective source of mechanical power, utilising for instance hydraulic motors of the 'swash-plate' type.
(f) Storage as mechanical energy in large flywheels. New designs of flywheel based upon the properties of new composite materials such as carbon fibre are now capable of storing energy for periods of some hours.

References to these and the other developments mentioned in this Appendix are given in the bibliography which follows.

BIBLIOGRAPHY

Shortage of space precludes the inclusion of a comprehensive bibliography containing references to all the relevant work published since Golding's original book – such a bibliography would of necessity contain several thousands of references. Fortunately, however, this is not necessary because bibliographies on the relevant subject areas are avaible from other sources as follows.

All aspects of wind power generation

Wind Energy Utilization, a bibliography with abstracts, Cumulative Volume 1944/1974, Energy Information Office, Technology Applications Centre, University of New Mexico, Alburquerque, 1975.

This is a comprehensive bibliography (with abstracts in many cases) of the entire subject of wind energy organised by subject area.

Measurements of wind structure

Counihan, J: Adiabatic Atmospheric Boundary Layers: A Review and Analysis of Data from the Period 1880-1972. *Atmospheric Environment,* **9**, 871-905, Pergamon Press, 1975.

A comprehensive survey of measurements in strong wind is given in this review paper which contains 570 references to measurements and measurement techniques with classification according to the nature of the terrain.

Wind effects on structures

Wind Effects on Buildings and Structures, Proceedings of the First International Conference held at NPL Teddington U.K. 1963. H.M.S.O., London, U.K.

Wind Effects on Buildings and Structures, Proceedings of the Second International Conference held at NRC Ottawa 1967, Univ. of Toronto Press, Toronto, Canada.

Wind Effects on Buildings and Structures, Proceedings of the Third International Conference held at Tokyo 1971, Saikon Shuppan, Tokyo Japan.

Wind Effects on Buildings and Structures, Proceedings of the Fourth International Conference held at Heathrow 1975, Cambridge University Press, U.K.

The proceedings of these four conferences provide a comprehensive account of measurements, wind tunnel simulation and mathematical treatment of the structure of natural wind and its effect on engineering structures. There are many hundreds of references to other source material.

TABLE A1

Designs for wind-driven generators

The table below gives details of the larger machines built in the post-war period up to about 1965, which were not included in Golding's Table XXII (p. 220). The writer has included details of all the machines of which he is aware but cannot guarantee that the list is exhaustive. The table only gives details of machines of a capacity greater than 20kW. With the exception of the 25kW Dowsett machine, of which a number were manufactured, all the other machines were one-off prototypes.

Machine	Rotor diameter (feet)	Rated wind speed (mph)	Output (kW)	Form of rotor	Speed or output control	Optimum tip speed ratio	Rotational speed of rotor (rpm)	Tower height (feet)	Method of weather-cocking
SEAS Denmark	80	33	200 kW induction generator	3 blades	Fixed Pitch Blade tip rotation spring controlled	–	30	76	Electric motor drive
E.d.F. France	70	28	132 kW Asynchronous	3 blades	Variable pitch hydraulic servo motor	6	56	56	Blades down-wind of tower + positioner
E.d.F. France	99	37	800 kW Synchronous alternator	3 blades	Fixed Pitch Mechanical and electrical brakes	4.5	47.3	105	Blades down-wind of tower
E.D.F. France	115	38	1000 kW Asynchronous alternator	3 blades	Variable pitch automatic regulator	–	–	98	Blades down-wind of tower weather vane
Hutter Germany	110	18	100 kW	2 blades	Variable pitch hydraulic servo motor	–	42	73	Vane actuated electric motor
Isle of Man Smith UK	50	41	100 kW induction generator	3 blades	Fixed pitch Automobile disc brake, Emergency spoiler flaps	–	75	35	Tail vane
Dowsett U.K.	42	25	25 kW induction generator or d.c.	3 blades	Variable pitch hydraulic operation	3.9	65	33	Twin fantails

SUBJECT INDEX

NAME INDEX